D1287013

The Justice Motive in Social Behavior

ADAPTING TO TIMES OF SCARCITY AND CHANGE

CRITICAL ISSUES IN SOCIAL JUSTICE

Series Editor: MELVIN J. LERNER

University of Waterloo
Waterloo, Ontario, Canada

THE JUSTICE MOTIVE IN SOCIAL BEHAVIOR: Adapting to Times of Scarcity and Change
Edited by Melvin J. Lerner and Sally C. Lerner

A Continuation Order Plan is available for this series. A continuation order will bring delivery of each new volume immediately upon publication. Volumes are billed only upon actual shipment. For further information please contact the publisher.

The Justice Motive in Social Behavior

ADAPTING TO TIMES OF SCARCITY AND CHANGE

Edited by
MELVIN J. LERNER
and
SALLY C. LERNER

University of Waterloo
Waterloo, Ontario, Canada

WITHDRAWN

PLENUM PRESS • NEW YORK AND LONDON

Hm
216
.J88

Library of Congress Cataloging in Publication Data

Main entry under title:

The justice motive in social behavior.

(Critical issues in social justice)
Based on papers presented at a meeting prior to the annual conference of
the American Psychological Association, 1978, Toronto, Canada.
Bibliography: p.
Includes index.
1. Social justice—Psychological aspects—Congresses. 2. Social
change—Psychological aspects—Congresses. 3. Adaptability (Psychology)—
Congresses. I. Lerner, Melvin J., 1929-   II. Lerner, Sally C., 1931-
        III. Series.
HM216.J88   303.4   81-10605
ISBN 0-306-40675-6   AACR2

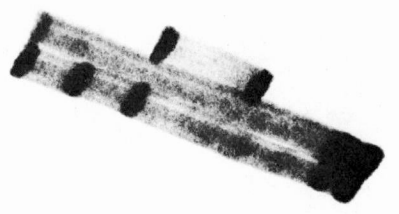

© 1981 Plenum Press, New York
A Division of Plenum Publishing Corporation
233 Spring Street, New York, N.Y. 10013

All rights reserved

No part of this book may be reproduced, stored in a retrieval system,
or transmitted, in any form or by any means, electronic, mechanical,
photocopying, microfilming, recording, or otherwise, without written
permission from the Publisher

Printed in the United States of America

SEP 2 0 1982

For MIRIAM and DANIEL

Contributors

Ellen Berscheid

Department of Psychology, University of Minnesota, Minneapolis, Minnesota 55455.

Philip Brickman

Department of Psychology and Research Center for Group Dynamics, University of Michigan, Ann Arbor, Michigan 48106.

Bruce Campbell

Department of Psychology, University of Minnesota, Minneapolis, Minnesota 55455.

Ellen S. Cohn

Department of Psychology, University of New Hampshire, Durham, New Hampshire 03824.

Thomas D. Cook

Department of Psychology, Northwestern University, Evanston, Illinois 60201.

William Damon

Department of Psychology, Clark University, Worcester, Massachusetts 01610.

Andre deCarufel

Faculty of Administration, University of Ottawa, Ontario, Canada K1N 9B5.

Morton Deutsch

Department of Psychology, Teachers College, Columbia University, New York, New York 10027.

Nicholas P. Emler

Department of Psychology, University of Dundee, Dundee, Scotland.

Michele A. Fagan

Department of Psychology, University of Chicago, Chicago, Illinois 60637.

Robert Folger

Department of Psychology, Southern Methodist University, Dallas, Texas 75275.

Erica Goode

Department of Psychology, University of California at Santa Cruz, Santa Cruz, California 95064.

Jerald Greenberg   Faculty of Management, Ohio State University, Columbus, Ohio 43210.

Robert Hogan   Department of Psychology, The Johns Hopkins University, Baltimore, Maryland 21218.

John G. Holmes   Department of Psychology, University of Waterloo, Waterloo, Ontario, Canada N2L 3G1.

Rachel Karniol   Department of Psychology, Tel Aviv University, Ramat Aviv, Israel.

Louise H. Kidder   Department of Psychology, Temple University, Philadelphia, Pennsylvania 19122.

Robert L. Kidder   Department of Sociology, Temple University, Philadelphia, Pennsylvania 19122.

Melvin J. Lerner   Department of Psychology, University of Waterloo, Waterloo, Ontario, Canada N2L 3G1.

Sally C. Lerner   Environmental Studies, University of Waterloo, Waterloo, Ontario, Canada N2L 3G1.

Dale T. Miller   Department of Psychology, University of British Columbia, Vancouver, British Columbia, Canada V6T 1W5.

Dean E. Peachey   Department of Psychology, University of Waterloo, Waterloo, Ontario, Canada N2L 3G1.

Barbara Pearlman   Department of Psychology, Northwestern University, Evanston, Illinois 60201.

Edward E. Sampson   Department of Sociology and Social Anthropology, Clark University, Worcester, Massachusetts 01610.

Yaacov Schul   Department of Psychology and Research Center for Group Dynamics, University of Michigan, Ann Arbor, Michigan 48106.

Carolyn H.Simmons   Department of Psychology, University of Colorado at Denver, Denver, Colorado 80202.

Neil Vidmar   Department of Psychology, University of Western Ontario, London, Canada N6A 5C2.

Preface

This volume was conceived out of the concern with what the imminent future holds for the "have" countries . . . those societies, such as the United States, which are based on complex technology and a high level of energy consumption. Even the most sanguine projection includes as base minimum relatively rapid and radical change in all aspects of the society, reflecting adaptation or reactions to demands created by potential threat to the technological base, sources of energy, to the life-support system itself. Whatever the source of these threats—whether they are the result of politically endogeneous or exogeneous forces—they will elicit changes in our social institutions; changes resulting not only from attempts to adapt but also from unintended consequences of failures to adapt.

One reasonable assumption is that whatever the future holds for us, we would prefer to live in a world of minimal suffering with the greatest opportunity for fulfilling the human potential. The question then becomes one of how we can provide for these goals in that scenario for the imminent future . . . a world of threat, change, need to adapt, diminishing access to that which has been familiar, comfortable, needed.

There is an answer to that question which has unexplored and potentially important ramifications. Social scientists have generated increasing documentation for the common observation that justice plays a singular and pervasive role in our culture. More specifically, they have shown that people can and do adapt successfully to changing circumstances, including those of scarcity and diminishing access . . . if they perceive their lot to be fair and just. More recently, research has provided considerable evidence that judgments of fairness and justness are not simply reflections of the prevailing normative system, but rather are well-grounded in psychological structure and processes. The implications of this are extremely relevant to the future, since the commitment to justice and the forms this commitment can take provide an as yet unexamined potential for creating the directions and the means for successful adaptation during times of change and scarcity.

Although the contributors to this volume conceptualize the psychology of justice in various ways, they have all demonstrated amply in their previous work the extent to which the commitment to justice guides important social behaviors. In this volume, they focus their attention on the establishment of guidelines for effective and constructive solutions to the set of human problems with which we must soon deal.

MELVIN J. LERNER
SALLY C. LERNER

Waterloo, Ontario
January 17, 1981

Acknowledgments

In the latter part of August, 1978, just prior to the American Psychological Association annual meetings in Toronto, a unique and important event took place on the campus of the University of Waterloo. For three days and evenings, twelve social psychologists, a sociologist, and an anthropologist pooled their expertise—through formal presentations, critical analyses, and informal discussions—in pursuit of answers to this question: how can the human concern with justice provide opportunities for constructive responses to future social dilemmas which may involve scarcity of resources and rapid change? As it turned out, this occasion was unique in many respects, not the least of which is that never before or since has there been such a gathering for the purpose of exchanging ideas. The setting and the topic provided a marvellous opportunity for this group to generate important contributions to our understanding of justice and of ways to meet the challenges which face our society.

This volume is one outcome of that 1978 conference. Although the chapters in the volume differ to some extent from what took place that August, the impetus for the volume came from those meetings and many of the individual contributions were stimulated and shaped by the dialogues that took place in the meetings.

This is the occasion to express our gratitude to those who made that unique conference possible. The encouragement and support of our colleagues in the Department of Psychology and the Faculty of Environmental Studies helped generate the conviction that not only was such a conference desirable, but feasible as well. Certainly, without the grants from the University of Waterloo Research Grant Program and the Department of Psychology, this estimate of feasibility would not have been transformed into reality. In a less direct but equally essential fashion, the Social Sciences and Humanities Research Council of Canada (File No. 410-77-0601-X2) and the Ontario Ministry of Energy provided generous support for the research which formed the conceptual background for much of this volume.

Contents

xiii

BASIC PROCESSES

Carolyn H. Simmons

INSTITUTIONAL SETTINGS

ENDNOTE

INTRODUCTION

1

Adapting to Scarcity and Change (I)

Stating the Problem

SALLY C. LERNER

1. THE NECESSARY TRANSITION

Many respected observers believe that North Americans are entering an inevitable and potentially dangerous 30- to 50-year period of transition, a transition marked by increasingly intense competition for a variety of scarce resources (e.g., Brown, 1978; Commoner, 1977; Ophuls, 1977; Schnaiberg, 1980). Representative of this outlook is political economist Robert Heilbroner's influential book, *An Inquiry into the Human Prospect* (1974). After outlining the limits to material growth posed by environmental constraints, Heilbroner points out that

> the difficulty of managing a socially acceptable distribution of income in the capitalist nations is that it will have to contend with the prospect of a decline in the per capita output of material goods. . . . The difficulties of a limited oil shortage have brought home to many Americans the hitherto unimaginable possibility that their way of life might not be indefinitely sustainable. If that shortage is extended over the next generation or two to many kinds of material outputs, a climate of extreme goods hunger seems likely to result. In such a climate, the large-scale reorganization of social shares would have

---

SALLY C. LERNER ● Environmental Studies, University of Waterloo, Waterloo, Ontario, Canada N2L 3G1.

to take place in the worst possible atmosphere, as each person sought to
protect his place in a contracting economic world. (p. 88)

In a new edition of his book ("updated and reconsidered for the 1980s"),
Heilbroner (1980) saw no reason to revise his previous conclusions as
to the problems we will face in an era of dwindling oil supplies, persistent
inflation or recession, structural unemployment, and precarious international relations. With regard to global population growth, a cause for
despair in the early 1970s, Heilbroner noted that a slowing of growth
rates during the past several years should produce a leveling off at 8–10
billions of people worldwide by about the middle of the next century.
Although this means a doubling of the present population, the trend
now appears to be toward a "manageable plateau" rather than toward
"hopeless and endless proliferation." But slower population growth will
do little, particularly in the short term, to relieve population pressures
on land and resources in many parts of the world. Nor is there much
real hope for a narrowing of the gap between rich and poor regions and
nations, particularly if this gap continues to be viewed as one that calls
for have-nots to play catch-up in the material goods sphere with primary
reliance on nonrenewable or polluting sources of energy. Heilbroner
concluded that while we cannot avoid a transitional period of diminishing resources and intensified competition in many of the conventionally
rewarding areas of human life, there is no compelling reason to despair
of finding long-term solutions; the Malthusian nightmare has given way
to a more challenging scenario.

The clearer understanding that we now have of the energy–
economy–environment meta-problem—and of the prospects for dealing
with it—should serve as an impetus for social researchers to focus increasing attention on questions of how people perceive and respond to
situations involving allocation of scarce resources. After decades of increasing affluence based in part on cheap energy, North Americans are
facing the possibility of a downturn in living standards for the great
middle majority. Social scientists have devoted relatively little attention
to the possible consequences of such a trend, since none has been evident
during the past 40 years. But, as Stretton (1976) pointed out, the era of
"environmental politics"—including problems of inflation and energy
supply—will make distributive dilemmas a central concern:

The questions which environmentalists ask make sure of that. What resources
should we use? What goods should we produce? Who should get
them? . . . Practical disagreements about environmental policy are rarely
simple conflicts between growth and conservation. They are conflicts about
what to use, what to produce and how to pay for it—conflicts between people
competing in familiar ways for rival values or for shares of scarce goods.
(p. 4)

Three factors, Stretton believes, will contribute to people's interest in questions of equality, inequality, and justice. First, such concerns will become more salient as slower economic growth translates into little or no improvement in lifestyle for middle- and working-class people. The inequality of both incomes and ownership of income-producing wealth, largely ignored or tolerated during a long period of rising affluence for the majority of North Americans, will attract sharper attention as the growth of the economic pie slows and tiny shares remain so or even diminish. Particularly provocative for the middle class will be the fading of hopes for a comfortable old age and for their children's educational and material advancement. Continued inflation, a second major factor, can contribute crucially to the dashing of such hopes and thus to the average person's feelings of being badly used, cheated, unfairly deprived of earned and planned-for rewards. Indeed, it is precisely this fear of falling behind, of losing out, that fuels the psychology of rising inflation and that also serves to make justice concerns more salient. Finally, altered perceptions of the nature of goods and opportunities (as limited rather than relatively unlimited) can be expected to intensify justice-related social conflicts:

> People who tolerate unequal sharing of ordinary goods feel differently about goods which they come to think of as absolutely limited and exhaustible. People also feel gains and losses with different intensity. . . . The division of gains which they accepted on the way up will not be the division of losses which they will want to accept on the way down. Besides particular goods like coal and oil, those feelings may apply to whole incomes if shortages of energy or other resources ever force a general fall in living standards. (Stretton, 1976, p. 6)

In this connection, Stretton made the very suggestive point that in a period of debate about how to distribute *exhaustible* goods (e.g., whether to ration gasoline by controlled program or by price), it is to be expected that more and more people will question the bases on which *all* necessary and highly valued goods are allocated. Thus, we would expect yet another pressure toward more concern on the part of the average citizen about questions of fairness.

2. TRAGIC CHOICES IN THE "CRUNCH"

To some extent, then, we face a period that will require societies to deal with what Calabresi and Bobbitt (1978) have termed "tragic choices." The nature of such choices is more complex than the simple phrase suggests, involving conflict between a culture's basic, highly prized values and the necessity of allocating certain goods and bads that involve extreme deprivation or even death. As the authors noted:

> When attention is riveted on such distributions, they arouse emotions of compassion, outrage and terror. It is then that conflicts are laid bare between, on the one hand, those values by which society determined the beneficiaries of the distributions, and (with nature) the perimeters of scarcity, and on the other hand, those humanistic moral values which prize life and well-being.
>
> In such conflicts, at such junctures, societies confront the tragic choice. They must attempt to make allocations in ways that preserve the moral foundations of social collaboration. If this is successfully done, the tragic choice is transformed into an allocation which does not appear to implicate moral contradictions. But unless the values held in tension have changed, the illusion that denies their conflict gives way and the transformation will have only been a postponement. When emotions are again focused on the tragic choice, action will again be required. (p. 18)

As this definition indicates, the choices that we can expect to be increasingly salient involve two kinds of decisions: how much of a scarce good will be made available ("first-order determination") and who shall get how much of what there is ("second-order determination"). These authors pointed out that scarcity is most commonly "not the result of any absolute lack of a resource but rather of the decision by society that it is not prepared to forego other goods and benefits in a number sufficient to remove the scarcity" (p. 22). Here, then, is the crux of the concept of tragic choice: these are choices that, at the level of first-order decisions, most often involve not absolute necessities but politically based priorities that must be presented (rationalized, publicized) as necessities if second-order allocation procedures and outcomes are to be accepted by those affected. To the extent that either second-order procedures or, especially, first-order decisions are perceived as unjust, the orderly functioning of the society may be threatened.

It seems clear that as North Americans enter a period that promises to focus public attention on a wide range of allocation situations that involve actual or threatened lifestyle changes, there is an imperative need for a better understanding of both the risks and the opportunities that confront us. We can speak of both risks and opportunities because such watersheds in a society's existence have historically given rise to a wide variety of system-level responses, ranging in type from draconian attempts to maintain existing power arrangements to basic social change embodying a more humane cultural paradigm. One challenge for social analysts is to develop further their understanding of these system-level responses so that we may hope for effective and humane societal guidance in the period ahead. Since system-level response is usually predicated on the assessment and often the manipulation of collective citizen response, another level of questions that must be addressed concerns how people respond to changes in circumstance that they perceive as having a crucial relevance to their own well-being.

It is in this context of precarious societal balance that we begin to see clearly the current need for a better understanding of the role of justice concerns in social behavior. One task is to develop a coherent, empirically based model of the functioning of justice concerns in the relatively neglected area of what might be termed behavior in the "crunch," that is, in situations that involve actual or perceived deterioration of the quality of one's life—including the fear or conviction that such deterioration is about to occur (Lerner, 1979). In particular, if we are to effect a peaceful transition to a more sustainable society, it is crucial to understand how people caught in a crunch will behave toward others. Historically, people in such situations have responded in a variety of ways, ranging from heroic self-sacrifice to vicious scapegoating. We already have substantial knowledge about many such episodes derived from attempts to codify data and develop generalizations about collective behavior. We also have useful data from areas of research that bear on the question such as those concerned with downward mobility, relative deprivation, status frustration, choice of reference groups, attribution of responsiblity and blame, adaptation to change, and response to various kinds of stress. Thus, we understand that complex variations in leadership, cultural tradition, the nature of the situation, and a multitude of other factors shape responses to crunch situations (see Figure 1.1). Yet, clearly, how a person or group *perceives* a situation or event is basic to the development of a response. If the perception leads to a definition of the situation highlighting absolute or relative deprivation vis-à-vis some valued good (or the fear of such an occurrence), then justice concerns almost inevitably become salient. Indeed, the centrality of such concerns can shape the perception of what constitutes a deprivation: an outcome or an anticipated outcome may not even be perceived as a deprivation *unless* it is felt or discovered to have been produced by unjust rules or rulers. Understanding how a person decides whether something is just or unjust, then, is basic to developing a better understanding of two processes: how people come to define a situation as a crunch and how they will behave, particularly toward others, in situations that they have so defined. Returning for a moment to the concept of tragic choice, it is clear that allocation choices that may be less than "tragic" for the large majority in times of plenty are likely, in times of declining stocks of limited resources, to approach tragic status insofar as *tragic* connotes choices that are likely to throw a spotlight on the value conflicts, class differences, and intergroup tensions in North American society. Herein lie the very real risks in the approaching transition from a period of consumerism based on cheap fuels and ignored

Mediating/conditioning variables

A. *Nature of the situation,* e.g.,
   timing (sudden, gradual);
   duration (brief/prolonged);
   duration (actual/expected)

B. *Differential impacts* by region,
   socioeconomic status, age, race,
   occupation, sex, etc.

C. *Reference points*
   Others as referent (reference group,
   social comparison processes)
   Own past as referent
   Mythic past (own/group) as referent
   Previously anticipated future as
   referent

D. *Nature of authorities' response,* e.g.,
   timeliness, effectiveness, equity

E. *Interpretations available and
   introduced,* e.g., existing
   stereotypes and antipathies; cultural
   traditions re violence, adaptation,
   compromise; extent and nature of
   previous experience with similar
   situations; conventional wisdom
   about similar events; entitlement
   ideologies; communications from
   leaders, media, significant others

F. *Perceived and actual opportunities
   to affect events,* e.g., access to
   political means; access to
   communication media; resources of
   various actors (to cope, profit,
   promote, or control violence, etc.)

G. *Types of (emergent) leadership,*
   e.g., motives, guiding beliefs, action
   preferences, intelligence, style

"Crunch" situation (perceived present/
potential deterioration of actor's quality
of life)

Cognitive-emotional responses
   Cognitive: perception (symbolization,
   framing), interpretation, explanation
   Emotional: fear, resentment, anger
   (relative deprivation); relief, optimism,
   joy
What is happening?
   Opportunity ←——→ Disaster
   Good ←——————→ Bad for me, those I
   care about?
   How to cope with change?
   How to minimize detriments?
   Who is responsible?
   Who is in control?

Behavioral responses
   Planning, sharing, organizing, etc.
   Withdrawal
   Passive resistance
   Acceptance/support of extremist
   measures and solutions;
   scapegoating, witch-hunting;
   physically violent "hostile" outbursts
   (riot, pogrom, lynching, etc.)

SPECIFICATION

Involving whom?	Scope, duration intensity?	Out-comes?
Choice of allies		
Identification of deviants, enemies		
Selection of objects of conversion, reeducation, discrimination, repression, aggression, extermination		

Figure 1.1. Behavior in the crunch: A paradigm

externalities to a more materially constrained society. As more and more people, middle and working class, take an intense interest in "who gets what and how much," there will be increased questioning not only of the fairness of allocation *procedures* but, undoubtedly, also of the justice (in terms of basic human needs) of the *priorities* expressed in current social, economic, and political arrangements (first-order matters). People today are considerably less likely to blame themselves for personal difficulties involving shoddy housing, unemployment, underemployment, and the like (Lambert & Curtis, 1979). The 50 years of education (and disillusionment) since 1930 have resulted in a much stronger tendency for people to place responsibility squarely in the lap of government—for its allowing such things to happen and for failing to remedy the problems once they exist. One serious risk in this state of affairs is exacerbated social conflict, in terms of political disaffection that leads to more negatively anarchistic behavior in some areas and/or in terms of sharply heightened hostilities between diverse groups in society. The latter danger, inherent in the scarcity situation in any case, is all the more likely to materialize if those with political and economic power turn to propaganda and symbol manipulation in order to deflect public criticism. The temptation to provide manufactured explanations, even scapegoats, has seldom been resisted by powerholders in crunch situations that lead the public to demand simple, instant "solutions."

Increased scrutiny of the procedures as well as the premises of allocations is also to be expected, of course. Indeed, it is likely that questions relating to the fairness or unfairness of procedures will be recognized and dealt with more openly than will questions about the "societal" priorities that set limits on the availability and the total pool of certain goods. In this context, a better understanding of people's justice concerns is a necessity if we are to develop the kinds of allocation procedures that minimize illegalities and intergroup hostility and that maximize cooperation and orderly social interaction.

3. CONCLUSION

In conclusion, it should be said that there are those who see the transitional period under discussion as an opportunity—to bring about fundamental social change in the direction of a more decentralized, sustainable, humane society (e.g., Schumacher, 1973; Solomon, 1978). Such advocates, many of them sincere and eloquent, would object to any approach to justice concerns that seems to be aimed at understanding them better in the service of manipulating or damping them down. In order for the social change that they envision to come about, in the form of new arrangements for work, governance, education, and family living that both reflect and stem from altered political and economic values, justice concerns of both the first- and second-order types must be made central to an ongoing critique of existing social arrangements. They would ask, Is it possible that in studying how people develop, articulate and express justice concerns, social scientists will deliberately or unwittingly furnish the data necessary for those currently in power to maintain their control and thwart desired social change? There is this possibility, of course. But it can be said truthfully that the risk that a worst-case situation will develop, one in which the inequalities and cruelties of the present society are magnified as well as visited on a much larger proportion of the population, compels one to urge the study of justice concerns as a top priority.

REFERENCES

Brown, L. Global economic ills: The worst may be yet to come. *The Futurist*, 1978, *12*, 157–168.

Calabresi, G., & Bobbitt, P. *Tragic choices*. New York: W. W. Norton, 1978.

Commoner, B. *The poverty of power: Energy and the economic crisis*. New York: Bantam, 1977.

Heilbroner, R. L. *An inquiry into the human prospect*. New York: W. W. Norton, 1980, 1975, 1974.

Lambert, R., & Curtis, J. Notes on a sociology for bad times. *Alternatives*, 1979, *8*, 2, 32–33, 51–53.

Lerner, S. C. (Ed.). *Behavior in the crunch*. Special topic issue of *Alternatives*, 1979, *8*, 2.

Ophuls, W. *Ecology and the politics of scarcity*. San Francisco: W. H. Freeman, 1977.

Schnaiberg, A. *The environment: From surplus to scarcity*. New York: Oxford University Press, 1980.

Schumacher, E. F. *Small is beautiful: Economics as if people mattered*. New York: Harper & Row, 1973.

Solomon, L. *The conserver solution: A blueprint for the conserver society*. Toronto: Doubleday Canada Ltd., 1978.

Stretton, H. *Capitalism, socialism and the environment*. Cambridge: Cambridge University Press, 1976.

2

The Justice Motive in Human Relations

Some Thoughts on What We Know and Need to Know about Justice

MELVIN J. LERNER

1. SETTING THE STAGE FOR WHAT WE KNOW: A CONCEPTUAL ANALYSIS OF JUSTICE

And so, as described in the previous chapter, the attempt to understand how people will react to conditions of scarcity leads directly to the issue of whether they experience their fate and the fates of those they care about as just or unjust. This statement is not at all simple nor obvious in its implications, as we shall see in a moment. After all, identifying the sense of justice as the key issue is equivalent to saying that in order to understand what will happen in our future, collectively and individually, we must solve one of the most enigmatic and complex problems that has preoccupied social analysts throughout the history of Western civilization: How does the theme of justice appear in people's lives?

As someone who has been involved in the task of trying to answer this question for the past decade, I will describe what I think we have learned about the theme of justice in human affairs. I must, of course, admit at the outset that I will not be able to provide a set of neat answers

MELVIN J. LERNER ● Department of Psychology, University of Waterloo, Waterloo, Ontario, Canada N2L 3G1 to the questions of when people will experience a sense of justice or injustice, and what they will do when these experiences are elicited— and thus, wind it all up with a flourish by providing explicit recommendations for what we must do to generate "constructive" adaptations to conditions of scarcity—whatever those might be. I am not terribly ashamed of that confession, however—mainly, because I think we have learned a great deal about justice in the last decade or so. For example, we are much further along in being able to recognize what "justice" is not, what the blind alleys and pitfalls are in the search for answers. Also we have some glimmers of the solutions to the most central issues associated with interpersonal justice. Conceivably, we do know enough to point other social scientists to what problems need to be solved and to provide social activists and political practitioners with good hunches about what social devices might be useful and which might be counterproductive. But these are no more than "educated" guesses, educated, because we know a great deal about the relevant processes involved in the experience of justice but not enough of the complexities of translating this information into social engineering. But, hopefully, others know more about how to do that.

1.1. Some Definitional Properties That Describe What We Have Learned

As a first step, let us take a stab at outlining some of the conceptual issues involved in the study of justice. The definitional problem is not as sticky as it might seem at first look. Social analysts and scientists have, for the most part, been fairly comfortable with recognizing that justice is a judgment with evaluative and emotional components. It is based on the recognition at some level of awareness that there is or is not an appropriate correspondence between a person's fate and that to which he or she is entitled—what is deserved. The sense of appropriateness derives from a judgment of the value, to the people affected, of their fate and the value of the fate to which they are entitled. If there is a discrepancy on the evaluative dimension between the desirability of the person's fate and that to which they are entitled, then an injustice has occurred. It can appear as a judgment that someone has received more or less than is deserved. And most social analysts would agree that there are degrees of injustice that correspond roughly to the magnitude of the perceived discrepancy. At a more molar level of analysis, the judgment of the degree of injustice extant is a function of the collective number and magnitude of these discrepancies that appear in a given population over a specified period of time.

I am not aware of anyone's having satisfactorily solved the conceptual issue of how much of a discrepancy is needed before someone decides that an injustice has, in fact, occurred, nor of how people add together various weightings or entitlements and outcomes in arriving at a composite judgment concerning the justness of their fate. (See Walster, Walster, & Berscheid, 1978, and Leventhal, 1976, for an attempt to deal with some of these issues, and Harris, 1976, for some of the complexities and ambiguities involved in that effort.) But there is solid agreement that people do continually make these judgments concerning their own fate and the fates of all others in their environment.

It also appears that most often this judgmental process occurs at a preconscious level and is revealed indirectly in the person's reactions to a given event. Of course, there are other times when these judgments are very salient to the person, especially when a great injustice has occurred. On many occasions, people can retrieve the bases for their past acts that reveal an implicit assessment of what was just or unjust about what has happened, how they have arrived at that judgment, and how they feel about it. But for the most part, as we shall see in the later discussion, people are not aware of the constant monitor they maintain on whether their fate and the fate of others correspond with their entitlements, what they deserve. There is, nevertheless, every reason to believe, given the sensitivity and responsiveness to incidents of perceived injustice, that the monitoring process is ever vigilant—for most people. Although there may be some slight differences in nuances in the way people have construed judgments of deserving fairness, entitlement, justice, rights, "equity," it is probably safe to consider them all equivalent in terms of social-psychological processes.

1.2. Paradoxical Properties of Justice

Once we get beyond the beginning step of arriving at a usable and acceptable definition of *justice*, then the conceptual issues become much more fascinating. An important next step is to recognize the provocative paradoxes inherent in the appearance of justice in human affairs. It is quite natural to be intrigued by these apparent contradictions, but at a more serious level, they provide the clue that there is something special, possibly unique, about the role that justice–injustice plays in our lives and some hints as to the relevant social-psychological processes. First, let us take a look at some of these paradoxical qualities.

1.2.1. *Justice as "Objective" but "Varying"*

For example, having recognized earlier that the sense of justice is at the core of a person's reaction to conditions of scarcity or unexpected change, we can begin with two related observations. On the one hand, people typically experience the sense of justice or injustice as immediately compelling. It is experienced as no less real or objective than the surrounding physical world. It is possible to find common examples of this reaction. To many television viewers, the scene of mounted police charging the peaceful marchers on the bridge at Selma, Alabama, aroused a vividly clear perception of an injustice. The same is true of most people when they see an adult abuse a child or a worker cheated out of wages. Readers can easily demonstrate this objective and immediately compelling quality by recalling their own experiences. Of course, we may at times experience "gray areas" where we do not have clarity in our view of what is just. I suspect, however, that those are temporary states of ambiguity, which are soon resolved with more relevant information or else are the occasions when we experience conflicts among more than one "objective" assessment of who deserves what. (The issue of conflicting entitlement-based demands will be dealt with in considerably more detail later in this discussion). In any event, the main point to be made here is that at its core, the sense of justice or injustice appears as an immediate and compelling reaction to the perception of what is experienced as objective reality.

On the other hand, the perception of what is objectively just seems to vary from person to person and from situation to situation for the same person. The history of our own lives as well as that of recorded civilization provides ample evidence of disputes and conflicts over who is entitled to what, each party absolutely convinced that, or at least acting as if, it has justice on its side. Also, if we focus on one person, we can see justice take many different forms. I am convinced on various occasions that it would be a gross injustice if each of us did not receive what we deserved by virtue of our effort, or the relative quality of our product, or our need for a given resource. And, in fact, there are occasions when I am convinced that I deserve everything I can get my hands on as long as I don't violate or get caught violating any specific laws.

How can the experience of justice be a reaction to objective reality and yet vary so greatly across people and situations?

1.2.2. *The Importance of Justice as It Varies from Supreme to Irrelevant*

Another set of seemingly contradictory observations can be generated in response to the question of how important the theme of justice is in people's lives. Granted that there probably are differences among people in the degree to which they are sensitive to issues of justice and the extent to which they are affected by these concerns, what is apparently true for most people is that on some occasions they are capable of being moved to extremes of self-sacrifice in order to promote justice.

There is also considerable evidence that people are opportunistic and greedy. The media are filled with evidence of tacit acceptance of the suffering of innocent victims by the majority of society, as well as reports of people's exploiting and harming one another. However responsive we are to evidence of injustices or however committed we are to deserving what we need and want in our own lives, we do precious little to come to the aid of those who are unjustly deprived and in need. Most of the time we try to get the best deal for ourselves when we are negotiating for the things that matter to us: money, status, power, etc. If we care deeply about deserving and justice, then why do we not act on behalf of all the recognized innocent victims and against those who harm and exploit? Sometimes we do. And sometimes we do not act out of raw self-interest but try to do what is fair and just, regardless of the "cost" to ourselves.

So how much do people care about justice for themselves and others? It appears that the answer is, at times, supremely, a great deal, somewhat, very little, not at all. How can we make sense out of this variation? What factors and processes could explain this great variability in the apparent concern with justice?

1.2.3. Justice as a Powerful Force for Both Social Change and Maintaining the Status Quo

Equally paradoxical is the role justice plays in the political arena. It is typical to find an appeal to the sense of justice in the rhetoric of those who demand radical social change. Revolutionary movements are often fired by the arousal or articulation of a sense of being treated unjustly. It is equally true that the procedures designed to guarantee justice and prevent the creation of undeserved victims provides a strong basis of resistance to radical or rapid social changes. The conservative forces in a society are often able to justify the status quo as having evolved out of "just" procedures and the functioning of politically legitimate social institutions. If radical change requires a reassignment of power and desired resources, one can always point validly to the risk of creating further injustices unless traditional modes of due process are followed in the use of power. Typically, due process takes time and requires economic and political power. Those are the very resources that usually are not available to those who experience injustice sufficiently to demand radical change. The commitment to justice: a strong basis for change, and a strong source of justification of the status quo and resistance to change?

1.2.4. Summary: The Paradoxes

How is it possible for justice to have the qualities of an intuitively compelling "objective" reaction and yet to vary so greatly across situations and people; to be so important in human events that it can legitimize the sacrifice of all other goals and yet often to be ignored for the sake of personal profit; to provide the sense of outrage and motivation for radical change and the compelling rationale to resist changes in the status quo; to extract extreme degrees of sacrifice from individuals or to enable people to ignore the suffering of their neighbors and require them to punish yet others? And so on.

What can we learn from these and other observations about how justice appears in human affairs? Two things we can say with certainty. *In our civilization, the theme of justice plays a central role in the public dialogues associated with the way people treat one another and go about acquiring and allocating resources. Also, it is clear that there may be some common elements in all the conceptions of justice, fairness, and deserving; the rules for deciding what is just are not invariant: justice takes many different forms.*

There is more that can be inferred, but not without a discussion of prior attempts to describe the way justice appears in our lives.

2. WHAT WE THOUGHT WE KNEW ABOUT JUSTICE: JUSTICE AS AN INSTRUMENTAL DEVICE FOR MAXIMIZING OUTCOMES

Until very recently, the social psychology of justice has been for all intents and purposes simply an extension of the familiar assumptions concerning the appearance of "normative" behavior, of any sort. The relevant psychological processes were those that entered into the formation and perpetuation of the "social contract."

2.1. The Appearance of the "Social Contract" and Its Tentative Acceptance

A basic assumption has been that people are motivated by the desire to "maximize their outcomes." It has also been assumed that people are intelligent enough to realize that it is to everyone's benefit to develop rules about how to get along with one another and to engage in jointly profitable endeavors in order for each to be able to maximize their outcomes in the long run. Thus, social rules, the "contract" in its various forms, have been developed and more-or-less effectively transmitted to members of subsequent generations through instruction, modeling, and conditioning. However, since each individual is presumably dominated by the private motive to maximize his or her outcomes, it is necessary also to create institutions, both formal and informal, that ensure through threat of sanctions that people will follow the rules.

Although it is assumed by "social contract" theorists that most people, certainly in our society, have an image of themselves as "just" people inculcated as part of their self-concept, nevertheless it is also commonly assumed that the allegiance to the social contract is dependent on its perceived utility to each person's attempt to get those things that he or she wants. For example, according to Walster *et al.* (1978) and most other social theorists, people follow rules of justice because (a) if they did not, their "consciences" might bother them; (b) they are afraid that others might punish them in some way if they are discovered violating the rules; and (c) they believe that it is to their advantage to go along with these rules in order to get others to join with them in a profitable venture. But it is also assumed that the persistent and overriding motivation in every situation is the attempt to maximize one's outcomes in terms of desired resources and that people will use rules of justice or abandon them as they are viewed to be more or less profitable in the outcome-maximizing process.

> Proposition I: Corollary 1: So long as individuals perceive they can maximize their outcomes by behaving equitably they will do so. Should they perceive that they can maximize their outcomes by behaving inequitably, they will do so. (Walster *et al.*, 1978, p. 16)

How do these theorists account for the fact that the rules for deciding what is just or deserved vary from situation to situation? The general solution is consistent with the previous assumptions. Individuals or people acting together use their intelligence to decide what rules of conduct would be most beneficial in accomplishing the desired ends. Within this consensus, there are some noteworthy differences among these theorists in terms of the underlying dynamics.

2.2. Justice as the Expression of Social Power: Contemporary "Equity Theory"

Walster *et al.* (1978), for example, elaborate their initial assumptions to achieve a radical perspective. For them, it is apparent that all people attempt to get as much as possible for themselves and that therefore, each tries to persuade the others to accept the rule of entitlement that would yield the persuader the greatest share. The most productive worker would try to persuade the others that pay should be allocated in proportion to productivity; the older worker would try to insist that it is only fair that pay should be allocated according to the number of years of service; the poorest worker with no seniority would try to convince the others that since they are all workers together, all pay should be shared equally; and the one with the largest family would insist that the amount of need in terms of the number of mouths to feed should determine the amount of money each worker is entitled to have. Finally, these theorists point out that, in fact, there is considerable evidence from the history of our society that those sectors of the social unit that have the most power do, in fact, get the most desired resources and that they are able to persuade the weaker and less well-off that the status quo is "justifiable" by virtue of some superior quality that they (the powerful) possess or some defect in the weak and relatively poor. Presumably, the impotent and thus deprived members of the society are motivated to perceive and accept the status quo as the natural of order of events.

According to Walster *et al.*, the "war of all against all" continues, but mainly covertly using the terms of a tentatively held and continually negotiated "social contract." All members of the society use their wits to maximize their own outcomes while using the rules of justice to give them every edge possible in this competitive struggle. If the manipulation of the particular rules of justice that apply in a given situation do not yield enough profit, then people will abandon those rules—give up the social contract.

2.3. Man as a More Rational Animal

There are other theorists who make similar assumptions about the human as a "rational animal," but they find greater evidence or at least more elaborate social structures that are attributable to the "rational" rather than the private pleasure-seeking motivation (Campbell, 1975; Deutsch, 1975). Deutsch (1975), for example, in common with Walster *et al.*, assumes that people employ rules of justice as means to accomplish their goals and express their values. However, he offers the hypothesis that the social unit as well as the individual adopts particular rules of entitlement or justice because it is *the* intelligent way to solve the problem. To Deutsch, "the essential values of justice are those which foster effective social cooperation to promote individual well-being." He assumes that when "economic productivity is a primary goal, equity . . . will be the dominant principle of distributive justice" (1975, p. 143). The reason is that the "equity" principle allocates resources according to productivity. It follows that since people want to maximize their outcomes, if one wants people to produce a great deal one makes their pay contingent on how much they contribute. On the other hand, as Deutsch reasoned, allocating resources on the basis of relative productivity might lead to envy or a sense of superiority. It is much wiser, then, if "enjoyable social relations is the common goal, equality will be the dominant principle of distributive justice" (p. 143). And finally, neither relative contributions nor an equal allocation makes sense when one is considering the needs of a child or someone who is stricken with illness. As Deutsch noted, when "fostering of personal development and personal welfare is the common goal need will be the dominant principle of distributive justice" (p. 143). In this manner, the resources (food, shelter, education, etc.) can be allocated in ways that would be most beneficial—for the recipient and, in the long run, for the social unit as well. Obviously underlying all these hypotheses are observations of the way some of the institutional arrangements in our society actually function. The norms for allocating resources in commerce, family, and among "friends" or teammates seem to reflect this adaptive, or functional wisdom.

2.4. Common Assumptions and Special Problems in These Perspectives

Both sets of assumptions, represented by Deutsch and Walster *et al.*, and those of others who present variants on the essential themes (e.g., Leventhal, 1976; Sampson, 1975) conceptualize justice as a social and personal device designed to facilitate the acquisition of other desired resources. The extent to which the dictates of justice are followed and the form that justice takes derive from this instrumental function. Justice prevails when it is perceived to be advantageous, and it is reshaped or abandoned when it is seen as a hindrance to other important goals. And there is evidence that one can cite, both experimental and observational, that would support both Deutsch's and Walster's assumptions. There are occasions when it is clear that the dialogue among participants that is couched in principles of justice, fairness, equity is a thin veneer over the underlying agenda of a contest in the self-serving use of power. Presumably, labor–management negotiations often fit this model as do international negotiations, and much of what occurs in legal encounters. The participants talk in terms of what is just and fair, but it is clear to all that each party will use whatever power is available to maximize their outcomes.

And there are numerous occasions when individual "power" seems to be of little influence and the allocations of resources do resemble Deutsch's observations of reasonable people assessing what would be the wisest way to proceed for all concerned. Most people do feel that children and the "stricken" are entitled to have their needs met. Also, we insist that each person is entitled to an equal share of and access to legal and political power.

As described earlier, however, whatever their differences both the Walster *et al.* and the Deutsch perspectives begin with the assumptions that justice, as a social device, is generated in the course of people's using their intelligence to gratify their appetites. Justice appears to be a rational way for people to coexist and at times to work together as they go about the business of getting what they want for themselves.

Walster's model of justice as the servant of the person's use of power to maximize private pleasure has considerable difficulty accounting for the patterned and pervasive appearance of the forms of justice and relatively selfless allocation patterns that Deutsch has described. Similarly, it is difficult to apply Deutsch's functional dominance of human intelligence to what occurs in the competitive world of commercial, political, and legal transactions. In these areas, justice does appear to be a servant of the exercise of power.

3.   NEED FOR AN ENCOMPASSING THEORETICAL PERSPECTIVE

What is most interesting, however, is that neither of these models, nor any of the others that have employed similar assumptions about the appearance of justice as an instrumental device designed to facilitate the acquisition of resources, is able to encompass the unique qualities of justice in human affairs.

3.1.   Justice in Social Institutions

No other value than justice can legitimize the intentional sacrifice of all other human ends and values, including liberty and human life. Every institutional arrangement in our society is legitimized and framed within rules of entitlements and rights: what is fair and just, who is entitled to what from whom under what circumstances. Justice plays a similarly central and dominant role in the lives of most individual members of our society. Certainly, the commonly available examples of the willingness of obviously sane people to risk their own welfare and conceivably sacrifice their lives in order to see to it that justice is served cannot be fit easily within the instrumental assumptions that the commitment to justice is dependent on its facilitating the acquisition of desired resources. More to the point, however, there is considerable evidence from systematic observation that, in fact, the most singularly "desired resource" is that people have what they deserve.

3.2. Justice as It Defines What Is a "Desired Resource"

According to the available evidence, there is no amount of desired resources—money, prestige, power, etc.—that is sufficient to be considered an acceptable fate, if it is less than people believe they deserve. And at the other extreme, there is no amount of deprivation and suffering that is unacceptable, or an undesirable outcome, if the person does not consider it an unjust fate (see Crosby, 1976). For example, one of the most striking findings that has been carefully documented recently is the extent to which people will alter their self-esteem in order to believe they deserve their fate, even when, by any objective assessment, that fate has been inflicted on them by external forces beyond their control (Apsler & Friedman, 1975; Comer & Laird, 1975; Lerner & Miller, 1978; Rubin & Peplau, 1973). Other investigators (Bulman & Wortman, 1977) have documented the way people inflicted with an "objectively" terrible fate (e.g., physically crippled as the result of an accidental injury) typically remove the experience of injustice associated with their fate by discovering fully compensating rewards, usually spiritual in nature, associated with the injury.

There is no doubt that people wish to maintain a positive self-image and pursue their self-interest in the most effective manner; however, what is also true is that these goals are framed within the more general commitment to deserving and justice. Rather than being an instrumental device to facilitate the acquisition of desired resources, justice appears as the guide for assessing what "resources" are desirable.

3.3. Centrality of Justice in the Acquisition and Allocation of Resources

If one examines the vast array of data generated by social scientists studying the behavior of children and adults in situations where they have been given the power to allocate resources generated by a common endeavor, the single dominant finding is that they use this power so that each worker gets what she or he deserves, even if that means allocating relatively little to themselves. There is also evidence that some people, when not under public scrutiny, will keep somewhat more for themselves than they "deserve." But this clearly is a weak response in both magnitude and frequency (see Walster *et al.*, 1978; Berkowitz & Walster, 1976; Leventhal, 1976). The experimental data also mimic familiar observations. For example, if the subjects are explicitly instructed to maximize their own outcomes, they will engage in self-interested tactics and use their power to bargain for the maximum share. But even this self-maximizing behavior occurs in normatively patterned ways. The contestants experience their self-interested acts as "justified" and "fair" (Leventhal, 1976).

Of course, it is possible to construe these findings as manifestations of the enlightened expression of self-interest as in the earlier models. But two important facts need to be emphasized. *The patently dominant behavior reveals a commitment to rules of justice. The interpretation of this pattern as a form of self-interest is an indirect theoretical inference* of the processes underlying that behavior. Second, *the evidence for the unilateral use of power to maximize one's outcomes is extremely weak,* both experimentally and in terms of observational evidence. Of course, people do at times engage in self-maximizing strategies, but this behavior appears typically in those situations that are defined as arenas of competition, where it is normatively legitimate–fair–just to use one's power to enhance one's own interests (Lerner, 1971).

If this analysis is valid, then we are led to the question of how to explain the uniquely central role that justice plays in people's lives. It is conceivable that through the accident of our Judeo-Christian history or the adaptive wisdom of our civilization (Campbell, 1975), rules of justice have come to play a central role in our normative system and are thus transmitted to and implanted in the members of each succeeding generation. There are, however, many reasons to doubt that there are specially effective devices for transmitting to our children the rules of justice—and especially the commitment to those rules—rather than any other set of moral norms. On the other hand, Piaget (1965) and many others have pointed to strong evidence for the natural invention of rules of fairness and justice, as well as a commitment to those rules, out of the interaction of the children's cognitive capacities with the contingencies in their environment, especially their experiences with others. Of course, much is adopted from the culture, but the commitment to the various forms of justice as they appear in our society is most probably an interactive effect of cognitive structure and environmental contingencies.

4.   WHAT SOCIAL-PSYCHOLOGICAL PROCESSES COULD ACCOUNT FOR THE WAY JUSTICE APPEARS IN PEOPLE'S LIVES

The most important step in developing an adequate theory of justice is to recognize that the traditional assumption that people are continually and centrally concerned with the process of maximizing their outcomes (call it drive reduction, or profit maximizing, or pleasure enhancing, or pain avoidance) must inevitably lead to a model that fails to capture the unique qualities associated with justice in human affairs. The main reason is that by beginning with that assumption of human motivation, justice must be construed as a more-or-less useful social invention. The commitment to justice remains continually dependent on its demonstrated utility to the person or the social unit, to be abandoned or shaped in the pursuit of maximizing profit. As a consequence, there is no way of encompassing the most obvious fact about justice: the central role it plays in our culture and in the lives of most individuals.

Once having arrived at that conclusion, we have achieved something of a breakthrough in the study of justice. It is now possible to take a fresh look at what social-psychological processes might be involved in explaining the way the theme of justice appears in people's lives.

4.1. Maturation of Cognitive Processes: The Understanding of "Impersonal Causation"

As it turns out, there are some encouraging possibilities. We can all probably agree that people are born with desires and wants and that something happens in the course of development that alters the way they go about meeting their desires. In terms reminiscent of the earlier discussion, it appears that the child becomes more "rational" and intelligent in the efforts to get what he or she wants. But we can say much more than that about these cognitive processes. Things of enormous relevance to the understanding of justice occur as the child becomes able to retain and symbolically represent ways of acting that are associated with different particular outcomes. As a consequence, the child understands that there are causal processes—ways of acting with predictable outcomes—that are impersonal in the sense that they are available to use by others. In other words, there are ways of operating on the environment that are more-or-less effective, and as the child becomes "wise," the child learns to analyze the possibilities presented in a given situation for the most effective alternative way of acting in order to achieve a valued goal. Typically, this analysis requires that the child view the situation not only in terms of immediate possibilities but from a longer time perspective in order to decide what course of action would be most beneficial in the long run.

4.1.1. People as Occupants of "Positions"

In becoming able to act on the basis of what we could consider more rational or "enlightened" self-interest, the child has accomplished two things that are of critical importance. One is having learned to view desired outcomes—or outcomes of any kind, for that matter—as the end product of "causal" processes that are impersonal, in the sense that anyone with the requisite abilities could engage in them. The implications are that children can view themselves and others as "actors" engaged in sequences of planned activities rather than simply as unique social objects. The reason that Tommy can hit the ball farther than Jimmy is not that Tommy is simply older or a better person than Jimmy; it is because Tommy has bigger muscles and better coordination or because he has practiced longer and developed his muscles and coordination. In fact, how far *anyone* can hit a ball depends on the extent to which she or he develops the requisite skills. Presumably, along with other factors, that practice requires willingness, the choice to invest the necessary amount of time and effort. So, the first important event that occurs is that as children learn about impersonal processes (sequences of cause and effect for operating on the environment), they are able to view themselves and others as actors or as occupants of positions in these sequences, people with more or less practice and skill at doing or preparing to do X.

4.1.2. The Development of the "Personal Contract"

The second event that occurs is that the child gains an important (but not the earliest) experiential referent for deserving and entitlement. As the child goes through the transition of becoming "wise," there is a conflict between the aspect that desires immediate gratification and the enlightened, intelligent, problem-solving aspect. What happens for most children in the normal course of events can be described metaphorically as follows: the intelligent aspect offers the promise to the part of the child that wants immediate gratification that if immediate self-interest is postponed and, in fact, further costs in terms of time and effort are incurred, there will be an appropriately compensating outcome at the end. In a sense, then, the child makes a "personal contract" with himself or herself based on the promise that the outcome will be appropriate to the investments or costs incurred. One can find in this commonly observed sequence of development, which underlies the "personal contract," the experiential prototype for the justice of "equity," in which people's entitlements are based on their relative investments.

4.2. The Basic Templates for Organizing One's View of the World

Prior to and concurrent with the development of the understanding of impersonal processes and the personal contract, something else is happening that is extremely important in understanding how justice appears in people's lives. The child seeks to generate an organized construction of the world. The patterns of early experiences, especially those associated with important and enduring consequences, provide the outline of this construction. Given the child's investment in creating order and continuity, these early outlines provide the integrative structure for processing new information. Thus, with additional experience, they will remain viable if they are sufficiently valid and effective so that their integrity does not disintegrate in the face of overwhelmingly "surprising" or contravening experiences.

4.2.1. The Three Prototypical Events

Interestingly enough, one can find plausible scenarios for the appearance of developmental analogues for three basic "kinds of relations" in the typical experiences of growing up in our culture. The earliest is the *identity relation:* the persistent and powerful empathic experience of sharing the emotional state of others. In particular, this reaction is most likely to appear when others transmit certain affective cues. There are even stronger associations of this "identity" reaction with others with whom one has been closely associated during this early period, such as members of one's family. That is one kind of relation with others.

The other two appear somewhat later, as the child interacts in various contexts with others with whom there is a perception of *similarities* and *differences:* the perception of a *unit* (similar) or *nonunit* (different) relation with someone. Among children's earliest and most common experiences are the organization and definition of important areas of their lives in terms of this kind of categorization. How they are treated by others in their world—parents, peers, etc.—is determined in patterned ways! They experience the same outcomes as those who are defined as the "same" in terms of age, sex, neighborhood, etc. Also, it is equally common and important that they are given "more or less," treated more or less well than those who are "different" than they. Other children who are "similar" to them are more likely to generate positive affect and to act cooperatively, while the "different" ones are typically competitors or hindrances to what one wants. It is quite plausible that these early experiences generate the commonly observed reaction among adults of favoring one's own kind versus "them." Apparently, almost any dimension of similarity–difference is capable of eliciting this preferential treatment of the "in-group" over those from the "out-group." Tajfel (1970) has referred to this phenomenon as the "generic norm."

Presumably, then, these three interpersonally contingent kinds of early experiences provide the basic templates for the way people define and view their relations to others. Initially in their lives, the various meaning elements appear together. The perception of others as "same" (identity), "similar" (unit), or "different" (nonunit), are associated with ways of treating one another: respectively, they are *vicarious dependent*, mutually facilitating *cooperative*, or hindering and *competitive*. And these activities are linked to patterns of outcomes. In identity relations, we increase "their–our" sense of well-being; in unit relations, we share equally; and in nonunit relations, we have more or less than they (see Table 2.1).

4.2.2. Separation and Resynthesis of "Relation" and "Process"

As children mature, they engage in more extended goal-acquisition with others. These experiences generate an important elaboration of the three basic "templates." The comprehension and use of impersonal processes in goal acquisition enable the child to separate the three kinds of activities associated with the perceived relation to others. In other words, the child learns that it is possible and may be desirable to work cooperatively with one of "them," or necessary to compete with one of "us," and, at times, to engage in "identity"-based activities with anyone. To recapitulate a bit, the identity activities are those in which the person has the ability to enable others to acquire a desired resource. Although the person does not have direct access to the resource in question, there is the expectation of vicariously experiencing the other's fate. An obvious example is parents' investing in and then vicariously sharing their children's accomplishments.

In any given situation, however, the person resynthesizes both elements—the perceived relation to the other and the task-relevant acquisition process—into an organized construction of the event. In the synthesis, either the relation or the acquisition process becomes dominant as "figure" (the focus of the activities), while the other becomes "ground"

Table 2.1. Prototypical Experiences

Associated cognitive elements	Identity	Unit	Nonunit
Person perception	Same "me"	Similar "us"	Different "them"
Activity relation to goal	Vicarious dependency	Mutually facilitative convergent	Hindering divergent
Outcomes	Need welfare	Equality equivalence	More less

(the setting of acceptable limits). Once the synthesis is constructed, the resultant cognitions, affect, and activities emerge. They appear in the stream of actions, having the experiential concomitant of a "good gestalt," a sense of requiredness and appropriateness. As portrayed here, the assessment of "who" is entitled to "what" is no less an inherent aspect of any encounter than the recognition of the activities that I must engage in to get what I want, and my perception of who the other person is and how he or she helps or hinders me.

The term emergent activities is intended to capture the elements of immediacy and requiredness associated with the person's organized construction of an interpersonal encounter. The construction includes acts, actors, and assessments of appropriate outcomes. "Who does what" and "with," "to," or "for whom" emerges out of the perceived relation to the other(s), the process involved in goal acquisition, and the relative salience of these elements (figure–ground). The resultant organized construction is experienced as a demand, an impulse to act and feel in a particular way.

4.2.3. Recapitulation: Justice as an Integral Part of Interpersonal Relations

The model presented here of the way justice appears in peoples lives integrates these concerns with the context of a more general conception of the social-psychological processes involved in interpersonal relations. In common with other models, it is assumed that people are self-interested in the sense that they are born with and generate desires and goals. And it is also assumed that the person's rational processes play a part in shaping behavior. However, a core set of assumptions in the present model is that the outcome-maximizing self-interested motivations are radically altered very early in the developmental process. Developing children, no less than adults, must organize their experiences not only in the historical sense but, even more importantly, in the dynamics of each encounter. They create meaning that provides continuity and stability to their activities.

Obviously, then, the person is supremely concerned with maintaining an organized structure of the environment. Given a structure in which the person has such a strong investment, new information is processed and fitted within that organizing framework. The central motivational force probably stems from the need to protect the integrity of this organizational structure. If one is looking for the functional bases of this force, it could easily be posited that in order to operate effectively on the environment, the person needs the order and stability generated by this organized structure.

These assumptions are not at all radical or unfamiliar, but if we are to generate a usable model, we need to provide more information about the nature of the psychological structures and processes that determine the way the person organizes the perceptual inputs. Our initial efforts to find patterns in social judgments about deserving and justice yielded evidence for three kinds of relations that appear in the way we view others. These were defined roughly as (a) an *identity* relation, in which we are psychologically indistinguishable from the other and we experience what we perceive they are experiencing; (b) a *unit* relation, in which we perceive ourselves as distinct from but similar to the other on relevant dimensions; and (c) a *nonunit* relation, based on the perception of being different from the other in ways that are meaningful and have value (Lerner, 1975, 1977).

We also described important developmental events whereby the child learns and is able to manipulate sequences of activities that generate a conception of impersonal "processes," or ways of acting and producing outcomes that are independent of any given individual. Not only does this conception enable the child–adult to develop a "personal-contract" orientation to deserving his or her outcomes, but, in addition, it provides a way of viewing his or her own and others' activities. People can be seen in "positional" terms, as actors located at one or another stage in an impersonal activity. Once the child understands and can make use of the information that there are alternative ways of acting in any given context, then the behavior of others can be assessed in terms of their choices in electing a given sequence. In addition, their behavior at any given moment can be viewed as reflecting their position in an elected or imposed process (Lerner, 1977; Piaget, 1965).

What is proposed here, then, is that in any encounter with others, people attend to and process cues of who they are in terms of "same," "similar," or "different," as well as the kinds of activities required to accomplish their goals: either they must experience their goals vicariously through the other, or their goals are convergent, or their goals are divergent, requiring competitive acts. These two elements (relation and acquisition process) are integrated so that the more salient aspect generates the focus of organization, and the other aspect provides the limiting conditions. The considerations of how one acts, how one thinks and feels, and what is an appropriate way to allocate resources emerge as a meaningfully organized pattern.

5. EXAMPLES OF THE "EMERGENT ACTIVITIES": HOW JUSTICE APPEARS IN INTERPERSONAL ENCOUNTERS

If there is any validity to this conceptual system, we should be able to take even these crude definitional statements and propositions and, by using some of the familiar meaning elements associated with these terms, generate recognizable descriptions of *emergent activities*. Unfortunately, we simply do not have an adequate language system for designating more precisely those processes subsumed under the labels of *vicarious dependent* or *unit-related–convergent goals*. Nor can we do any better at this point with the perceptual categories that frame how we perceive others: *identity, unit,* or *nonunit*. Fortunately, we do have enough understanding of these structural elements and the processes to begin generating meaningful descriptions of each of the ways in which they can appear in our lives (see Table 2.2).

In scanning Table 2.2, the psychological set of the person is derived from the perception that I and the "other" are not distinct psychologically, we are the "same;" or we are "similar" in all essential respects; or we are "different." Thus, I feel impelled to arrange matters so that our needs are met (identity), or that we have the same outcomes (unit), or that our fates are different (nonunit) and they correspond to our relative status (typically, I get more than you). These are not rational considerations; they are simply felt as appropriate, natural ways for things to happen.

At the same time, I perceive the activities required to gain the desired resource: the expected outcomes in the encounter. The activities, however, involve one of three general modes of responding. Either I have the power to enable you to get the "desired" resource, or what is required is that we integrate our activities to gain a commonly desired resource, or in order for either of us to gain and possess the resource, we must do so at the cost of what the others want in the situation.

Depending on which set of considerations is most important to the person in a situation (the relation to the others, or the gaining of the resource), the person will perceive others and respond to them within a problem-solving set, viewing them in relatively impersonal terms as occupants of positions in an acquisition process, or as "kinds" of people who merit certain kinds of outcomes regardless of the particular acquisitional activity in which they happen to be engaged.

5.1. Conditions of Divergent Goals

Let us look at some of the examples in Table 2.2. Those of particular relevance to this volume are the ones often associated with conditions of "scarcity," in which there are mutually divergent goals, and people perceive that they must compete for the same scarce resource. If one looks down the third column (nonunit), it is clear that there are six modes of responding—what I have termed *emergent activities*—that may appear. If one perceives an "identity" with the others, and if that is the dominant concern, one will find quite naturally the appearance of apparently self-sacrificial activities. In fact, psychologically, for each it does

Table 2.2. Summary: Emergent Activities (EmAc)

Process

Relation to others	Identity vicarious dependency	Unit convergent goals	Nonunit divergent goals
Identity "same"	Nurturant[a] Os ultimate welfare *(upper)* / Meets[b] Os needs *(lower)*	Collective oriented *(upper)* / Individual oriented commune *(lower)*	Utilitarian decisions *(upper)* / Heroics martyred self-sacrifice *(lower)*
Unit "similar"	Mutual responsiveness *(upper)* / Reciprocity *(lower)*	Cooperative relative contribution equity *(upper)* / Team effort parity *(lower)*	Parallel competition Justified Self Interest *(upper)* / Formal contest *(lower)*
Nonunit "different"	Evaluate Os acts correspond with Ps goals *(upper)* / Judge Os personal worth correspond with Ps values *(lower)*	Contractual relations mock equity *(upper)* / Status-consistent division of labour *(lower)*	Regulated conflict maximize legal outcomes *(upper)* / "Fight" maximize differential outcomes *(lower)*

[a] Process is dominant in the upper right half of each cell.
[b] Relation to O is dominant in the lower left half.

not matter who actually has the "resource" in the sense that they all share in it vicariously. When the resource itself becomes the dominant concern, then one finds "utilitarian" decisions appearing. Tacitly or explicitly, people elect ways to minimize the total amount of deprivation experienced over a given time period by all members concerned. Given the problem-solving set, the concern is the ultimate welfare of the members, not their immediate needs.

Where the bond between the members is not one of "identity" but is a "unit" relation, involving the perception of being "similar" or equivalent in essential respects, then under conditions of divergent goals, a form of "competition" takes place. If the dominant concern is maintaining the perception of being a "unit," then one finds procedures designed to minimize the divergent consequences of the competition—as in "contests," where at the end there are rituals that portray each of the contestants as equally important. Where the resource in question is the dominant concern, then what emerges is a form of interaction based on the assumption of "equal opportunity" or access in the acquisitional process. Typically, this interaction takes the form of "parallel" or "indirect" competition as students or employees—often friends—vie for the relatively rare fellowship, promotion, etc. Norms of "justified self-interest" apply here, where each is allowed and expected to gain what she or he can within the rules of fair competition. This is experienced as a "fair" process for allocating scarce resources among people who are entitled to be treated equally.

When the others are perceived as "different," and that nonunit perception dominates the situation, then the appropriate way to respond is to attempt to maintain one's superior status over "them" at whatever cost. If, however, the resource acquisition is of primary concern, then simply doing as much as possible on one's own behalf to gain power over the resource appears appropriate. In both situations, the issue of dominance and subordination shapes what emerges. In the former, the attempt is to maintain dominant status and control, while in the latter, the focus is dominance over the resource in question. In either case, any set of outcomes or acts that would lead "them" to have more of either the status or the resource would be construed as unfair, unjust, and inappropriate—an "outrage" to the sense of decency and fairness.

5.2.   Conditions of Convergent Goals

There are, of course, other ways in which people can respond to the objective conditions of decreasing availability of desired resources. It is conceivable that the members could perceive these occasions as ones in which they have common interests rather than divergent goals, and thus the process would be one of mutually facilitating interaction. In that case, the emergent activities that appear in the second column of Table 2.2 ("Unit Process") are pertinent. If the members perceive each other in "identity" terms, then, depending on whether resource acquisition or the relation is dominant, one will find either a collective or a communal way of responding. The distinction is that in the former, the environment is viewed and the participants are defined in terms of understanding the relevant processes required to gain the generally desired resources. In the latter, the common endeavor is to facilitate each member's uniquely defined goals.

When there is the perception of similarity or equivalence rather than difference, what emerges are cooperative activities. Where the members' similarity is dominant, they act as a "unit," a team, and naturally allocate resources equally. Where the focus is on the acquisitional activity, then they view each other in positional terms: each member's distinct contribution in the common enterprise is emphasized. Naturally, then, the commonly acquired resource is allocated according to each member's relative contribution.

Finally, when the perception of "nonunit" relations or difference is maintained, there can still be cooperative or mutually facilitating activities. But these differ from the others in that dominance and inferiority are maintained and generalized to the roles played by each member in the common activity (when relation is dominant). However, if the acquisitional activity is the primary concern, then there are bargaining and contractual agreements designed to provide sufficient organization to be effective on at least a temporary basis. The participants nevertheless believe that they are entitled to more than "them." Typically, in our society, one finds a form of "mock" equity appearing in these contractual agreements. In order to achieve the amount of cooperation necessary to be an effective unit, the "different" participants agree that all members' outcomes should correspond to their relative contributions to the common enterprise. But, of course, each member feels that he or she is entitled to more than that because he or she is working with "them."

5.3. Conditions of Vicarious Dependency

It is also conceivable that under conditions of economic scarcity, many people will find themselves on one or the other end of an "identity" process, involving one person's dependence on the other for a desired resource. There is no mutuality or direct interdependence of their fates, at least in that situation. When the others are perceived in identity or unit terms, the powerful member's behavior is easily identifiable from the labels in the first column of Table 2.2. However, when the powerful, as is often the case, see the dependent as "them" (in nonunit relations), then what appears is the use of power to shape the behavior or reward the attributes of "them" that resemble the values with which the powerful identify. For the powerful to act otherwise would be an unjust failure to do their duty as decent citizens.

6. SUMMARY OF WHAT WE "KNOW" AND NEED TO KNOW ABOUT THE JUSTICE MOTIVE IN A COMPLEX SOCIETY

So far, what has been presented is a way of understanding the motivational and cognitive bases for the commitment to justice. In this model, the perception of who is entitled to what appears as an intrinsic part of the way people organize their experiences and respond to their environment. The common observation that what is considered "just" can vary radically is derived from an analysis of the "templates" people use to organize their experiences in terms of "kinds" of people and processes involved in acting on the environment.

If this analysis is valid, then we now understand something more about the motivational and cognitive bases underlying the way people treat, or could react to, one another. For various reasons, I will not discuss in any systematic manner those conditions—personal–environmental, situational, and historical—that elicit the perception of a particular kind of relation to others (identity, unit, or nonunit), or the relevant acquisition process, or what determines which will be dominant (relation or acquisition process).

The main reason is that in a very real sense, the rest of this volume is designed to speak to these issues in one way or another, for example: Under what conditions are people likely to perceive an identity relation with one another or a nonunit relation? What will determine when the anticipation or experience of decreased availability of resources leads to the perception of convergent and cooperative efforts rather than divergent goals and competitive if not combative activities? Although there is a great deal that is already known about those factors and has been described elsewhere (Lerner, 1978; Lerner & Whitehead, 1980), the chapters in this volume examine these issues within the context of various social institutions and arrive at important and often highly creative contributions to our understanding.

Before turning to these specific contributions, however, there is one other theoretical issue that needs to be touched on. Assuming that we do know a great deal about the occurrence of the sense of justice in its various forms and that in each instance it is experienced as a compelling aspect of how the person reacts in that situation, what we have not discussed are the "facts of life" that people in our society must build into their interpersonal relations.

One of these is that the consequences of any important activity have implications for many people other than those directly involved on that occasion. Typically, the person has prior commitments and/or antici- pated future commitments that generate a demand for a given level of resources. For example, in order to meet the "needs" of the family, the worker must obviously generate a certain level of pay or job security. This demand will shape the way the person is able to interact with others on the job. As a result, the person may employ "defensive" or functional ways of viewing others, so that the way she or he must treat them does not violate the sense of what is just and fair.

As a result, in our society one often finds people engaging in in- stitutionalized or privately prearranged strategies to ensure that they will perceive others in "positional" terms, where process is dominant, and as having a nonunit relation with them rather than a unit or identity relationship. The nonunit relation enables people to see others in ways that legitimize working cooperatively with them on a "contractual" basis. But then one can act unilaterally to maximize one's outcome whenever possible. This is a way of reducing the possibility of failing to meet the felt obligations to those with whom one has prior and enduring "identity" relations. It also prevents the development of additional iden- tity or unit relations that might create further demands.

As a result of this common strategy, people recognize that in their work activities and economic exchanges, all are out to maximize their own gain, and to look out for their own private interests. Obviously, though functional in some ways for the individual, this strategy for enabling one to avoid violating the sense of what is just and fair leads to a feeling of estrangement, if not distrust, of most other members of the society. This attitude could be most problematic under conditions of an economic crunch, where genuinely cooperative endeavors and even utilitarian decisions are required if all are to adapt successfully. In fact, the stage is set for the emergence of the normative response that everyone is entitled to engage in the self-interested defense of one's own kind against "them."

It is most important to remember that people remain capable of perceiving a strong sense of identity with people other than their own immediate family and that there are many occasions on which people feel a sense of similarity (unit relation) with others. Although our eco- nomic institutions and traditional social arrangements may require that people carefully control and tailor the occasions on which those percep- tions are likely to arise, they remain a psychological resource of enor- mous power for eliciting alternative ways of coping with conditions of scarcity. On the basis of what we now suspect to be true about the central role of justice in people's lives and the conditions under which people experience various forms of justice in their relations with others, there is every reason to believe that *there are many ways of adapting to conditions of scarcity that are available within the human potential. It would be just as natural for people to work cooperatively for one another's welfare as to fight over the dwindling resources. It just depends. . . .*

REFERENCES

Apsler, R., & Friedman, H. Chance outcomes and the just world: A comparison of observers and recipients. *Journal of Personality and Social Psychology*, 1975, *31*, 884–894.

Berkowitz, L., & Walster, E. (Eds.). Equity theory: Toward a general theory of social action. *Advances in Experimental Social Psychology*, Vol. 9. New York: Academic Press, 1976.

Bulman, R. J., & Wortman, C. B. Attributions of blame and coping in the "real world": Severe accident victims react to their lot. *Journal of Personality and Social Psychology*, 1977, *35*, 351–363.

Campbell, D. On the conflicts between biological and social evolution and between psychology and moral tradition. *American Psychologist*, 1975, *30*, 1103–1127.

Comer, R., & Laird, J. D. Choosing to suffer as a consequence of expecting to suffer: Why do people do it? *Journal of Personality and Social Psychology*, 1975, *32*, 92–101.

Crosby, F. A model of egoistical relative deprivation. *Psychological Review*, 1976, *83*, 85–113.

Deutsch, M. Equity, equality, and need: What determines which value will be used as the basis of distributive justice? *Journal of Social Issues*, 1975, *31*, 137–149.

Harris, R. J. Handling negative inputs: On the plausible equity formula. *Journal of Experimental Social Psychology*, 1976, *12*, 194–209.

Lerner, M. J. Justified self-interest and the responsibility for suffering: A replication and extension. *Journal of Human Relations*, 1971, *19*, 550–559.

Lerner, M. J. The justice motive in social behavior: introduction. *Journal of Social Issues*, 1975, *31*, 1–19.

Lerner, M. J. The justice motive in social behaviour: Some hypotheses as to its origins and forms. *Journal of Personality*, 1977, *45*, 1–52.

Lerner, M. J. The justice motive in social behavior: Hypotheses as to its origins and forms II. 1978. Research Grant Proposal, the Canada Council (Social Sciences and Humanities Research Council) 410-77-0601-X.

Lerner, M. J., & Miller, D. T. Just world research and the attribution process: looking back and ahead. *Psychological Bulletin*, 1978, *85*, 1031–1051.

Lerner, M. J., & Whitehead, L. A. Procedural justice viewed in the context of Justice Motive Theory. In G. Mikula (Ed.), *Justice in social interaction*. Bern: Huber, 1980. New York: Springer Verlag, 1980.

Leventhal, G. S. Fairness in social relationships. In J. W. Thibaut, J. T. Spence, & R. C. Carson (Eds.), *Contemporary topics in social psychology*. Morristown, N.J.: General Learning Press, 1976.

Piaget, J. *The moral development of the child*. New York: Free Press, 1965.

Rubin, Z., & Peplau, A. Belief in a just world and reaction to another's lot: A study of participants in the national draft lottery. *Journal of Social Issues*, 1973, *29*, 73–93.

Sampson, E. E. On justice as equality. *Journal of Social Issues*, 1975, *31*, 45–64.

Tajfel, H. Experiments in intergroup discrimination. *Scientific American*, 1970, *223*, 96–102.

Walster, E., Walster, G. W., & Berscheid, E. *Equity: Theory and research*. Boston: Allyn & Bacon, 1978.

BASIC PROCESSES

Origins and Development of the Sense of Justice

Since justice concerns will play an increasing role in shaping how people treat one another in the years directly ahead, it is important to assess what we know about the origins of such concerns and the influences that shape their variable development. The three articles in this section address these interests directly.

Simmons provides an overview of the issues and lines of research in the general area of "development of the sense of justice." Her discussion of the major theoretical approaches to the problem—social learning, psychoanalytic, and cognitive—serves as a background for consideration of current issues in the area. These include questions about the functions of role-taking ability and experience in the evolution of moral reasoning, the ways in which justice-based decision-making is culturally and situationally malleable, and the problems of interpretation posed by findings obtained in hypothetical and/or "scrutinized" research situations. By delineating the major unresolved issues pertaining to the development of justice concerns, Simmons sharpens our understanding of what is significant and innovative in research strategies such as those described by Damon. As well, awareness of the ongoing debates in this problem area points up the need for an integrative framework, such as that provided by Karniol and Miller, if we are to evaluate the adequacy of existing research and identify future research needs.

Damon's report of his studies of children's allocation behavior details one researcher's response to the challenges of finding out when and how justice concerns develop. His focus is the positive aspects of children's justice conceptions, those related to sharing and the fair distribution of valued goods. Reviewing his work of the past several years, Damon describes a program of research that has addressed many of the relevant theoretical and methodological issues. His most recent studies, involving "real-life" situations, contribute to our understanding of the complex interrelationships between what children say they would do and what they do in various allocation situations. In conclusion, Damon offers a model of "justice development" that incorporates a number of currently recognized factors, including self-interest and social experience.

Karniol and Miller undertake first the complex task of describing the conditions of arousal of the sense of injustice from a developmental perspective. They then analyze the process by which the individual's concern for his own just treatment evolves into "morality," the generalized concern about social justice. Distinguishing among situations involving commutative, distributive, and procedural justice, they construct an integrative framework within which to view diverse research findings. The child's attribution of responsibility, understanding of intentionality, need for justification, and like concerns are discussed with the aim of determining what we currently know about the interactive influence of such basic variables as socialization, cultural frame, developmental stage, situational variation, and role-taking opportunity on the evolution of a sense of injustice.

3

Theoretical Issues in the Development of Social Justice

CAROLYN H. SIMMONS

1. INTRODUCTION

Interest in the origins and development of social justice principles would appear to be an obvious starting point for the construction of theoretical perspectives on social justice. Everyday observation confirms that children in various cultures demonstrate sharing, assistance, and equity— in short, perceptions of fairness in their relationships with others. Indeed, some of the earliest psychological theories on moral judgment (Piaget, 1932/1965) as well as early research studies on the knowledge of justice norms (Hartshorne & May, 1929) were based on the behavior of children. However, this promising beginning was not followed up in any systematic way; only within the past dozen years or so has there been intensified and expanded investigation into the ways in which children judge and react to each other's fate, develop and refine their notions of various forms of justice, and behave in ways that demonstrate a commitment to deserving in their relationships with others.

It is clear that social justice is a powerful theme in human civilization that has attracted attention from several different theoretical perspectives. Common sense and observation tell us that a child's sense of justice changes with age. But what are the developmental processes involved in the emergence of concepts of justice?

CAROLYN H. SIMMONS ● Department of Psychology, University of Colorado at Denver, Denver, Colorado 80202.

2. SOCIAL LEARNING THEORY

Although social learning theory has not dealt directly with social justice concepts, much of the research on social justice behaviors has used a social learning approach, investigating the effects of modeling, role playing, socialization influences, and other social reinforcers on behavioral demonstration of fair or just behavior. Traditional learning theory tended to focus on overt behaviors in specific situations (see, e.g., Skinner, 1974), while social learning theory as developed by Bandura (1969) and others has taken a broader stance, emphasizing the importance of social context and incorporating such factors as delayed gratification, internalized self-reward, and modeling as variables affecting just behavior (Mischel, 1961, 1966; Aronfreed, 1968; Bryan & Walbek, 1970a,b; Harris, 1971; Midlarsky, Bryan, & Brickman, 1973; Yarrow, Scott, & Waxler, 1973).

The importance of modeling and imitation of significant adults who provide learning cues for children is well-covered by Staub (1975) in his work on the effects of socializing influences on children. As Staub pointed out, children can learn to behave in conformity with the expectations of society, yet also reflect in their behavior societal conflicts among such values as personal ambition, gaining the approval of others, and taking responsibility for the welfare of others. Staub's discussion of nurturance effects on imitation points out that nurturance may have a negative effect on imitation when there is a conflict between the child's self-interest and the demand for sharing limited resources, since the child can interpret adult warmth as a cue that punishment will not be a consequence of failing to follow society's expectations. Therefore, nurturance unaccompanied by firm expectations for socially approved behavior would not be expected to explain learning-based just behavior in many situations. In her influential studies on the effects of parental socialization on child development, Baumrind (1971a,b) reported research illustrating the connection between parental style and the child's developing response to social demands for fairness. According to Baumrind, parental style characterized by warmth, clear expectations of maturity, consistent control of the child, and encouragement of responsible behavior does result in teacher ratings of responsible behavior in preschool children as they interact with peers.

Review of the considerable research literature based on principles of social learning theory leads to the conclusion that learning theorists have creatively studied many kinds of social cues, both direct and indirect, that affect behaviors related to justice. Of concern to some psychologists, however, is the emphasis on observable behavior and the neglect of possible underlying general principles of justice. From a social-learning-theory perspective, social justice becomes only one of many possible reasons for behavior transmitted to children via effective learning situations. As will be discussed later in this chapter, there are some promising indications that concepts of social justice are a dominant theme in child development rather than one of several possible explanations for just behavior.

3. PSYCHOANALYTIC THEORY

A second theoretical approach to the emergence of justice principles derives from psychoanalytic theory. Freud believed that the child developed from a pleasure-seeking organism to a rational individual capable of weighing reality in arbitrating between the demands of raw self-interest and the equally irrational prohibitions of society and parents internalized through the "cultural superego." According to Freud (1961), the first requirement of civilization is a sense of justice. Erikson (1963) has elaborated on this theme, theorizing that the capacity for fairness requires a sense of basic trust in the environment and the child's reliance on the justness of the world about her/him. From Erikson's perspective, the groundwork for a sense of social justice is laid well before the emergence of the superego and thus is based not only on the internalization of social prohibitions but also on rational cognitions about the nature of the social environment and social interaction or transactions.

Despite the paucity of controlled research to specify and confirm the clinical observations on which psychoanalytic theory is based, the model has had a significant effect on current thinking about social justice origins. Of particular importance is the generation of theory based on psychoanalytic concepts of identification and delay of gratification and on Erikson's stages of ego development, which unfold in interaction with influential experiences mediated by significant adults. Although increased understanding of child development from a cross-cultural approach has led us to question the applicability of psychoanalysis to non-Western cultures, the theory does offer insight into the understanding of behavioral dynamics. The challenge for current researchers interested in using this approach to study social justice in children is to specify and operationalize the processes affecting ego and superego development in varying cultural and religious environments.

4. COGNITIVE DEVELOPMENTAL THEORY

4.1. Piaget's Perspective

A third theoretical model for the study of the origins of social justice in children uses underlying processes of cognitive development as the basis for investigating moral development. Jean Piaget was the first to assume a relationship between four cognitive stages and the ability of children to develop a mature sense of social justice. According to Piaget, children in the preoperational stage of cognitive development view rules as externally given, unchallengable, and immutable. Their sense of justice is based on behavioral consequences rather than on the motives or intentions of the other person; and—because children at this stage are unable to consider the perspective of others—they assume that their judgments are shared by all others. The child's egocentrism during this stage of cognitive development is thus assumed to parallel his/her sense of absolute justice in Piaget's first level of moral judgment, called *moral realism*.

As the school-age child begins to apply logic to an understanding of the physical world and to acquire the capability of viewing the world from another's perspective, her/his view of moral decision-making also undergoes change, incorporating a sense of reciprocity. At about the same time that the child is capable of formal operations, moral decisions begin to include *moral relativism*. Judgments of fairness and justice can now take into account the intentions of others as well as the behavioral outcomes of actions. Rules for social behavior are understood to be socially determined and alterable when fair play and majority rule require adjustments in them.

The child is now capable of role taking; considerations of peer perspectives and the demands of social equity supercede the all-or-none justice concepts used earlier when the child had only limited social interaction and experience. Just as the processes of assimilation and accommodation affect cognitive schemata, so are they presumed to affect developing judgments of fair outcomes and equity of consequences (Piaget, 1932/1965).

Having postulated some important assumptions about the relationship between cognitive and moral development, Piaget left it to others to provide expanded empirical support for his ideas and to clarify related issues; included in these are questions of the effects of cognitive conflicts on justice decision-making and the influences of culture and varying peer experiences and relationships on developing concepts of social justice. Hoffman (1970) provided an effective review of such research, confirming a sequence of moral development for children in Western cultures. As research techniques have become more appropriate for studying young children, however, there is increasing evidence that children may be capable of making subtle discriminations, including an awareness of intention, at younger ages than Piaget supposed. Wellman, Larkey, and Somerville (1979) found that awareness of intention could be discriminated from accident by children as young as 3 years old (see also Surber, 1977).

4.2. Kohlberg's Stages of Moral Reasoning

Kohlberg was the next to pick up the challenge of possible relationships between the development of moral reasoning and concepts of
justice in children. Starting in the late 1950s, he developed a series of
moral dilemmas that formed the core of interviews used to study social
justice concepts in children and, later, adults. Each dilemma presents
a hypothetical conflict between the expectations of an established authority and the rights or well-being of individuals. Kohlberg elicited a
choice of hypothetical behavior from his subjects for each dilemma, but
he was primarily interested in searching for the reasoning process from
which the behavior decision was derived. Although Kohlberg's data
base, like Piaget's, was the verbal responses of children rather than
behavior in a real situation, this theory has been highly influential in
generating empirical evidence for the justice perspectives of children.

Kohlberg (1971) postulated three levels of moral reasoning, each
consisting of two stages. This justice structure becomes progressively
more differentiated at each stage as new logic is incorporated in the
individual's reasoning. At *Stage One*, the child's reasoning reflects the
egocentrism and self-interest of Piaget's moral realism stage. A child at
this stage expresses concerns about rewards and punishments as the
definers of just behavior. In *Stage Two*, the child's sense of justice is
tempered by a desire for reciprocal fair exchange and quantitative equality.

Older children (and a majority of adults) function at Kohlberg's
second level. Moral dilemmas are resolved at this level by conformity
to the social order and reference to social laws and norms governing
behavior. The ability for mutual role-taking or "ideal reciprocity" in
specified relationships thus becomes the defining characteristic for *Stage
Three* just behavior. *Stage Four* extends the concept of justice by recognition of a more general social order encompassing individual relationships but maintaining uniform rights and obligations.

Only a minority of adults, Kohlberg believes, reach the third level
of moral reasoning. Here the individual is capable of generating just
social contracts *(Stage Five)*. At *Stage Six*, moral decisions are based on
internally developed universal principles of justice that recognize the
dignity of human beings apart from fixed rights and contracts (Kohlberg,
1963).

Research by Kohlberg and his associates has demonstrated that a
similar order of moral development occurs in such disparate cultures as
Mexico (Kohlberg, 1969), Kenya (Edwards, 1975), and Turkey (Turiel,
Edwards, & Kohlberg, 1978), and among different social classes (Kohlberg & Kramer, 1969). Although the rate and extent of stage responses
varied in these different environments, Kohlberg (1971) has concluded

that the major source of this variation is developmental: the result of role-taking experience in social interaction.

4.3. Social Justice and Role-Taking Ability

More recently, Damon's research has shown a strong parallel between the levels of reasoning used by children in making decisions about sharing and allocation of rewards and the child's ability to express Piagetian stages of cognitive reasoning. According to this perspective, children moving from intuitive problem-solving with no role-taking ability toward concrete operational problem-solving with the ability to take the perspective of others demonstrate a parallel shift in moral reasoning. Using moral dilemmas adapted for young children, Damon (1975) found a progression from self-serving allocations through allocation based on strict equality to considerations of merit and deservingness in allocation decisions.

The relationship between role-taking ability and progression to higher levels of moral reasoning has also been demonstrated by Selman (1971), who concluded that role taking is a necessary condition for the development of Kohlberg's Stage Three moral reasoning. Eisenberg-Berg and Mussen (1978) have extended our understanding of the relationship between just decision-making and role-taking ability by looking specifically at affective role-taking and the responses of adolescents to hypothetical moral dilemmas in which conflicting individual needs and desires could not be resolved by reliance on external laws or authorities. They found that a progression from "immature hedonistic reasoning" to reasoning based on internalized values was correlated positively with both written and behavioral measures of empathy.

4.4. Social Justice and Transformation of Thought

Concurrently, Turiel's research has focused on sequential stages and the increasing capacity for transformation of thought as the child interacts with the environment (rather than changes as a function of the learning of social values as postulated by, e.g., Berkowitz and Daniels, 1964, and Aronfreed, 1968). Using boys and girls from 6th, 9th, and 12th grades, Turiel (1976) administered both a "moral knowledge test" and an interview using Kohlberg dilemmas. He found that the younger children earned higher scores for their knowledge of norms and values regarding social justice, while the older children demonstrated higher levels of moral reasoning. Turiel interpreted these results as an indication that moral knowledge may be more clearly related to early stages of moral reasoning, when society's values are accepted as the standards for behavior, rather than to higher levels of moral reasoning, when internal standards of ethics have developed.

We find, then, that the social justice motive is a central theme in children's behavior that has engendered much attention and debate, and that a sense of social justice and entitlement changes with age.

5. PARITY AND EQUITY

The use of a "justice of parity" as the basis for behavior apparently comes early and requires minimal cognitive complexity (Damon, 1975; Hook & Cook, 1979). At this level, children distribute punishments and rewards so that "we all get the same"; presumably, the perspectives of others are not incorporated into such decision making. As children experience expanded interactions with the environment and develop more sophisticated cognitive schemata and role-taking abilities, a "justice of parity" emerges for team relationship allocations and a "justice of equity" occurs in situations where relative contributions and efforts of independent workers can be specified (Walster & Walster, 1975; Anderson & Butzin, 1978; Hook & Cook, 1979).

How early equity and parity considerations occur in children's allocation decisions is currently under debate. Research has yielded mixed results, with some investigators claiming that kindergarten children are not capable of comparing relative contributions to reach equitable distributions (Masters, 1973; Lane & Coon, 1972), while others interpret their results to conclude that children at this age do use a justice of equity in allocating rewards (Leventhal & Anderson, 1970; Leventhal, Popp, & Sawyer, 1973).

Lerner (1974) has demonstrated that first-grade children and kindergarteners are capable of considering the relative contributions of workers in distributing rewards, whether they are sharing the reward with another worker or acting as impartial supervisors. The significant factor determining whether equity or parity was followed in making distributions was the description of a fictitious other worker as either a team member or simply an independent worker. Under team conditions, children took about half the reward regardless of their relative contribution (reflecting parity), while under nonteam conditions, they took rewards closely matching their input to the work effort (reflecting equity).

A recent study by Nelson and Dweck (1977) shows that situational variables such as instructions for reward distribution are of significance in influencing the behavior of even younger children. Thus, 4-year-old children in their study allocated rewards on the basis of their own contribution without regard to the input of another independent worker when they were instructed to divide the reward any way they wanted to. When the instructions to the children made equity salient by requiring division so that each child got the "right amount" for the work he or she had done independently, equity theory predictions were substantiated whether the child was sharing a reward with a supposed co-worker or was acting as a supervisor dividing a reward between two workers.

6. JUSTICE MOTIVE THEORY

6.1. The Personal Contract

Lerner has proposed that as the child develops, she/he refines the capacity to calculate costs and benefits in anticipating alternative outcomes. If the child has experienced a stable environment that allows for the development of basic trust and the assumption of a just world in which outcomes generally match deserving (Lerner & Simmons, 1966), then the child is willing to postpone immediate gratification of personal desires in exchange for long-term benefits. In other words, the child becomes committed to a *personal contract* that presupposes that people deserve their outcomes and receive—in the long run—what they are entitled to receive, no more and no less. The personal contract made by the child at an early age and reaffirmed throughout adult life thus represents an inner dialogue between the child's wish for immediate rewards and the rational weighing of greater possible rewards that are promised by self-commitment to deserving, entitlement, and social justice.

6.2. The Social Contract

A *social contract* develops as the child comes to recognize the obligation and necessity for other people in his/her world to operate by similar personal contracts. In order to be able to deserve one's own outcomes, the child must adopt procedures that both enable and require the others in the world to do likewise. From this perspective, the concern with justice for others and the forms it takes are derived from and shaped by everyone's enlightened self-interested desire to organize his or her goal-seeking around deserving what is desired.

6.3. Identity, Unit, and Nonunit Relationships

Research using school-age children reveals that children are indeed sensitive to situational cues that indicate when their personal and social contracts are most appropriately served by, for example, a justice of parity or equity or justified self-interest (Long & Lerner, 1974; Braband & Lerner, 1975). These cues include recognition of three kinds of relationships among people that affect the choice of appropriate and just reactions: *identity, unit,* and *nonunit* relationships. Identity relations arise out of an empathic perception of another person, motivating a *justice of need* to reduce the distress of and provide assistance to another. A *justice of entitlement* develops instead when we perceive an identity relationship with the circumstances in which the other is involved (see, e.g., Eisenberg-Berg & Mussen, 1978). Weighing the long-term effects of the situation from a more objective position, we act on the basis of social obligations and the other's deserving.

A *unit* relationship among people assumes either the perception of similarity with the other (yielding a *justice* of *equality* or *parity*) or a similarity with their circumstances and current contributions (resulting in a *justice of equity*). *Nonunit* relations with others derive from either the perception of conflict and competition with another person (thereby eliciting *Darwinian justice*) or the perception of another in a competitive position relative to one's own situation. This conflict can be resolved by *competitive justice,* regulated by applicable laws or rules that allow self-interested behavior (Lerner, 1975, 1977).

Once the child has the ability to understand that there are alternative ways of acting in response to given situational cues, the child can assess his/her relationship to others in selecting an appropriate response to those specific circumstances. Justice motive theory thus assumes the child's need and wish to have predictable bases for reconnoitering the world and for continuing integration of new experiences and circumstances. Thus, a determination of *who* is entitled to *what* depends directly on the perceived relationship among the participants; and there is a clear pattern to the appearance of these perceptions in specific encounters. Identity relations with others apparently occur first, as the child develops a sense of trust in a stable environment and in relationships with caretaking adults. The other two relationships develop later as the child learns to interact in a broader social context and becomes capable of recognizing similar and competing perspectives of peers as well as the identical perspectives shared with a few close others. The child's treatment of "identical," "similar," and "different" others therefore comes to reflect the appropriateness of nurturant assistance, cooperative sharing, or competitive confrontation with others. Further sophistication of the child, of course,. leads to the adaptive realization that relationships can change and therefore alter appropriate behavior toward another; this realization allows the child to compete with a similar other and even to cooperate with someone who is different from the child in some significant ways (Lerner, Miller, & Holmes, 1975).

There is also good reason to believe that children learn the social norms for desired behavior in their culture, and that they recognize the power of adults to enforce those norms, quite early in life. It is therefore not surprising that under conditions of social scrutiny, children delineate the appropriate cues for applying equity or parity or self-interest in making just decisions. But many experimental and real-world situations require that decisions be made and acted on under anonymous conditions, where the child's behavior apparently will never be known to others. Under these circumstances, what theoretical explanation accounts for decision making based on understanding of social justice?

A series of experimental studies on reward allocation has provided some insight into this intriguing question. A typical paradigm involved third- and fourth-grade children divided on the basis of their general willingness to delay gratification (DG), using a technique developed by Mischel (1973). Some days after this measure was taken, the children went singly to another classroom where the experimenter explained that the child would be constructing checkerboards for needy children. Each child believed that another child of the same age and sex was working at the same task in another room. Half of the time, the other worker was referred to as the subject's partner (unit relationship); the other half of the subjects believed that the other worker simply happened to be there at the same time as the subject (nonunit relationship). In reality, there was no other worker.

The experimenter explained to the child that the more checkerboards made as partners (or as independent co-workers), the more reward there would be to divide between them. Relative productivity was created by stopping the child's work when either two or four checkerboards had been completed and informing the child that the other worker had completed either four or two checkerboards in the same time (2:4 and 4:2 conditions). The experimenter then left six dimes for the child to divide with the other worker under conditions in which it was made clear that no one would know how the child actually allocated payment. After making the allocation in private, the child was asked to rate for the experimenter his/her own and the other worker's productivity, effort, and deserving.

Results indicated that both relative work contribution and the defined relationship between the children (unit or nonunit) affected the allocation by children with a commitment to a personal contract. Almost all the "high DG" children rated greater productivity by an independent co-worker as deserving of higher reward. When the other worker was their "partner," however they were likely to see the other as deserving the same reward regardless of relative contribution. Children with low DG scores generally indicated that they and the other worker deserved the same amount of pay regardless of productivity or relationship to each other. Actual allocations of money closely paralleled the ratings of deserving for all children (Lerner & Meindl, 1979).

In a later variation of this study, where there was clear adult surveillance of reward allocation, "low DG" children were more likely to follow the justice-of-equity rules in rating a more productive other worker as deserving more than they and in dividing the money. Only under this public condition did the low DG children report and allocate money on an equity rather than parity basis when it was costly to them to do so (Lerner & Meindl, 1979).

These findings reveal that children at this age, regardless of their preference for delay of gratification, know that there are two rules of deserving and that in a co-worker situation, the equity rule is more appropriate, while in a team relationship, the parity rule is more appropriate. But only the more mature (high DG) children are committed strongly enough to a personal contract to use a justice-of-equity allocation under anonymous as well as public conditions when it lessens their own reward. Future research will explore more fully some related hypotheses regarding predicted patterns of behavior that are elicited by children's perceptions of the interaction of relationship and situation.

7. QUESTIONS FOR FUTURE STUDY

Although the first consideration of children's perceptions of social justice occurred approximately 50 years ago, there are still many unanswered questions for interested researchers to pursue. Greater understanding of social justice development in children appears close at hand with emerging interest in three research questions.

7.1. How Early Are Children Capable of Different Types of Justice-Based Decisions?

Adaptation of research techniques to make them appropriate for younger children is still needed. We do not yet have a clear answer to this question, although it is becoming apparent that even preschool children can discriminate intention from accident and evaluate consequences of behavior (Surber, 1977; Wellman, Larkey, & Somerville, 1979). Other justice perceptions may occur earlier than previously thought, also: Miller and McCann (1979) have found that even first-grade children are capable of using discriminations between intentional and accidental perpetrators of injustice in making decisions regarding deserved compensation for victims of those injustices.

7.2. What Range of Responses Are Children Capable of in Expressing Their Ideas of Social Justice?

Midlarsky has often urged the development of experimental designs that would allow children to demonstrate a broader range of skills and resources in responding to the needs and deservingness of others (see, e.g., Midlarsky & Midlarsky, 1970). Mischels and Mischel (1975), too, have urged research to investigate not only what the child *would* or *could* do under varying circumstances, but also what the child *does* do when faced with a nonhypothetical situation. A great deal of the published literature on social justice in children and adolescents has required only that children respond verbally to hypothetical situations or that they behave while under the observation of adults. Several researchers have creatively used employment–work paradigms to investigate justice rules for reward allocation by children and/or willingness to donate rewards to anonymous other children of varying deservingness. However, the development of new experimental strategies would be useful in the investigation of other forms of justice and of closer parallels to the complexities of justice demands in the real world. It seems especially appropriate in a society that expects and assumes just behavior in adults without constant surveillance, that more research involving children use experimental situations requiring behavior under apparent nonsurveillance circumstances.

7.3. What Are the Effects of Prevalent Strategies in Different Cultures on Children's Social Justice Behavior?

With increased sensitivity to the importance of socialization influences on children's behavior has come the need to investigate the effects of values and institutions of other cultures and religious traditions on children's formulations of personal and social contracts. Too, as resources become scarcer and the world more interdependent, we need to consider what forms of justice will become most necessary to children and adults in Western Judeo-Christian cultures. Is there, as Kohlberg (1971) claimed, a single set of universal ethical principles, including justice, that differ in form but not content from culture to culture? If so, how can we specify the optimal cultural conditions for the full development of social justice? Or, as Riegel (1978) claimed, is it not really possible to remove psychological research and theory from the specific cultural and historical contexts in which they develop? There are exciting challenges ahead for researching the expression of social justice in children and contributing to this central theme in human behavior.

REFERENCES

Anderson, N., & Butzin, C. Integration theory applied to children's judgments of equity. *Developmental Psychology*, 1978, *14*, 593–606.

Aronfreed, J. *Conduct and conscience*. New York: Academic Press, 1968.

Bandura, A. Social learning through moral judgments. *Journal of Personality and Social Psychology*, 1969, *11*, 275–579.

Baumrind, D. Current patterns of parental authority. *Developmental Psychology Monographs*, 1971, *4*, 1–101. (a)

Baumrind, D. Harmonious parents and their preschool children. *Developmental Psychology*, 1971, *4*, 99–102. (b)

Berkowitz, L., & Daniels, L. Affecting the salience of the social responsibility norm: Effects of past help on the response to dependency relationships. *Journal of Abnormal and Social Psychology*, 1964, *69*, 275–281.

Braband, J., & Lerner, M. A little time and effort . . . who deserves what from whom? *Personality and Social Psychology Bulletin*, 1975, *1*, 177–181.

Bryan, J., & Walbek, N. The impact of words and deeds concerning altruism on children. *Child Development*, 1970, *41*, 747–757. (a)

Bryan, J., & Walbek, N. Preaching and practicing generosity: Children's actions and reactions. *Child Development*, 1970, *41*, 329–353. (b)

Damon, W. Early conceptions of positive justice as related to the development of logical operations. *Child Development*, 1975, *46*, 301–312.

Edwards, C. Societal complexity and moral development: A Kenyan study. *Ethos*, 1975, 505–527.

Eisenberg-Berg, N., & Mussen, P. Empathy and moral development in adolescence. *Developmental Psychology*, 1978, *14*, 185–186.

Erikson, E. *Childhood and society* (2nd ed.). New York: W. W. Norton, 1963.

Freud, S. *Civilization and its discontents*. New York: W. W. Norton, 1961.

Harris, M. Models, norms and sharing. *Psychological Reports*, 1971, *29*, 147–153.

Hartshorne, H., & May, M. *Studies in the nature of character*. New York: Macmillan, 1928–1930.

Hoffman, M. Moral development. In P. Mussen (Ed.), *Carmichael's manual of child psychology*. New York: Wiley, 1970.

Hook, J., & Cook, T. Equity theory and the cognitive ability of children. *Psychological Bulletin*, 1979, *86*, 429–445.

Kohlberg, L. The development of children's orientations toward a moral order. I: Sequence in the development of moral thought. *Vita Humana*, 1963, *6*, 11–33.

Kohlberg, L. Stage and sequence: The cognitive–developmental approach to socialization. In D. Goslin (Ed.), *Handbook of socialization theory and research*. Chicago: Rand McNally, 1969.

Kohlberg, L. From is to ought: How to commit the naturalistic fallacy and get away with it in the study of moral development. In T. Mischel (Ed.), *Cognitive development and epistemology*. New York: Academic Press, 1971.

Kohlberg, L., & Kramer, R. Continuities and discontinuities in childhood and adult moral development. *Human Development*, 1969, *12*, 93–120.

Lane, I., & Coon, R. Reward allocation in preschool children. *Child Development*, 1972, *43*, 1382–1389.

Lerner, M. Justified self-interest and the responsibility for suffering: a replication and extension. *Journal of Human Relations*, 1971, *19*, 550–559.

Lerner, M. The justice motive: "Equity" and "parity" among children. *Journal of Personality and Social Psychology*, 1974, *29*, 539–550.

Lerner, M. The justice motive in social behavior: An introduction. *Journal of Social Issues*, 1975, *31*, 1–20.

Lerner, M. The justice motive in social behavior: Some hypotheses as to its origins and forms. *Journal of Personality*, 1977, *45*, 1–52.

Lerner, M., & Meindl, J. *The justice motive and self-interest: Some factors influencing the emergence of "equity" and "parity" among children*. Unpublished manuscript, University of Waterloo, 1979.

Lerner, M., Miller, D., & Holmes, J. Deserving vs. justice: A contemporary dilemma. In L. Berkowitz (Ed.), *Advances in experimental social psychology*, Vol. 8. New York: Academic Press, 1975.

Lerner, M., & Simmons, C. The observer's reaction to the "innocent victim": Compassion or rejection?" *Journal of Personality and Social Psychology*, 1966, *4*, 203–210.

Leventhal, G., & Anderson, D. Self-interest and the maintenance of equity. *Journal of Personality and Social Psychology*, 1970, *15*, 312–316.

Leventhal, G., Popp, A., & Sawyer, L. Equity or equality in children's allocation of reward to other persons? *Child Development*, 1973, *44*, 753–763.

Long, G., & Lerner, M. Deserving the "personal contract" and altruistic behavior by children. *Journal of Personality and Social Psychology*, 1974, *29*, 551–556.

Masters, J. Effects of age and social comparison upon children's contingent and non-contingent self-reinforcement. *Child Development*, 1973, *44*, 111–116.

Midlarsky, E., Bryan, J., & Brickman, P. Aversive approval: Interactive effects of modeling and reinforcement on altruistic behavior. *Child Development*, 1973, *44*, 321–328.

Midlarsky, E., & Midlarsky, M. Aiding under stress: The effects of competence, status and cost to the aider. *Proceedings of the Annual Convention of the American Psychological Association*, 1970, *5* (Pt. 1), 439–440.

Miller, D., & McCann, C. Children's reactions to the perpetrators and victims of injustice. *Child Development*, 1979, *50*, 861–868.

MIschel, W. Preference for delayed reinforcement and social responsiblity. *Journal of Abnormal and Social Psychology*, 1961, *62*, 1–7.

Mischel, W. Theory and research on the antecedents of self-imposed delay of reward. In B. Maher Ed.), *Progress in experimental personality research*. New York: Academic Press, 1966.

Mischel, W. Processes in delay of gratification. In L. Berkowitz (Ed.), *Advances in experimental social psychology*, Vol. 7. New York: Academic Press, 1973.

Mischel, W., & Mischel, H. Moral behavior from a cognitive social learning viewpoint. Society for Research in Child Development meeting, 1975, Denver.

Nelson, S., & Dweck, C. Motivation and competence as determinants of young children's reward allocation. *Developmental Psychology*, 1977, *13*, 192–197.

Piaget, J. *The moral judgment of the child*. Glencoe, Ill.: Free Press, 1965. (Originally published 1932.)

Riegel, K. E. *Psychology mon amour: A countertext*. Boston: Houghton Mifflin, 1978.

Selman, R. The relation of role taking to the development of moral judgment in children. *Child Development*, 1971, *42*, 79–91.

Skinner, B. *About behaviorism*. New York: Knopf, 1974.

Staub, E. *The development of prosocial behavior in children*. Morristown, N.J.: General Learning Press, 1975.

Surber, C. Developmental processes in social inference: Averaging of intentions and consequences in moral judgment. *Developmental Psychology*, 1977, *13*, 654–665.

Turiel, E. A comparative analysis of moral knowledge and moral judgment in males and females. *Journal of Personality*, 1976, *44*, 195–208.

Turiel, E., Edwards, C., & Kohlberg, L. Moral development in Turkish children, adolescents and young adults. *Journal of Cross-Cultural Psychology*, 1978, *9*, 75–85.

Walster, E., & Walster, G. W. Equity and social justice: An essay. *Journal of Social Issues*, 1975, *31*, 21–44.

Wellman, H., Larkey, C., & Somerville, S. The early development of moral criteria. *Child Development*, 1979, *50*, 869–873.

Yarrow, M., Scott, P., & Waxler, C. Learning concern for others. *Developmental Psychology*, 1973, *8*, 240–260.

4

The Development of Justice and Self-Interest during Childhood

WILLIAM DAMON

1. INTRODUCTION

Anyone who spends time around young children eventually will encounter the ringing refrain of "That's not fair!" Beginning as early as the preschool years, fairness plays a central role in every child's social life. Particularly in its breach, it is a notion of which even young children are acutely aware. The outraged reaction of a 2-year-old to being denied a turn on a swing or a share of a cookie will attest to this early sense, at least as evoked in personally experienced injustices.

Fairness is important in children's social lives for the same reason that it is essential to us as adults: it is the best means of resolving conflicts concerning rights and claims between persons. When parties agree that a certain resolution of a conflict is fair, there is no longer a conflict. In contrast, when one or more of the parties considers the resolution unfair, the conflict will continue brewing, however submerged it may remain for necessary pragmatic purposes. It is in this sense that "fair" resolutions are the best ones. In their interactions with peers and adults, children soon come to realize the unique power of fairness in regulating social conflict between persons and therefore in facilitating the maintenance of cohesive social relations.

WILLIAM DAMON • Department of Psychology, Clark University, Worcester, Massachusetts 01610. The research described in this paper was supported (in part) by a grant from Carnegie Corporation of New York.

Childhood social relations, of course, are quite different in character than adult social relations, despite their common reliance on fairness as a social regulator. For this reason, there is a change in children's use and understanding of fairness as they grow into adulthood. In fact, there are a number of changes, because the social relations of children, along with the children themselves, are continually developing all through childhood and beyond. Therefore, although justice fulfills an essential social function for human social life in any form, its meaning and specific nature undergo systematic changes throughout the various phases of human social development.

For the past several years, I have studied the development of children's understanding of justice (Damon, 1973, 1975, 1977). Some of this research has focused on the punitive or retributive aspects of justice (Selman & Damon, 1975); but my major interest has been in the "positive" aspects of children's justice conceptions, in particular those aspects related to sharing and the fair distribution of resources. Strangely, the major theories of moral development (Piaget, 1932/1965; Kohlberg, 1963; Rawls, 1970) have overlooked childhood manifestations of justice like sharing, and consequently, they have mistakenly placed the developmental roots of childhood morality in a stage of unilateral obedience to adult authority, as if justice were a notion confined to the world of the adult. As I noted above, any caretaker who has witnessed children arguing about fairness could attest to the adult-centered blindness of such formulations. My hope has been that the study of childhood sharing and resource distribution would reveal more accurately the origins of morality in humans. These origins can be observed in the rich understanding of fairness, kindness, and other prosocial concepts that children display throughout the course of their earliest attachments and peer relations (Hoffman, 1975).

My initial studies of children's positive-justice conceptions explored subjects' verbal responses to hypothetical stores and dilemmas. This proved to be a fine means of determining the cognitive basis for the development of justice in children, but it left unanswered the question of how cognition about justice relates to children's everyday social behavior. In later studies (Damon, 1977; Gerson & Damon, 1978), I extended the scope of my investigations to include children's social conduct during real-life, peer-group situations centering on problems of fairness. The data from these studies quickly revealed that there are complexities in children's justice development that go beyond children's cognitive understanding of fairness. One of these complexities is children's understanding of their own self-interest, another social-cognitive phenomenon that interacts and sometimes conflicts with their understanding of justice. This interaction may take a variety of shapes, each of which has a particular and striking influence on children's social development and conduct. Other complexities include the nature of the physical and social context in which the child operates, the child's interpretation of that context (another cognitive phenomenon related to justice and self-interest), and the child's general developmental capacity and potential. In this paper, I shall describe a model of justice development that tries to take into account the interaction of these various factors. I shall make reference to a previous description of this model (Gerson & Damon, 1978) and to some data from previous studies (Damon, 1977); and I shall also introduce some new findings that some further reanalyses of the data from the earlier studies have yielded.

2. THE DEVELOPMENT OF CHILDREN'S HYPOTHETICAL REASONING ABOUT JUSTICE

In order to investigate the general developmental course of children's justice conceptions, I have constructed a number of verbal justice problems that are appropriate for children in the age range between 4 and 12. These problems present issues of fairness to children in a story context that is familiar and comprehensible to the children. Picture cards illustrating the various characters in each story are also shown to children as comprehension aids. A sample positive-justice story, along with probe questions, follows:

> All these boys and girls are in the same class together. One day their teacher let them spend the whole afternoon making paintings and crayon drawings. The teacher thought that these pictures were so good that the class could sell them at the fair. They sold the pictures to their parents, and together the class made a whole lot of money.
>
> Now all the children gathered the next day and tried to decide how to split up the money.
>
> 1. What do you think they should do with it? Why?
> 2. Kathy said that the kids in the class who made the most pictures should get most of the money. What do you think?
> 3. Andy says the kids who made the best ones should get the most. What do you think?
> 4. There was a lazy kid in the class, Rebecca, who didn't draw very much in comparison with the others. What about her?
> 5. Jim says that the best-behaved kids should get more than the rest.
> 6. Melissa says the girls should get more than the boys. Why would she say that? Is she right? John says that the boys should get more. Is that fair? Why/why not?
> 7. Lisa says that the poor kids should get the money, because they don't have much.
> 8. Billy, here, comes from a very poor family and doesn't get an allowance. What should the class do about him?
> 9. Someone said the teacher should get the money, because it was her idea to sell the pictures. What do you think?

10. What's the fairest way to divide up the money?
11. What should the kids do?
12. Should anyone get more than anyone else?

Remember Billy, he's the poor kid who never gets any allowance. Sometimes he asks Miss Townsend, the teacher, for a free candy bar at school. Some of the kids tell Miss Townsend that she shouldn't give Billy a free one, because they have to pay a dime for theirs, and it's not fair that he should get one free.

13. What should Miss Townsend do?
14. Someone else suggests that Billy should do a chore or some extra work to earn the candy bar. Is that fair?
15. What about having a contest to give him a chance to win it?
16. What's the fairest thing to do about Billy? Is that fair to the other kids who have to pay? To Billy? To the teacher?

Remember there was a lazy kid in class: Rebecca. Now, Rebecca is very smart. She never studies or does her homework, because she is so lazy. But she always gets all the answers right on tests [or for younger subjects, when Miss Townsend asks her questions]. Peter, here, is just the opposite. He works very hard, but he's not as smart and usually makes lots of mistakes.

17. Miss Townsend has to decide whom she should give the best mark to for schoolwork [or for younger subjects, whose paper she should put the most stars on], Rebecca or Peter. What should she do?
18. Why does a teacher give marks [put stars on papers]? Is that a good reason?
19. What should Miss Townsend do? What's fairest to Rebecca and Peter? How should Miss Townsend decide?

In administering this and other interviews to children, we follow the tenets of the "clinical method" (Piaget, 1928). Since the interviewer's goal is to "test the limits" of the child's understanding, the wording and order of the questions may vary somewhat from child to child. The youngest subjects in the 4–12 age range sometimes have difficulty with concepts like money, and so we make appropriate substitutions where necessary (for example, using scoops of ice cream as the reward).

In a number of studies (Damon, 1973, 1975, 1977), we have found children's responses to this and similar interviews to conform to the following sequence of developmental levels. These levels are closely age-related, as has been confirmed by both cross-sectional and longitudinal data. The age norms noted below are for middle-class North American populations and are only approximate. Fuller descriptions of the levels, as well as more extensive empirical data, are given in Damon (1977).

3. BRIEF DESCRIPTIONS OF EARLY POSITIVE JUSTICE LEVELS

1. *Level 0-A* (age 4): Positive justice choices derive from the wish that an act will occur. Reasons simply assert the choices rather than

attempting to justify them ("I should get it because I want to have it").

2. *Level 0-B* (ages 4–5): Choices still reflect desires but are now justified on the basis of external, observable realities, such as size, sex, or other physical characteristics of persons (that is, "We should get the most because we're girls"). Such justifications, however, are invoked in a fluctuating, after-the-fact manner and are self-serving in the end.

3. *Level 1-A* (ages 5–7): Positive justice choices derive from notions of strict equality in actions (that is, that everyone should get the same). Equality is seen as preventing complaining, fighting, "fussing," or other types of conflict.

4. *Level 1-B* (ages 6–9): Positive justice choices derive from a notion of reciprocity in actions: that persons should be paid back in kind for doing good or bad things. Notions of merit and deserving emerge.

5. *Level 2-A* (ages 7–10): A moral relativity develops out of the understanding that different persons can have different, yet equally valid, justifications for their claims to justice. The claims of persons with special needs (that is, the poor) are weighed heavily. Choices attempt quantitative compromises between competing claims.

6. *Level 2-B* (ages 8–12): Considerations of equality and reciprocity are coordinated, so that choices take into account the claims of various persons and the demands of the specific situation. Choices are firm and clear-cut, yet justifications reflect the recognition that all persons should be given their due (though, in many situations, this does not mean equal treatment).

Sample quotes from six children responding at the six developmental levels are offered below as a means of illustrating the respective modes of justice reasoning. Each quote, of course, represents only one of many possible manifestations of the various reasoning levels.

0-A: Suppose you and Sammy are playing together and you have these (five) toys? Would you give him any? *I would give him these two.* Why those two? *Because I got to keep three. These are the ones I like.* Suppose Sammy said, "I want to have more"? *If he took one then I would take one back from him.* Why is that? *Because I want three.* What will Sammy do then? *He will say that's OK because he likes these (toys).*

0-B: (Keeps four blue chips and gives Jenny two white chips). You would keep all these blues for yourself? *Yes, because I like blue. And then I'd play with them.* Let's pretend Jenny said, "I like blue." Would you give her any blue? *Never, because I have a blue dress at home.* So you wouldn't give her any blue at all. Is it fair to do it this way? *Ah, ha, I've got it. I'd give her two of the white: I'd give her those two because she's younger than me, and I get four because I'm four.*

1-A: Do you think anyone should get any more than anyone else? *No, because it's not fair. Somebody has 35 cents and somebody has 1 penny. That's not fair.* Andy said that he thought the one who made the best stuff should get the most money. *The best person who makes the best stuff is not polite, because you should make them have the same alike—give everything the same. The same, because it's not polite when you give people the most and they [the rest] don't have one. It's fair to the other children that they have to get it, too.*

1-B: But Kathy thought that she should get the most because she made the most things. Do you think that's fair? *Well, if she made more things, then she'd get more money.* What if she didn't. *If she didn't want to make anything, then she wouldn't get as much as all of these kids.* Well, what about the poor kid here, who doesn't have any money to begin with? *Well, the poor kid should make some stuff, then he'd get more.* What about the lazy kid? *Well, he shouldn't get as much if he didn't work as much, if he didn't do his work.*

2-A: What if Kathy made more stuff? Should she get more money? *Oh, about 7 more pennies. It depends on what she made—If she made something easy or hard.* What if she made something hard? *About 10 cents more.* What about Andy, who made the best stuff. Should he get more? *Well, maybe he should. But since she made more, she may have some good ones, so then maybe he can get around 5 cents more.* What about Billy, who doesn't get any allowance? *He should only get about 3 cents more because—if he got a lot more, he might even have more than anybody adding up their allowance.* What about these others? *No, because they don't have such a big reason.*

2-B: Should the kids who were most cooperative get bigger shares? *No! Because that really doesn't make much sense. They are not in a contest about attitude and how you share with other people. They don't care about that. They just want to have people do good stuff, they don't care about that.* How about giving bigger shares to the poor kids? *No. They [the class] don't care if they [the poor kids] are poor or not. Well, we might feel a little sorry for them. But they don't care about that. They just want the ones who did the best to get the most money.* And why is that? *I just said, that way they'll all try to do better next time.*

The sequence of justice levels described above are clearly related to children's general intellectual development. In reasoning about positive justice, children call on their abilities to catagorize the world, to order the world, to establish reciprocal relations between events in the world, to quantify goods and rewards. In this manner, children's logical thinking is reflected in their justice reasoning; and in one investigation (Damon, 1975), it was found that children's logical abilities do indeed develop side-by-side with their justice conceptions. That is, subjects who were more advanced in logical thinking (as measured by Piagetian tasks) tended also to express justice conceptions reflective of the higher levels, and vice versa for subjects who were less advanced in logical thinking. Of course, both logical and justice reasoning were also closely associated with age, but significantly less so than with each other.

Because children's justice reasoning, as measured by hypothetical stories and verbal interviews, is so closely related to children's general cognitive abilities, one wonders how important this type of reasoning is to children's real-life feelings and behavior. Does a child's verbal response to a hypothetical justice story indicate anything important about the child's conduct or thinking in actual social encounters, or is it merely an intellectual exercise with no real-life implications or influence? This has long been a problematical question for moral-development research, since the main body of work (e.g., Kohlberg, 1969) has focused solely on subjects' verbal conceptions; and those who have considered conduct in addition to judgment (e.g., Hartshorne & May, 1928, 1930) have generally failed to establish meaningful relations between the two. In order to extend my investigation of justice development during childhood into the realm of real-life reasoning and conduct, I have turned to laboratory studies of children engaged in actual peer-group interactions.

4. CHILDREN'S PEER-GROUP BEHAVIOR IN A DISTRIBUTIVE JUSTICE SITUATION

A total of 144 boys and girls, 36 at every other age from 4 through 10, were placed in the following "real-life" distributive justice situation (Damon, 1977). It should be noted that each subject participated in this situation with two same-age peers (who were also subjects in the study) and one younger child (who was not a subject of the study). The younger child was younger by two years in the 6-, 8-, and 10-year-old groups, and by one year in the 4-year-old group. Attempts were made to ensure that the younger child was brought from a different population than the three subjects and was hence unknown to them. Also, care was taken to ensure that each experimental group consisted of two boys and two girls.

The four original children were brought into a room with two adults. In the room was a table with an array of beads and strings on it, and six small chairs were set around the table. The children were told by one of the adults that he liked bracelets made by little children. He then asked the children if they would make some for him.

The four children worked for about 20 minutes making bracelets. The two adults helped the children string and tie up the bracelets upon request. At the end of about 20 minutes—or when the group had made an assortment of bracelets—the adults stopped the bracelet making and took the children from the room. The children were separated, and the younger child was thanked for his/her efforts and taken home. The younger child in all cases had made the fewest, or the least-completed, bracelets.

The remaining three children were brought back into the room, and each was asked to place the bracelets he/she made in front of him/herself. The child who had made the most was then congratulated and was told that the experimenter also thought that his/hers were very pretty. This child's name was then written on a blackboard next to the words "pretty and most." A second child was then picked out as being the biggest boy or girl present (whichever was appropriate). This child's name was then written on the board next to the words, "biggest boy (or girl)." The third child was also thanked and was told his/her bracelets were nice, too. His/her name was written on the board next to the word "nice."

The three children were again taken from the room and were brought back one at a time. As each was brought back, he/she was told that the experimenter was grateful for the necklaces that the children had made. As a reward, he said, the children were to receive 10 candy bars, but the children themselves had to decide how to split them up between themselves. Each of the three subjects in turn was then asked individually for a decision and interviewed about that decision. The three children were then brought back as a group and asked to reconsider their individual decisions. They were then asked to reach a group consensus. Each child's distribution decisions were considered his/her "conduct." Accordingly, there were three choice points at which children's justice reasoning and conduct were recorded: before the child was placed back in the group, at the beginning, and at the end of the group session.

There were three experimental conditions in this situation, though in a sense, the conditions were self-assumed by the children. The three conditions were those of "pretty and most," "biggest boy or girl," and "nice"; each of these criteria represented a different kind of claim. The truest claim to merit was considered "pretty and most," since this claim represented payback for meritorious achievement.

Candy bars were distributed to the three subjects—and to the non-present younger child at a later time—according to the group's decision. In those cases where the group was unable to arrive at a consensus, a compromise was arranged by the experimenters, although this compromise was not recorded as part of the children's own behavior.

In addition to the basic experimental condition, there were a number of control groups, some of which were run subsequently to the initial write-up of the study. The most important of these for the present account was a group of 36 children who, instead of being placed in the experimental real-life situation, were asked hypothetically what they would do in such a situation and then were asked to distribute cardboard candy bars to imaginary fellow participants in the hypothetical story. For both the experimental and control subjects, our standard positive-justice interviews were administered either one month before or one month after the bracelet-making experience (real or imaginary), in a counterbalanced design.

In some respects, the results from this experimental study showed that children's hypothetical reasoning about justice is indeed related to their social behavior in actual justice situations. First, it was found that children's justice reasoning in the experimental situation was age-related in much the same way as described in the sequence of developmental levels derived from hypothetical positive-justice measures. In the experimental situation, the youngest subjects reasoned at 0-A or 0-B, the oldest ones at 2-A and 2-B, and so on throughout the age levels. The order, sequence, and quality of the conceptual development was therefore identical in hypothetical and real-life contexts.

One difference, however, was striking. Subjects' reasoning in the experimental situation was often lower (by one or two levels) than was their reasoning on the hypothetical situation. Rarely was it higher; in many cases (50%) it was the same, but in those 50% of cases in which there was a difference, it was almost always in the direction of the hypothetical reasoning's being higher.[1] Yet this "lag" between the two did *not* obtain in the control condition in which the subjects were asked only hypothetically about the bracelet-making situation. We may conclude, therefore, that there was something about the "real-life" nature of the experimental context that was responsible for the reasoning lag (rather than, for example, a difference in story complexity or other extrinsic factors).

To explain this lag between hypothetical and real-life justice reasoning, I suggested the notion of self-interest, defined cognitively as the subjects' conception of how to maximize one's own advantage (meaning, in this case, one's own share of the reward). In a positive-justice situation, it is in a child's self-interest to use a lower rather than a higher level of justice reasoning, since the lower levels tend to be more egocentric and self-serving. For this reason, it is not surprising that children demonstrate more advanced positive-justice reasoning while considering imaginary rewards than real ones. But it is also interesting that children's conceptions of their self-interest did not manifest themselves in an unmitigated way in our experimental situation.

The children who showed lower reasoning in the real-life setting did so in a modified way, with levels only one or two levels below their hypothetical reasoning. And, as noted above, one-half of the subjects showed no lag at all. Therefore, there was some constraint operating on their notions of self-interest, a constraint in the form of their conceptions of justice. Indeed, there was direct correlational evidence ($r = .78$, $p < .001$) that children's hypothetical justice reasoning was associated with their reasoning in the experimental situation, despite the lag be-

[1] The only age group that did not show this "lag" in the experimental condition was the 10-year-old group, the oldest in the study. The contrast between the 10-year-old group and the other subjects was significant in this regard ($p < .05$). As will be noted later, this markedly greater consistency between hypothetical and real-life reasoning in the oldest group may indicate that the interaction between justice and self-interest conceptions in children changes dramatically with development.

tween the two. It seems clear from the pattern of results that children's reasoning in the experimental justice situation was the product of an interaction between their conception of justice on the one hand and their conception of self-interest on the other.

In this study, further evidence of the justice–self-interest interaction can be seen in the relations between children's hypothetical reasoning and their conduct in the experimental situation. By "conduct" is meant the children's actual choices about how to distribute the 10 candy bars among the four bracelet-makers. Let us first consider the findings relating to subjects' *initial* choices at the beginning of the group session (prior to the peer debates). The main findings were as follows: (1) All children in the study gave *some* candy bars (at least one) to each of the four participants. (2) Nevertheless, there was a significant tendency at all ages for children to favor themselves. Overall, subjects gave an average of 29% of the candy bars to themselves and an average of 23% to each of the other participants. (3) The reverse was true of the control group: children in the imaginary bracelet-making–cardboard-candy condition gave themselves an average 18% of the fake candy and others an average of 27% each. (4) Although the tendency to favor the self in the experimental condition was present at all ages, it was significantly greatest ($p < .001$) among children whose hypothetical justice reasoning was 0-A or 0-B. 0-A subjects gave themselves an average of 39% of the 10 candy bars, with some of these children trying to take as much as 7 candy bars for themselves. (5) Unlike lower-level subjects, children whose hypothetical reasoning was 1-B or higher tended to favor themselves most when they were in the "merit" condition (having been the participant who made the prettiest and the most bracelets). Conversely, the lower-level hypothetical reasoners favored themselves most when they were in the more diffuse condition of being called "nice." These contrasting relations were all found to be significant ($p < .001$) on a series of analyses of variance. (6) Higher-level children (2-A and 2-B) were the ones most likely to give the nonpresent, "incompetent" younger child a share of the reward nearly equal to that of the other participants.

In these results, we can once again see the interplay between justice and self-interest. That justice reasoning had an influence on children's distribution choices is demonstrated by the finding that 0-A and 0-B children were the ones who chose the most self-oriented distributions, and by the finding that the most advanced justice reasoners were the ones who were most considerate of the special circumstances surrounding the younger participant's poor bracelet-making performance. It is also noteworthy in this regard that no subject ever neglected to give at least one candy bar to each participant. This is even more remarkable when it is remembered that these findings pertain to subjects' *initial* choices, before their actual confrontations with their peers.

Yet, the clear influence of children's self-interest on their conduct must also be acknowledged. The striking contrast between the experimental and control group performances demonstrates the instigator of self-interest in the experimental condition: it seems obvious that the presence of real candy bars made all the difference. Again, however, self-interest had its constraints. Although children of all ages tended to favor themselves in the experimental condition, there was a moderation in this self-favoring. This moderation is evident not only in the finding that subjects never tried to give everything to themselves, but also in the revealing interactions that were uncovered by our analyses of variance.

In particular, two interactions between reasoning level and position of the self stand out. The first was the finding that children at 1-B and above, unlike other subjects, were particularly likely to favor themselves when they were in the merit position; the second that 0-A and 0-B children, unlike others, showed a particularly strong propensity to favor themselves when they were in the "nice" position.

Taken together, these two findings indicate that, though self-interest does not disappear with development, its nature changes. The claim of merit (payback for having made pretty and most) was more accessible to children at 1-B and above than to younger children, and therefore, it was used disproportionately at the higher levels in constructing self-serving distributions. This is not surprising, since it is at level 1-B that notions of merit, earning, and reciprocal payback are first truly understood. The 0-A and 0-B children, on the other hand, seemed to be particularly influenced by the claim of "nice," a more general and diffuse form of praise. Probably because "nice" did not represent real merit or investment, the older children did not take it seriously, even when it was in their self-interest to do so. For the younger children, however, it represented the one comprehensible and believable criterion with which they could distinguish themselves favorably from others. It seems that justice reasoning, though never abrogating the self-interest factor, does determine the kinds of self-interested choices that a child is likely to make, or, phrased differently, that advanced justice reasoners find a different way to favor themselves than do less advanced ones—a way that is partially constrained by their more sophisticated notion of fairness.

The explanatory model proposed here is one of a developmental differentiation and subsequent interaction between justice and self-interest conceptions in the child. Early in development—at levels 0-A and 0-B—there is radical confusion between the child's conception of what is fair and what serves his/her own interest. Choices at these levels are therefore self-oriented in an exaggerated way, although still not totally so. As the child's conceptions of both justice and self-interest develop in their own distinct directions, the child realizes that fairness often means denying the self. At this point, self-interest becomes external to the child's sense of justice, and the child's distribution choices may reflect a more articulated moral orientation. But, although now distinct, self-interest still has some influence on justice reasoning and choices, as we have witnessed in the present set of findings. This influence may take the form of retarding justice reasoning in a real-life setting (as in the first findings described above), or it may take the form of encouraging a child to support a self-serving reward-distribution with higher-level reasoning. In either case, the influence of self-interest seems to become increasingly modified by justice constraints as the child's conception of each develops. Here we may recall the finding that only the 10-year-old group did not show the hypothetical–real-life reasoning lag, as well as the interactional findings regarding children's initial choices in the experimental situation.

Further insights into the concomitant development of justice and self-interest may be culled from our experimental subjects' *final* distribution choices at the end of the group session. This was the choice that each subject finally settled on after discussion and debate with peers. Again, the most revealing aspects of subjects' final choices were the differences between younger and older subjects' conduct. First, one very simple finding may be noted: there was an increasing ability, with age, of children to reach agreement with their peers. In the oldest groups (the 8- and 10-year-olds), all children reached some sort of consensus by the end of the group session. This consensus, in some cases, resulted from all three children's coincidentally agreeing from the start, and in other cases, it resulted because one or more children in the group changed their decisions for the sake of agreement. In either case, for none of the older groups was the experimenter forced to arrange a compromise for the children. On the other hand, 8% of the 6-year-old groups and 22% of the 4-year-old groups failed to reach a consensus among themselves, no matter how long they argued or disagreed. It should be noted that all groups were given as much time as they needed to come to an agreement, and the experimenter encouraged agreements by refusing to distribute candy bars until the group was in accord about how the bars were to be given out. Only after it was clear that there would be no progress toward a consensus did the experimenter step in. In spite of all this encouragement, a sizable number of younger children failed to arrive at a group agreement.

One is immediately led to ask how our older subjects were able to establish a consensus more easily than the younger subjects. We may perhaps infer from this finding that children in general do seem to "get along better" with age, if we take *getting along* to mean agreeing with each other more easily. But we still have not learned how they come to accomplish this. Of course, we might speak broadly of increasing "social sensitivity," or of developing "role-taking ability." But I think that a closer look at children's specific final choices in this distributive justice situation may provide us with another explanation for the developing ability of these children to reach agreements. Table 4.1 shows the raw number of children who constructed equal versus unequal distributions, broken down by age group (4, 6, 8, and 10) and by choice point (before the group session, at the beginning of the group session, and at the end of the group session). An equal distribution was one in which 25% of the candy bars were given to each of the four original participants; an unequal distribution was one in which at least one child was given more than 25% of the candy. (Percentages rather than raw number of candy bars was used as a measure because some children refused to distribute all 10 candy bars).

As can be seen in Table 4.1., there are two similar and dramatic movements in the proportion of equal to unequal solutions among the children in this study. The first movement coincides with the progressive increase in age represented by the four age groups: the older children construct equal solutions more often than do the younger ones, at each of the three choice points. Similarly, there is a continual increase in equal distributions that coincides with the amount of time the children spend in the group session. This increase occurs at each of the four age levels. In other words, the longer the child spends in the group, the more likely she or he is to distribute the candy equally, regardless of age. The two movements toward equal distributions reinforce each other, so that all 72 of the oldest two ages of children constructed equal distributions by the end of the group session.

Table 4.1. Number of Subjects Constructing Equal and Unequal Distribution at Each of Four Age Levels, and at Each of Three Choice Points

	Choice point					
	Before group session		At start of group session		At end of group session	
Age	Equal	Not equal	Equal	Not equal	Equal	Not equal
4	7	29	6	30	15	21
6	19	17	22	14	24	12
8	22	14	32	4	36	0
10	28	8	34	2	36	0

To understand what is happening here, I think that we must consider for a moment the function of equality. Equality is, of course, one of the fundamental aspects of justice, though it certainly does not, in itself, represent the most advanced form of justice principle. Many of the older children in this study showed themselves capable, prior to the group session, of constructing more sophisticated kinds of justice solutions than simple equality, for example, 1-B solutions that employ the notion of merit (reciprocal payback for investment of talent or work) and 2-A solutions employing the notion of benevolence (inequality in the service of special need, or to remedy prior deprivation). Although the use of equality as a means toward fairness is a notion that must develop (that is, it is more advanced than the purely selfish 0-A or 0-B solutions), children by the age of 5 or 6 normally have consolidated this notion and are ready to construct more sophisticated justice principles. Therefore, the age-related movement toward equality as the final solution, as shown in Table 4-1, may seem somewhat puzzling at first glance.

But even though strict equality is fairly primitive as a justice principle, it nevertheless is an excellent way to reach agreement in a group. Equality is a perfect leveler of differences, a means of compromise *par excellence*. In an equal solution, a person's full "deserving" may not be recognized; and yet, no one's claims are to any great degree ignored. In most peer-group distribution situations—as in the one used in this study—it is to everybody's advantage to reach some consensus, since only then may the disputed reward be distributed. The accomplishment of the older children in our study was to recognize that, whatever their personal beliefs about fairness, it was equality that would work best in this group situation. Their selection of a primitive justice principle for the sake of compromise and group consensus was their means of "getting along" with their peers in this situation.

Clearly, not all groups, in childhood or elsewhere, are regulated by principles of equality. In other types of social interaction, other kinds of regulating principles are more appropriate. Youniss's work (1975, 1980) on kindness and affirmation in relation to friendship shows how various types of reciprocity are primary as social regulators in most peer interactions. But what the experimental findings reported above do suggest, I think, is that children continuously develop organizing principles to regulate their social lives, and that they use these organizing principles with increasingly greater effectiveness as they experience social interactions. The principles may be drawn from the conceptions of justice (which are particularly appropriate in a distribution situation), or from conceptions of more practical social considerations (like affection and social status). In either case, the principles may serve either justice or self-interest, or some interaction between the two. The role of social experience in encouraging a child's construction and effective use of such organizing principles is illustrated in the finding (shown in Table 4-1) that participation in the group session itself instigated a greater proportion of equal solutions across all age groups. Here we see influence of social experience on children's developing abilities to understand and organize their interacting conceptions of justice and self-interest.

5. DESCRIPTION OF FURTHER RESEARCH

My current use of experimental positive-justice situations is to explore more systematically the influence of social experience—an influence that we only glimpsed during the study described above. I am currently placing a new group of subjects in the bracelet-making situation, but now the purpose is to observe the dynamic influence of children on each other, rather than to examine the relation between hypothetical and real-life moral judgment. Accordingly, I have added some new features to the basic experimental design. Most importantly, the constitution of the groups is arranged in certain ways prior to the group experience. Based on the results of a pretest, children are matched with children at their own hypothetical justice level, or with children either more or less advanced than themselves. During the actual group debate, we look for various kinds of social conflict between the children that may be expressed. For example, children may disagree about their choices concerning how to distribute the candy bars. Or there may be differences between the levels of justice reasoning used by children to justify their choices. Some groups are rife with all kinds of conflict, whereas others come to an agreement almost immediately. The question under investigation is, What effects (if any) do the various kinds of social experience (nonconflictual versus conflictual in one or another way) have on children's justice conceptions and conduct?

My operating hypothesis is that it is exactly through such peer experiences that children work out new and better ways of dealing with social concerns like justice. Now, in the course of this one minor experience, I do not expect to find great changes in children's justice knowledge or behavior; I would be surprised, for example, if many of my experimental subjects showed changes in justice reasoning as great as a developmental level (say, from 1-A to 1-B) beyond the average changes showed by control subjects. But more subtle shifts from pre- to posttests are measurable, and it is these that I am particularly looking for. For example, I shall determine whether experimental subjects are using a greater proportion of higher-level notions than matched control subjects, even though the experiment subjects may not have completely reorganized their thinking by a whole stage. Similarly, I shall look to see whether vestiges of lower-level thinking may be dropping out of experimental subjects' reasoning disproportionately to control subjects. In addition, it is interesting not only to look for change between a pre- and a posttest, but also to measure change from the beginning to the end of the group session itself. Here, we may find small but potentially important changes in subjects' orientation to the justice problem, for example, an increased awareness of contradiction, a greater self-questioning, or a tentative attempt at new solutions. To capture these partial and tentative movements, I have had to develop some new observational and interview measuring-techniques that focus more on the dynamic nature of children's developing justice knowledge than solely on stable "stagelike" behavioral patterns. Although results from my current study are not yet in, it is this general research direction, I believe, that is most likely to capture the active and ever-growing character of children's senses of justice and self.

REFERENCES

Damon, W. *The development of children's conceptions of justice*. Paper presented to Society for Research in Child Development, 1973.

Damon, W. Early conceptions of positive justice as related to the development of logical operations. *Child Development*, 1975, *46*, 301–312.

Damon, W. *The social world of the child*. San Francisco: Jossey-Bass, 1977.

Gerson, R., & Damon, W. Moral understanding and children's conduct. *New Directions for Child Development*, Vol. 2. San Francisco: Jossey-Bass, 1978.

Hartshorne, H., & May, M. S. *Studies in the nature of character*, 2 vols. New York: Macmillan, 1928–1930.

Hoffman, M. The development of altruistic motivation. In D. Depalma & J. Foley (Eds.), *Moral development: Current theory and research*. New York: Wiley, 1975.

Kohlberg, L. The development of children's orientations toward a moral order. *Vita Humana*, 1963, *11*, 1–32.

Kohlberg, L. Stage and sequence: The cognitive-developmental approach to socialization. In D. Goslin (Ed.), *Handbook of socialization theory and research*. Chicago: Rand McNally, 1969.

Piaget, J. *The child's conception of the world*. New York: Littleton, Adams, 1928.

Piaget, J. *The moral judgment of the child*. New York: Free Press, 1965. (Originally published 1932.)

Rawls, J. *A theory of justice*. Cambridge, Mass.: Harvard University Press, 1970.

Selman, R., & Damon, W. The necessity (but insufficiency) of social perspective taking for conceptions of justice at three early levels. In D. Depalma & J. Foley (Eds.), *Moral development: Current theory and research*. Hillsdale, N.J.: Lawrence Erlbaum, 1975.

Youniss, J. Another perspective on social cognition. In A. Pick (Ed.), *Minnesota Symposium on Child Psychology*, Vol. 9. Minneapolis: University of Minnesota Press, 1975.

Youniss, J. *Parents and peers in social development*. Chicago: University of Chicago Press, 1980.

5

Morality and the Development of Conceptions of Justice

RACHEL KARNIOL and DALE T. MILLER

1. INTRODUCTION

Social philosophers from Aristotle to Rawls have argued that justice is the primary value underlying all morality. Nevertheless, our understanding of how the concern with justice guides human behavior is far from complete. The explication of this relationship is made especially difficult because it is not easy to define justice. In fact, it often appears that *justice* is only definable in contrast with *injustice*. When we speak of a just act, we generally mean that the act has remedied or prevented an unjustice. From this perspective, the pursuit of justice can best be defined as the active process of remedying or preventing what would arouse the sense of injustice (Cahn, 1949). In this chapter, we attempt to describe how the sense of injustice is aroused and how the pursuit of justice becomes a moral value.

Our exposition is guided by two assumptions. First, it is proposed that the sense of injustice is not innate but is acquired socially. Our second assumption is that the sense of injustice first manifests itself in reactions to one's own experiences and that it is only with experience

RACHEL KARNIOL ● Department of Psychology, Tel Aviv University, Ramat Aviv, Israel. DALE T. MILLER ● Department of Psychology, University of British Columbia, Vancouver, British Columbia, Canada V6T 1W5. The writing of this chapter was completed while the authors were summer participants at the Center for Advanced Study in the Behavioral Sciences.

and cognitive maturation that the sense of injustice becomes aroused by the experiences of others. The transition from self-concern to concern about social justice is the *sine qua non* of the process of becoming moral. Justice and morality, then, are intimately intertwined.

In our discussion, we distinguish among three types of justice. Commutative justice (Smith, 1759) is the act of refraining from doing harm to someone's person or property. Distributive justice involves the fair allocation of the goods and services that are available. Related to distributive justice is procedural justice, which involves the means by which principles of allocation and rule making are decided on. We will discuss each of these types of justice in turn.

2. COMMUTATIVE JUSTICE

We indicated that there are two spheres in which commutative justice applies: (1) harm to one's person and (2) harm to one's property. We will first examine the arousal of the sense of injustice in the case of personal harm. Harm to one's person may be caused by the infliction of either physical or psychological pain. For the sake of exposition, we will focus on the experience of being subjected to physical pain. The experience of physical pain always arouses negative affect. Such an experience may frequently lead to a sense of injustice, but it does not always do so. It becomes important, therefore, to determine the conditions under which the negative experience of physical pain is accompanied by a sense of injustice.

The present analysis identifies four factors as affecting the degree to which a sense of injustice is aroused by the act of physical harm: (1) the degree of responsibility for the act attributed to the harm doer; (2) the legitimacy of the harm doer's authority; (3) the comparative consistency of the harm doer's treatment of us with that accorded to others similar to us in behavior or some other characteristic; and (4) the consistency of this treatment with past experiences in similar situations.

2.1. Responsibility and Commutative Justice

The question of responsibility attribution in harm infliction is one that has been extensively discussed within both social-psychological and Piagetian frameworks (e.g., Rule & Nesdale, 1976). In his discussion of naive psychology, Heider (1958) differentiated five levels at which people assign responsibility for events: responsibility by association with the deed or the doer, responsibility by simple commission of the act, responsibility through the foreseeability of the consequences, responsibility through intentionality, and, finally, responsibility through valid justification of intended acts.

Cultures and individuals differ greatly in the levels they use in assigning responsibility. For example, in contrast to the customary practice in our culture, there appear to be a number of cultures that endorse the principle of blame by association (Kelsen, 1946), a tendency against which the biblical injunction to refrain from blaming sons for their fathers' sins appears to be oriented. Similarly, Shaw and Sulzer (1964) have shown that children tend to use different criteria than adults to attribute responsibility for actions. The attribution of responsibility both by association and by commission are examples of attributional tendencies found to be much more prevalent in children than in adults.

Perhaps the most widely discussed difference between younger and older children in this regard has to do with their sensitivity to the distinction between well-intended, ill-intended, and unintended acts of harm. According to Piaget (1932/1965), children in the stage of heteronomous morality weigh consequences more heavily than intentions in assigning blame, whereas the opposite pattern appears in older children and adults. The greater use of consequences in assigning blame may not be as highly dependent on developmental stage as is suggested by Piaget, however. It is known, for instance, that by the early elementary-school years, children readily distinguish between accidental and intended acts, though they still have difficulty with intended acts that vary in both valence of intent and severity of outcome (Karniol, 1978). Moreover, as Miller and Vidmar noted in their chapter, even adults often hold individuals responsible for unintended acts with severe consequences (see, e.g., Walster, 1966). Despite these complexities, it is still fair to propose that, ordinarily, a sense of injustice is less likely to be aroused if the act is judged to have been unintended.

Note, however, that irrespective of the intentionality of the act, negative affect attends the experience of physical harm, and that even an act of accidental physical harm may lead to the arousal of a sense of injustice. This happens if the harming agent does not apologize or try to make amends for his actions. In our culture, a harming agent is expected to apologize or to try to make amends even if his actions were accidental. Furthermore, the greater the harm done, the greater is the expectancy or sense of obligation.

2.2. Authority and Commutative Justice

If the act is judged to have been intended, the question of the harm doer's legitimate authority to inflict such harm becomes of issue. Through socialization, we come to learn who is regarded as having the legitimate authority (French & Raven, 1959) to administer physical pain. In our society, it is considered legitimate and hence not unjust for some persons in authority to inflict pain on us in order to try to change our undesirable behavior (e.g., parents and sometimes teachers). Indeed, children often justify their parents' use of physical power by citing legitimate parental authority (e.g., "parents are supposed to bring their kids up," Kohlberg, 1958, p. 432). It is similarly considered legitimate for a person to inflict harm as part of his aim to improve our personal welfare (e.g., a dentist or a surgeon). It is not considered legitimate in most instances, however, for a peer or a person not in authority to inflict pain; nor is it legitimate for someone to inflict pain in order to improve his own welfare (e.g., rape, violent crime).

Norms about the legitimate use of physical power vary across social groups (Bernstein, 1971). For instance, children from position-oriented homes appear to hear more status-oriented legitimations for pain infliction than do those from person-oriented homes (e.g., "He can hit you because he's older than you"). Socialization experience can modify not only beliefs pertaining to which agents have legitimate authority to inflict physical pain but also beliefs concerning the conditions under which legitimate authorities may exercise their power.

2.3. Consistency and Commutative Justice

Assuming that the authority in question is deemed legitimate, the question of consistency of treatment becomes relevant. If the agent of harm is perceived as having legitimate authority *and* if consistency of treatment is evident (i.e., all cases of A are given treatment X), it is unlikely that the sense of injustice will be aroused, irrespective of the treatment. This state of affairs is exemplified in the words of an individual who played football for the notoriously harsh coach Vince Lombardi: "Is he fair? Sure he's fair; he treats us all alike, like dogs." According to Perelman (1969) and others, *the* formal principle of justice is that people be treated equally and that similar situations be handled in similar ways. If inconsistency in treatment is evident, it is virtually certain that a search for possible explanations of divergent treatment will be launched by the participants (Golding, 1963).

In short, even if the harm doer is a legitimate authority, the sense of injustice with respect to physical pain will be aroused if the dictum "Treat like cases alike" is violated. The importance of this principle is reflected in the fact that agents generally attempt to justify differential treatment by pointing out differences in cases (e.g., "He's too old to get a spanking; you're not"). Socialization factors become of critical relevance here since the dimensions considered important for differentiation vary greatly across social groups. The most important point to be made, however, is that once expectations for the bases of divergent treatment are established, failure to satisfy these expectations will lead to perceptions of injustice, irrespective of the socialization experiences of the individual.

To recapitulate, we see that the act of physical harm does *not* arouse a sense of injustice when (1) the agent is not held responsible for his actions; (2) the agent is seen as having legitimate authority to be administering such harm; and (3) the harm delivered is similar to that given to other people considered similar to the recipient.

The second sphere in which commutative justice operates is in the removal or damaging of someone's property. As in the case of physical harm, the question exists as to when the removal or damage to one's property arouses not only negative affect but also the sense of injustice. In answering this question, we find once again that the social context is of critical importance. When we learn to use possessive pronouns, we learn that they imply not only that we own the item but also that this ownership entails certain additional benefits and/or duties. If we own an item, we are expected to take care of it, prevent its loss, enjoy its benefits, and perhaps let others "share" its benefits as well. The value of possession is that it allows the owner to deny access to the item to anyone he chooses (e.g., "I don't think he should give it up if it's his," Kohlberg, 1958, p. 432), especially to anyone who might damage it (Furby, 1978). As well, anyone who would like to have access to another's possessions may gain such access only by asking permission, and he must do so with the understanding that access to the item does not entail a change of ownership. The socialization process determines who is permitted access to another's property. In our society, parents and siblings are ordinarily considered the only ones who have legitimate access to children's possessions (Furby, 1978). Other cultures appear to allow a much broader range of people legitimate access to one's property (Read, 1955).

Violation of the cultural norms pertaining to property access will very likely arouse the sense of injustice, particularly when the other has no intention of returning the item in question. If someone has gained illegitimate access to the item but has returned it in its original state, the sense of injustice is reduced (e.g., "If he paid it back after he stole it, it wouldn't be wrong," Kohlberg, 1958, p. 444).

The issue becomes quite different, though, when we turn to damage to one's possessions. Essentially, no one has the legitimate right to inflict harm on someone else's property. A sense of injustice is certainly aroused by such intentional harm. But a sense of injustice may also attend accidental harm of this type. We indicated earlier that reparatory action for accidental harm to one's person is culturally obligatory. In the case of harm to one's property, reparatory action is much more culturally prescribed. In most instances of property harm, unlike physical or psychological harm, the harm can be rectified. One can often replace the item with a new one or make the damaged item appear to be undamaged (e.g., wash a soiled garment, glue a broken vase). Not to attempt to rectify harm done to another's property, even if it was accidental, produces a sense of injustice and condemnation of the harming agent. Not surprisingly, children are taught very early about such cultural demands for reparative action (e.g., "Jill, you tell Susie right now you're sorry you broke her doll! Don't you worry, Susie, we'll fix your doll when we get home").

In discussing both aspects of commutative justice, it appears that notions of legitimacy, intentionality, and inconsistency are critical in the reaction to physical harm, while legitimacy and reparation are focal in the determination of the sense of injustice that attends harm to one's property.

3. DISTRIBUTIVE JUSTICE

What arouses the sense of injustice when allocation of goods and services is involved? In most instances, the arousal of the sense of injustice in this context follows the violation of the princple of "Share and share alike." If two people are categorized as being similar, it is expected that they will be treated in a similar fashion (Lerner, 1974). Dissimilar allocation to individuals who perceive themselves as being similar will very likely provoke a reexamination by one or more of the participants of the categories being employed. For instance, if two children receive a different allocation of resources (e.g., cake), it is likely that one or both of the children will feel that the situation is unfair unless differences between them are cited and accepted as meaningful differences. A sense of injustice is not necessarily aroused in a situation where the child who received the most cake was a boy and the other a girl. In this circumstance, shared attitudes about sex differences may mute the sense of injustice (e.g., "Boys get more cake because they grow faster").

3.1. Similarity and Distributive Justice

The central issue here, of course, is how it is decided what *alike* means. Through the years, different societies have employed different definitions of similarity in this context. Achievement, effort, ability, and need have all been used as principles for defining *alike* and *not alike*.

Aristotle speaks of three main theories of distributive justice: (1) oligarchical, based on wealth; (2) aristocratic, based on merit; and (3) democratic, based on essential equality and humanity.

Whatever definition of *alike* one employs, distributive justice appears to have three defining characteristics. First, decisions about the fairness of a given distribution are comparative in nature. That is, one decides if he has received a fair share only after finding out what others have received (Festinger, 1954). The comparative nature of decisions about the perceived fairness of allocation is illustrated in the reasoning of a young child in a 1977 study by Damon, who said that "if they gave somebody else two pieces and they only gave me one," that would not be fair (p. 82). Second, in order for one to be able to compare one's own share meaningfully with that of another, the other must be perceived as being similar on the dimension implied by the relevant principle. If the principle of allocation is based on need, one cannot argue "unfair" unless one is comparing himself with others who are alike in need. There is considerable research evidence that comparisons are, in fact, based on similarity in categorization. The following statement by a 10-year-old is illustrative of the type of response that Damon (1977) found in his study of young children's sharing responses: "He's a child too, so he should get even" (i.e., the same amount) (p. 128). The importance of categorization in the context of perceived fairness was emphasized many years ago by Krech and Crutchfield (1948) when they argued that one of the deadliest weapons in propaganda is the strategy of making some segment of a group feel that they are suffering more hardships or are gaining fewer benefits than are other similar segments.

The third defining characteristic of distributive justice is that whenever departures from absolute equality in distribution are noted, the allocator is expected to justify the deviation. If his justification for differential categorization and allocation is deemed valid in the situation, then the sense of injustice is reduced. We should reemphasize that one should not equate the sense of injustice with negative affect. Deprivation always arouses negative affect, but only when such deprivation is considered unjustifiably discriminatory is this negative affect forged with a sense of injustice.

3.2. Justifying Unequal Distributions

The question of what constitutes valid justification for unequal allocation of resources is an interesting one. It appears that there is a developmental sequence in the kinds of justification used by children in explaining deviations from equality. At the earliest level, young children tend to focus on the characteristics of the goods to be distributed (i.e., shape, color, quantity) since these are the most salient aspects of the situation. At a later age, children focus on relationships between the allocator and the individual to whom allocation is to be made. Last to develop are justifications based on qualities that are intrinsic in the individuals to whom allocation is to be made. Here, children first focus on status factors, probably because these are often cited by socialization agents. Need and productivity appear to be fairly late to develop as justificatory principles, and productivity, in particular, may be the most dependent on cultural context (cf. Furby, 1978, 1979). It appears that the developmental sequence from object-intrinsic to recipient-intrinsic justification may be universal, but the specific types of intrinsic characteristics focused on are culturally dependent.

Support for this conceptualization of the development of justifications for unequal distributions is found in the data collected by Damon (1977). The youngest children in Damon's research appear to use justifications that refer to the intrinsic qualities of the object being distributed (e.g., "I like blue," p. 80; or "I want 7," p. 118). Later, there is a focus on the relationship between the allocator and the recipient (e.g., a child who said that he'd give a certain group more because "I like them," p. 119, or because "I hate girls," p. 121). Last to emerge in Damon's research are references to characteristics of the recipient. References to age or status appear first (e.g., a child who would give a friend less because "I's bigger than him," p. 80), and need (e.g., "He really needs it because he doesn't get money too often," p. 85) and productivity (e.g., "We should get more because we did more," p. 134) tend to appear only in somewhat older children.

The foregoing examples suggest that children apparently learn not only that deviations from equality must be justified but also what types of justifications are culturally accepted as reasonable (e.g., Eckhoff, 1974). Interestingly, there is often a lack of consensus about which justifications of unequal distributions are most valid in any given situation (e.g., "We'd have everybody say all their reasons, then we'll pick the 3 best reasons," Damon, 1977, p. 85). As well, justifications often conflict with each other (see Brickman, Folger, Goode, & Schul, this volume) and cannot all be satisfied in the same situation (e.g., need and desert). Consequently, the decision made ordinarily is for an equal distribution, a decision that avoids having to prove the validity of one justification over any other. Damon, in this volume, has also noted the conflict-resolving nature of equal distributions.

Another aspect of distributive justice that is important is reciprocity. In distributing goods, it appears to be a universal norm that one reciprocate when one is able to do so (Gouldner, 1960). A number of studies with children attest to the power of this norm in determining allocative behavior (e.g., Furby, 1978). In fact, in a recent study (Libby & Garrett, 1974), reciprocity was found to be more important in guiding grade-school children's allocative decisions than whether the other person's behavior was guided by benevolent or malevolent intentions or whether or not it was coerced by an external agent.

When it is not possible for the individual to reciprocate the same resource, there is an expectation that similar resources will be reciprocated, and that until such reciprocity has occurred, the recipient will have a sense of obligation toward the giver. Violations of reciprocity of either kind or deed arouse the sense of injustice.

In summarizing this section on distributive justice, we note that the sense of injustice is often aroused by intentional violations of legitimate expectancy and is frequently dependent on comparison with others who are considered similar. Whenever we are subjected to behavior that arouses our sense of injustice, we search for possible explanations for the violations. If the explanations for differential categorization and allocation are considered justified (a standard that is acquired through socialization), we may continue to feel deprived, but our sense of injustice will be muted. On the other hand, if we do not find valid excuses for the violation, we will desire to have justice restored.

4. PROCEDURAL JUSTICE: ESTABLISHING FAIR PROCEDURES

How a principle of allocation is decided on, how a structure of authority is determined, and how rules of harm infliction are established are all issues that fall under the rubric of procedural justice. If individuals do not feel that the prevailing principle of justice or fairness has been decided on by the proper procedures, a sense of injustice can arise independent of the particular principle that was selected. The appropriate procedures for establishing rules and principles of allocation and interaction appear to change over the developmental span and may depend on socialization experiences.

Young children appear to perceive authority as ruling by fiat (e.g., "When one boy says that you must play roulette, everyone plays that way," Piaget, 1932/1965, p. 43) and the rule espoused by the authority is presumed to be more fair than other rules, even when they are not democratically arrived at (e.g., captains of baseball teams should be big so they can "mush you up if you don't listen," Damon, 1977, p. 189). Such infliction of physical harm is seen as legitimate because the captain is "the boss of the team and you have to listen to him" (Damon, 1977, p. 188). One does not question the procedures by which captains are selected or by which the rules are arrived at. Essentially, the child is unable to conceive of alternative procedural rules. This inability is probably due to the young child's lack of experience as well as his heteronomous moral orientation.

The same type of trend in reactions to authority is seen in children's perceptions of how legal rules are arrived at. Young children argue that the president can change any law by fiat, by simply deciding to do so. Later, they realize that he must obtain approval for changes in law from his constituents (Hess & Torney, 1967). Similarly, young children believe that judges *are* the legal system rather than merely being its selected representatives (Hess & Torney, 1967; Kohlberg, 1958).

Later in development, there is general agreement among children that rules should be established democratically (e.g., "sometimes people play differently. Then you ask each other what you want to do. . . . We scrap a bit and then we fix things up," Piaget, 1932/1965, p. 72). In other words, it is recognized that rules are required to avoid disagreements and that rules are arrived at by consensus. It is because of the consensual nature of rules that they are now seen as obligatory for the individual (e.g., "All the kids have agreed to have him as captain and that means they have all agreed to play where he tells them to," Damon, 1977, p. 198). The sense of injustice at this stage is manifested whenever there exist nondemocratic procedures for rule determination, resource allocation, and leadership selection.

Unfortunately, we do not yet understand how socialization succeeds in teaching children the consensual nature of fair procedural rules. At the very least, however, this would appear to be a culture-bound phenomenon in that children in dictatorships would not be expected to espouse views on procedural justice similar to those espoused by children in democracies. Finally, in contrast to the case of commutative justice, acts that constitute procedural injustice are not intrinsically unpleasant. In fact, very young children may never experience a sense of procedural injustice. They may simply accept any procedure—especially adult-administered procedures—as fair. It is only with considerable experience and cognitive development that the child comes to believe that some procedural rules are more fair than others.

5. TRANSFORMING JUSTICE CONCERNS INTO MORAL AND LEGAL GUIDELINES

We now turn to a discussion of the ways in which the experience of injustice becomes translated into moral guidelines for our own behavior and into moral judgments of the behavior of ourselves and others.

5.1. Role Taking and Morality

There appears to be consensus among ethical philosophers that the most important process in becoming moral is one that we will call *projective role-taking*. This process has been called by many names: "imaginative interchange" (Cahn, 1949), "dramatic rehearsal" (Dewey, 1922), "imaginative projection" (Hare, 1962), "imaginative pretending" (Harman, 1977), "universalizing" (Hospers, 1970), and "imaginative sympathy" (Smith, 1759). Whatever label the process is given, all references to the phenomenon cite the process by which the individual imaginatively puts himself into another person's role.

Projective role-taking is presumed to facilitate a number of tasks that are considered critical for moral judgment: (1) the ability to judge one's own behavior from a spectator's point of view (Hare, 1962; Harman, 1977); (2) the ability to judge one's own or another's behavior from the point of view of the recipient of the action (Baier, 1958; Hospers, 1970); and (3) the ability to judge what one would do himself in another's situation (Smith, 1759).

While they may disagree as to the most important aspect of role taking, it is clear that most moral philosophers acknowledge the critical importance of this process. More specifically, philosophers and psychologists alike generally believe that it is through successful role-taking that the individual learns that all "nonreversible behavior" (i.e., behavior we would not like to be the recipient of) is unjust and morally wrong (Baier, 1958).

The implications of projective role-taking for justice considerations are important and complex. Since we are presumed to engage in such a process in order to see how *we* would react to the other's predicament, it may be speculated that our concern with justice for others is based on our desire to avoid injustice for ourselves. In line with this, Plato argued that "those who denounce justice do so not because they are afraid of doing it but of suffering it." La Rochefoucauld (n.d.) similarly argued that "the love of justice is, in most men, merely the fear of suffering injustice" (p. 580). In his theory of justice, Lerner (Lerner, Miller, & Holmes, 1976) suggested as well that concern for others arises because of the implications of their suffering for one's own well-being.

This line of analysis may be taken to imply that only egotistical concerns guide people, but this is not necessarily the case. Specifically, there is general agreement (e.g., Hare, 1962; Kohlberg, 1976; Metcalf, 1971) that through the process of projective role-taking, egotistical concerns become transformed into social ones. By projective role-taking, we abstract guiding moral principles against which events in the real world are assessed. Events that do not measure up to these principles violate our sense of injustice, irrespective of whether we are the recipients of the violation. Consequently, the process of projective role-taking yields not only personal guidelines (i.e., "I ought to do . . .") but more general action principles (i.e., "All people ought to do . . .").

This shift can be observed in the developmental sequence. As young children, we must be told to imagine what others feel when we treat them unjustly; later, we do so of our own accord. As adults, however, we have higher-order principles that have emerged through projective role-taking, and we perceive that violation of these principles arouses the sense of injustice. In short, it is suggested that the sense of injustice gradually becomes independent of our own experience and becomes extended into a moral orientation that guides our own actions toward others, as well as our reactions to both victims and perpetrators of injustice.

This moral orientation also guides conceptions of legal obedience. Children learn that legal rules guide other people's behavior and hence serve to protect us from the caprice of others (Tapp & Kohlberg, 1971). As Schopenhauer (1942) stated, the purpose of law is to see that *"everyone should have justice done to him,"* while morality is concerned that *"everyone should do justice"* (p. 103). It is because the law serves to guarantee that one will have justice done to him that it becomes morally binding for the individual to obey the law. Moreover, it is on the basis of this connection between law and justice that we come to view the violation of rules as unjust acts, irrespective of whether the violation directly affects us.

5.2. Moral Development and Legal Obedience

Because obeying the law does not always promote justice at the microlevel (see Brickman, Folger, Goode, & Schul, this volume), a critical issue is the manner of resolving conflicts between justice concerns and legal obedience. This question has provoked debate for many centuries. There are essentially three views on this issue, each differing in its implications for human action and justice. The first view, typified by the Old Testament, is that the moral and legal laws coincide because moral law, as pronounced by God, dictates human law. In this instance, pronouncement by God is perceived as sufficient reason to prohibit action without need for further justification.

The second view, natural law, as espoused by Kant, is that moral law is a higher law and any conflicts between legal and moral rules are to be resolved in favor of moral law.

The third view is that of legal positivism, according to which law and morality are separate spheres, and though conflicts between them may raise moral dilemmas, such conflicts do not alter the duty of legal obedience. Following this tradition, the Greeks believed that if legal rules conflicted with moral rules, it was the citizen's undeniable duty to obey the law while endeavoring nonetheless to change the law (Lloyd, 1964). In other words, one has a *moral* obligation to obey the law (Gross, 1973; Wasserstrom, 1970). President Kennedy exemplified this view when he said, "Americans are free, in short, to disagree with the law, but not to disobey it." Because legal positivism encourages a disposition to obey the law, it has often been argued that as a philosophy it was psychologically responsible for the unquestioning obedience of Nazi officers (cf. Shuman, 1963). Natural law, on the other hand, is often thought to characterize the philosophy of the draft resistance movement in the 1960s. The debate about which philosophical position is more valid is a lively one, with each side having its tenacious advocates (e.g., Fuller, 1940; Ross, 1974; Perry, 1976; Shuman, 1963). It does appear, however, that the manner in which such conflicts are resolved depends on one's philosophical orientation and cannot easily be predicted through a psychological analysis of reactions to injustice.

5.3. Specious Morality

We have characterized the transition from concern with one's own experiences of injustice to concern with the injustices experienced by others as being the crux of moral development. We do not mean to imply, however, that this transition occurs in everyone or at least to the same extent in everyone. Individuals differ greatly in projective role-taking ability (Hoffman, 1977), and they tend to differ accordingly in their moral judgment and behavior (Kohlberg, 1976).

The more sophisticated the moral principles a person cites, the greater is his presumed ability to role-take and abstract moral principles of action. We also do not wish to imply that all behavior that is morally relevant has concern for others as its impetus. It is entirely possible to engage in *specious* moral behavior, behavior that is engaged in for reasons that are not ordinarily considered "moral" ones. Such reasons may include self-interest and prudence (e.g., "It's bad to hit people cause they hit you back"), punishment avoidance (e.g., "It's bad to steal because they put you in jail"), and divine retribution (e.g., "God will punish you for that"). The belief in immanent justice (Piaget, 1935/1965) may also lead to specious moral behavior (cf. Karniol, 1980), since, in this as in the other instances, the person is not using justice concerns as the basis for establishing the morality of the act. It is usually argued that only behavior and principles that arise from justice concerns are truly moral (e.g., Kohlberg, 1958). Moreover, it is assumed that there is a developmental sequence in the types of principles that individuals abstract to help provide their moral behavior. Only with experience and cognitive maturity do individuals acquire the ability to use justice-based principles to guide moral behavior.

6. SUMMARY

In this chapter, we have attempted to illuminate the conditions that give rise to a sense of procedural, distributive, and commutative injustice. In doing so, we have taken a developmental perspective and have endeavored to trace how the cognitive and social development of the child shape and give definition to both the experience of injustice and the understanding of morality. Our analysis has been far from complete, however.

Perhaps most noticeably absent from our analysis has been any discussion of the behavioral and psychological reactions that individuals have when their sense of injustice is aroused. Elaborate discussions of this issue can be found elsewhere (Lerner, Miller, & Holmes, 1976; Miller & Vidmar, this volume; Walster, Berscheid, & Walster, 1976), but a brief discussion of possible developmental differences in these reactions may be appropriate here.

In the face of the perception of commutative injustice, a number of differences between younger and older children or adults may emerge. First, on the basis of the work of Piaget and others, it appears that young children are more likely to recommend expiatory punishment (analogous to the *lex talionis*) than are older children and adults. One reason for children's being more likely to endorse expiatory or severe punishment may be that they are quicker to infer a "bad character" from "bad behavior" than are adults. This tendency may well be a partial function of parents' socialization techniques, which tend to label the individual (e.g., "You're a bad boy, Johnny") rather than his undesirable behavior (e.g., "You did something bad, Johnny").

Additionally, it appears that young children, more than adults, believe that punishment of the harm doer eliminates the state of "injustice," whether the victim has been compensated or not (Miller & McCann, 1979). It is also possible that young children have different beliefs about the deterrent value of punishment, though this hypothesis has not yet received empirical attention.

Developmental differences are important also in the context of distributive and procedural justice. For instance, one of the most natural reactions to both procedural and distributive injustice is to attempt to change the situation so as to make it "just." The accomplishment of such change by the child depends first on the child's power within the situation. If he is a highly popular or influential child, his chances of instituting changes are much greater than if he holds a low power position. Second, instituting such changes depends to a large extent on the child's ability to persuade others of the value of adopting alternative procedural or distributive rules. To the extent that children are less persuasive in such contexts than adults (Weinstein, 1970), their ability to institute more fair procedures and distributions are limited as well.

Perceptual reactions to injustice have been extensively studied in adult populations (e.g., Lerner & Miller, 1978; Walster, Berscheid, & Walster, 1976), but we still know very little about the developmental course of these reactions. This and other issues pertaining to the development of justice must be addressed in the future if we are to have a comprehensive understanding of the psychology of justice.

REFERENCES

Allen, C. K. *Aspects of justice.* London: Stevens, 1958.

Armstrong, K. G. The retributionist hits back. In S. E. Grupp (Ed.), *Theories of punishment.* Bloomington: Indiana University Press, 1971.

Baier, K. *The moral point of view.* Ithaca, N.Y.: Cornell University Press, 1958.

Bernstein, B. *Class, codes and control.* London: Routledge & Kegan Paul, 1971.

Cahn, E. *The sense of injustice.* Bloomington: Indiana University Press, 1949.

Damon, W. *The social world of the child.* San Francisco: Jossey-Bass, 1977.

Dewey, J. *Human nature and conduct.* New York. 1922.

Eckhoff, T. *Justice: Its determinants in social interaction.* Rotterdam: Rotterdam University Press, 1974.

Festinger, L. A theory of social comparison processes. *Human Relations,* 1954, 7, 117–140.

French,, J. R., & Raven, B. The bases of social power. In D. Cartwright, (Ed.), *Studies in social power.* Ann Arbor, Mich.: Institute for Social Research, 1959.

Fuller, L. *Law in quest of itself.* New York: Foundations Press, 1940.

Furby, L. Sharing: Decisions and moral judgments about letting others use possessions. *Psychological Reports,* 1978, 43, 595–609.

Furby, L. Inequalities in personal possessions: Explanations for and judgments about unequal distributions. *Human Development,* 1979, 22, 180–202.

Golding, M. P. Principled decision-making and the supreme court. *Columbia Law Review,* 1963, 63, 35–58.

Gouldner, A. The norm of reciprocity: A preliminary statement. *American Sociological Review,* 1960, 25, 161–178.

Gross, H. Comments: Some moral perspectives on defying legal authority. In N. S. Care & T. K. Trelogan (Eds.), *Issues in law and morality.* Cleveland: Case Western University Press, 1973.

Hare, R. M. *Freedom and reason.* Oxford: Clarendon Press, 1962.

Harman, G. *The nature of morality.* New York: Oxford University Press, 1977.

Heider, F. *The psychology of interpersonal relations.* New York: Wiley, 1958.

Hess, R. D., & Torney, J. *The development of political attitudes in children.* Chicago: Aldine, 1967.

Hoffman, M. Empathy, its development, and prosocial implications. In C. B. Keasey (Ed.), *Nebraska Symposium on Motivation, 1977.* Lincoln: University of Nebraska Press, 1977.

Honderich, T. *Punishment: The supposed justifications.* London: Penguin, 1969.

Hospers, J. *Human conduct.* London: Rupert Hart-Davis, 1970.

Karniol, R. Children's use of intention cues in evaluating behaviour. *Psychological Bulletin,* 1978, *85,* 76–85.

Karniol, R. A conceptual analysis of immanent justice responses in children. *Child Development,* 1980, *51,* 118–130.

Kelsen, H. *Society and nature.* London: Paul, Trench and Trubner, 1946.

Kohlberg, L. The development of modes of moral thinking and choice in the years ten to sixteen. Unpublished Ph.D. thesis, University of Chicago, 1958.

Kohlberg, L. Moral stages and moralization. In T. Lickona (Ed.), *Moral development and behavior.* New York: Holt Rinehart, 1976.

Krech, D., & Crutchfield, R. W. *Theory and problems of social psychology.* New York: McGraw-Hill, 1948.

La Rochefoucauld, F. *Maximes.* Paris: Larousse, no date.

Lerner, M. The justice motive: "Equity" and "parity" among children. *Journal of Personality and Social Psychology,* 1974, *29,* 93–146.

Lerner, M. J., & Miller, D. T. Just world research and the attribution process: Looking back and ahead. *Psychological Bulletin,* 1978, *85,* 1030–1051.

Lerner, M. J., Miller, D. T., & Holmes, J. G. Deserving and the emergence of forms of justice. In L. Berkowitz and E. Walster (Eds.), *Advances in experimental social psychology,* Vol. 9. New York: Academic Press, 1976, pp. 133–140.

Libby, W. L., Jr., & Garrett, J. Role of intentionality in mediating children's responses to inequality. *Developmental Psychology,* 1974, *10,* 294–297.

Lloyd, D. *The idea of law.* Hammersmith, U.K.: Penguin, 1964.

Metcalf, L. E. *Values education.* Washington: National Council for the Social Studies, 1971.

Miller, D. T., & McCann, D. Children's reactions to the victims and perpetrators of injustices. *Child Development,* 1979, *50,* 861–868.

Perelman, C. Justice and reasoning. In G. Hughes (Ed.), *Law, reason and justice.* New York: New York University Press, 1969.

Perry, T. D. *Moral reasoning and truth.* Oxford: Clarendon Press, 1976.

Piaget, J. *The moral judgment of the child.* New York: Free Press, 1965. (Originally published 1932.)

Read, K. E. Morality and the concept of the person among the Gahuku-Gama. *Oceania,* 1955, *25,* 233–282.

Ross, A. *On law and justice.* London: Stevens, 1974.

Rule, B. G., & Nesdale, A. R. Moral judgment of aggressive behavior. In R. G. Geen & E. C. O'Neal (Eds.), *Perspectives on aggression.* New York: Academic Press, 1976.

Schopenhauer, A. On human nature. In *Complete essays.* New York: Wiley, 1942.

Shaw, M. E., & Sulzer, J. L. An empirical test of Heider's levels in attribution of responsibility. *Journal of Abnormal and Social Psychology,* 1964, *69,* 39–46.

Shuman, S. I. *Legal positivism.* Detroit: Wayne State University Press, 1963.

Smith, A. *A theory of moral sentiments.* London, 1759.

Tapp, J. L., & Kohlberg, L. Developing senses of law and legal justice. *Journal of Social Issues,* 1971, *27,* 65–91.

Turiel, E. Distinct conceptual and developmental domains: Social-convention and morality. In C. B. Keasey (Ed.), *Nebraska Symposium on Motivation, 1977.* Lincoln: University of Nebraska Press, 1977.

Walster, E. Assignment of responsibility for an accident. *Journal of Personality and Social Psychology*, 1966, *3*, 73–79.

Walster, E., Berscheid, E., & Walster, G. W. New directions in equity research. In L. Berkowitz & E. Walster (Eds.), *Advances in experimental social psychology* (Vol. 9). New York: Academic Press, 1976.

Wasserstrom, R. A. The obligation to obey the law. In R. S. Summers (Ed.), *Essays in legal philosophy*. Oxford: Blackwell, 1970.

Weinstein, E. A. The development of interpersonal competence. In D. A. Goslin (Ed.), *Handbook of Socialization Theory and Research*. Chicago: Rand McNally, 1969.

New Perspectives on the Social
Dimensions of Justice

The chapters in this section provide highly insightful, creative contributions to our understanding of the way the theme of "justice" appears in people's lives. I think it is safe to say that each contributor has reacted to what have appeared to be weaknesses in prior conceptions of the social psychology of justice. However disparate their conceptual tools may appear—"macrojustice versus microjustice" (Brickman, Folger, Goode, and Schul), or "frame of address" and the negotiated basis of justice (Sampson), or "retribution theory versus equity theory" (Hogan and Emler; Miller and Vidmar), they are all generated as a "versus" response to what had gone on before. In particular, they all, in one way or another, are attempting to make the social psychology of justice more "social" than have prior theories.

For example, in his chapter, Sampson portrays justice as a distinctly "social" phenomenon. Its origins, dynamics, and forms in human affairs are a direct expression of the social-historical factors operating at a given time. He construes justice as "part of a conversation taking place between people" imaginally or in vivo. And thus, he introduces the term frame of address *to emphasize that what occurs in that conversation is a reflection of the socially patterned and determined ways in which people come to view one another. His overriding point is that "our conceptions of what is just and fair are not motives that serve as springs to action from within but are socially sanctioned, public procedures that serve as the grounds for behavior." He sees the famous* Bakke *case as an interesting example of changes in social patterns that have generated a conflict between a universalistic (frame of address?) and a particularistic way of viewing others. Of course, it is all more complex than a simple dimension, as Sampson recognizes, but he sees in the* Bakke *case the conflict between applying general rules of merit (universalistic) for the acquisition of a scarce resource—that is, admittance to medical school—and the use of quotas or special rules for members of a particular category (particularistic), that is, white versus black, men versus women. His vision of the future is one of increasing self-interested behavior as scarcity leads to the elaboration of special interests and to fragmentation of the social structure. Sampson also assumes that guiding the use of various frames of address and negotiated terms of justice is the familiar self-interested model of human motivation. Man's thoughts and ways of seeing people may be shaped by the culture, but his motives are assumed by Sampson to be as simple as they are familiar: to negotiate the best deal for oneself using the rules or language of justice. In that sense, Sampson's assumptions resemble those of Walster (1978), but his emphasis on the power and shape of the "social contract" is a highly inventive contribution.*

Hogan and Emler appear to be even more impressed than Sampson with the central role of the "social contract" in people's lives. As a contrast to the presumably limited assumptions and applicability of "equity theory," Hogan and Emler offer what they portray as the more basic and universal model of "retribution theory." Their reasoning and the examples are delightfully persuasive. Their argument is that what is common to all societies and what underlies the most prevalent form in which justice appears is the knee-jerk reflex, the sense of moral outrage, that occurs when people see a norm violated (by someone else). So in a very real sense, all important concerns with justice are socially dependent on the reaction to and the upholding of the rules of conduct accepted in the society. Hogan and Emler also introduce the "social dimension" into their motivational system: Why do people behave this way? As they state:

> We believe that the members of any human group are in more-or-less constant competition for status, no matter how carefully the competition is disguised . . . retribution exists to punish those who take unfair advantage in this competitiveness.

I suspect, however, that Sampson might observe that their assumptions reflect the analysis, unintentional as it may have been, of what they see happening in their own society. And it does appear that a common feature of our society, especially in economic endeavors, is that people are in a "nonunit" relation (competing) with all others. What prevails, then, as described earlier, is a form of "mock equity" in which people engage in contractual forms of cooperative endeavors but in which essentially, each person is in the business of getting the most for himself or herself. Although the common focus of the endeavors is particular kinds of generally desired resources, it is clear that in any encounter with one of "them" (nonunit relation), the issue of superior and inferior status is always at least a tacit consideration. Norm violation (i.e., violating the rules of the "contract") invariably makes salient issues of superiority–inferiority, or status: "We can't let someone like that, one of them, get away with. . . ." Hogan and Emler's analysis provides a rather elegant elaboration of one of the "strat-

egies" people employ in the service of the justice motive (see the introductory chapters).

It is interesting, in this respect, that the cross-cultural examples presented by Hogan and Emler to illustrate the universality of the "retribution" response seem to fit the descriptions of "peasant" cultures that led Foster (the anthropologist) to develop his theory of the "psychology of limited good." Apparently, in these cultures, there is the common assumption that there is a finite and fixed amount of desirable resources available: only so much land that will produce only so much grain or support only so many fruit trees. As a result of living in this kind of static society, people develop a supreme concern with the issue of their relative status. Any change in one member's well-being is attributed to norm violation—an illegal act—and is taken as a threat to the status of all the other members. As a result, any positive change in circumstance, however it occurs, is met with the kind of outrage and vengeance portrayed so vividly in Hogan and Emler's discussion of the desire for retribution. Is it true, however, as Hogan and Emler are suggesting, that our more complex, pluralistic, secular society is equally dominated by this preoccupation with relative status?

Miller and Vidmar provide a superb overview of other private and public motives that are associated with a punitive response to rule violation. Along with documenting a variety of relevant motives and factors, they, too, find that the issue of relative status can be an important consideration in the punitive response (desire for retribution) to a harm doer. However, their intriguing hypothesis is that the motive for the punishment is to provide a public device for lowering the relative status of the offender in comparison with that of the victim.

The "social context" dimension appears in the Brickman et al. contribution in a most interesting form. Generalizing from the conceptual distinction that has appeared in other areas, such as economics and sociology, these authors suggest that there is an entire domain of macrojustice that has been ignored or at least insufficiently treated in prior efforts. For the most part, they find that the earlier models of justice focused on the individual not only as the origin but as the sole object of justice considerations. Justice was determined by assessing whether someone was entitled to his or her fate. That assessment, in terms of what these authors call microjustice, is based on applying one of the three micro rules of entitlement: ability, merit, or need. But in macrojustice considerations, people assume the posture of the social or political analyst and view the system in aggregate. As a result, they arrive at principles of macrojustice that "specify a priori constraints on the allowable form or pattern for an overall distribution of resources." The primary concern of this social perspective is with "the appropriate order of society or the goodness of society." Common examples of macroprinciples, the authors believe, can be found in some forms of procedural justice or in the common use of lotteries to distribute scarce resources.

In the pilot investigations that Brickman et al. report, they generate com-

pelling examples of classes of macroprinciples that are probably familiar to the reader. One of these they term the "minimum principle," whereby the smallest allocation cannot be too much less favorable than the others. And that makes sense. We would all probably agree that there are certain minimal conditions to which everyone is entitled, regardless of their ability or merit, and without careful assessment of individual need. The authors' "subgroup principle" is reminiscent of the Bakke case, in which it had been assumed by some administrators that there should be an appropriate balance between the average allocations to different subgroups. There should be an appropriate representation of women and minorities in the highly desired occupations and professions, even if that representation required setting aside considerations of individual entitlement based on ability or merit. Bakke argued his case for a place in medical school on the basis of rules of both procedural justice and microjustice.

The Brickman et al. chapter raises some key questions: To what extent are micro and macrojustice actually two different types of justice that operate according to different rules? How are they related to one another? From one perspective, it appears that these macroprinciples emerge from the recognition that there is a need for a more "just" procedure in a certain area of the society. This assessment of the need for a more just procedure and the form it should take is based quite explicitly on microprinciples of justice, illustrated vividly in the legislative and legal dialogues that formed the background for the Bakke case. For example, the successful advocates of quotas and different standards of admittance to medical schools argued that certain categories of people (women and members of ethnic minority groups) did not have equal access to the medical profession, because of what our society had imposed on them. They were hindered by "society" in their rightful pursuit of what they were entitled to by virtue of their individual ability, effort, need, etc. (microprinciples). Setting the quotas and altering the standards of admission were designed to prevent such injustices being done to these deserving candidates. At the other end, it was also reasoned that the genuine health needs of members of the minority communities were not being met, because there were not enough physicians willing and able to care for them. Again, that argument was based on the assessment that legitimate health needs were not being met—a microprinciple. It appears, then, that the macroprinciple, in the form of quotas for appropriate distribution of the allocation of this desired resource (admittance to a medical career), arose from the awareness of injustices as assessed by the typical microprinciples. Bakke, in arguing and winning his case, made it clear, as well, that each of these macroprinciples is continually evaluated in terms of the extent to which it permits people to gain or inhibits people from gaining what they deserve according to their need, merit, etc.

A singular contribution of the Brickman et al. chapter is that by introducing the concept of macroprinciples of justice, it highlights the common dilemmas and unavoidable problems created by the concern with justice in a complex society.

There will always be competing demands based on different principles of entitlement (need versus merit, effort versus investment) and the need to compromise these demands so that a minimum number of injustices are perpetuated in the society. It will be exciting to see how other investigators are stimulated to examine the processes that people employ in generating a given macroprinciple for a given area of activity.

REFERENCE

Walster, E., Walter, G. W., & Berscheid, E. *Equity: Theory and research.* Boston: Allyn & Bacon, 1978.

6

Social Change and the Contexts of Justice Motivation

EDWARD E. SAMPSON

In this chapter, our aim is to develop an understanding of justice and social change by locating this understanding in the context of an emerging, alternative model of social psychology. We open by examining the major thrust of the recent critiques of contemporary social psychology. The writings of Mead and Vygotsky on mind, Mills on motivation, and Sullivan on personality are then introduced as offering a distinctly alternative version of social-psychological inquiry. These serve as the basis for introducing a concept, *address frame*, that is responsive to the critiques and offers a useful new perspective for understanding social behavior, including justice motivation. We examine justice motivation reconceptualized in terms of the concept of address frame. Finally, we turn to how changing sociohistorical factors have constituted different contexts of justice and what some of the future issues might be.

1. THE CRITIQUE OF SOCIAL PSYCHOLOGY

The last decade has seen a resurgence of critiques directed toward the discipline of psychology, social psychology in particular (e.g., Cartwright, 1979; Gergen, 1973, 1978; Rosnow, 1978; Sampson, 1977, 1978).

EDWARD E. SAMPSON • Department of Sociology and Social Anthropology, Clark University, Worcester, Massachusetts 01610.

These critiques have taken several different forms. Some focus on the experimental methodologies that dominate the field and that often seem ill suited either to increase our enlightenment about significant human issues or to protect the rights and dignity of our human subjects (e.g., Kelman, 1967, 1972). Others, however, have focused directly on the model of science that dominates the field (e.g., Gergen, 1973; Gibbs, 1979; Levine, 1976; Riegel, 1976; Sampson, 1978).

In an earlier paper (Sampson, 1978), I referred to this dominant model as the natural-science paradigm; it is characterized by a pursuit of universal principles of human behavior that are "independent of any particular society or historical period" (p. 1332). A similar view has been expressed by several others: Gergen (1978) writes of the positivist–empiricist approach; Levine (1976) and Mannheim (1971) write of natural-science thought.

A close reading of the recent collection of paradigmatic critiques reveals two dominant and interrelated themes: the first emphasizes the individualistic, acontextual, and ahistorical nature of our science; the second, the value biases that thereby inhere in such a scientific paradigm. Basically, a science that ignores the particular context of its subject matter in the pursuit of some transhistorical certainties of human psychological functioning tends to be blind to the value biases that infuse its conceptions. The existing shape of social reality is treated as though it were the inevitable or necessarily ordained shape. Alternatives are either not sought or are sloughed off as aberrant cases not proper for scientific study.

Although I did not then develop the details of the paradigmatic case, this is the argument that I suggested in examining the popular conception of justice as equity (Sampson, 1975). When equity is treated as a universal property of the human mind, we fail to see the ways in which the equity principle is more meaningfully understood to be a sociohistorically created and sustained ideal. Equity motivation, therefore, is not the way human justice inevitably must be served. The social context and value that equity represents—a competitive business world involving exchange relationships among strangers—are ignored because the equity motive is treated as something fundamental about the way in which the individual mind and the motive system work. Since people are assumed "naturally" to be equity theorists, we fail to observe the connection between equity theory's view of human behavior and the sociohistorical context that is required for its view of justice to be realized.

The preceding suggests that many of the flaws in the dominant social-psychological paradigm stem from its pursuit of intrapsychic, natural, and universal fundamentals of human cognition and motivation.

This view is shared by many who have recently commented on the problems inherent within our existing formulations.

Writing on the concept of "selfhood," Brewster Smith (1978) saw the existing Newtonian forms of our scientific paradigm as blinding us to the ethnocentric bias of our understanding of behavior: "We psychologists, with our preoccupation with universal processes, have a strong bias against historical-cultural modes of thought and therefore risk being culture- and history-bound" (p. 1056).

In his thoughtful piece introducing the perspective of Vico to American social psychology, Ralph Rosnow (1978) commented on our existing paradigmatic assumption "that human nature . . . is fixed across time and space" (p. 1327), and our further assumption "that motive explanations of human behavior such as social comparison, cognitive consistency, and reciprocity are universals of social behavior that are fixed across time and culture" (p. 1327).

Gergen (1978) echoed Rosnow's view in writing about the aleatoric nature of human activity "that . . . may largely reflect contemporary contingencies" (p. 1353) and may not be as stable or replicable as the subject matter of the natural sciences. It is because we search for something fundamental within the psychology of the individual—Gergen sees this effort as descriptive of dissonance theory, balance theory, integration theory, attribution theory, and so forth—that we fail to note how these may simply represent "the transient peculiarities of contemporary life" (p. 1353).

Riegel (1976) captured a similar outlook in noting

> that only a conception in which the individuals are seen in their developmental context, and in which, moreover, individual developmental changes are seen in their changing cultural-historical context, can lead to a comprehensive understanding of human activities. (p. 354)

Therefore, Riegel recommended a dialectical formulation of psychological processes to replace the existing paradigm and its pursuit of stable, universal, and timeless entities.

1.1. Elements of an Alternative Model

Needless to say, it would be possible though not helpful to continue to outline the many critiques that have been introduced. What would be helpful, however, would be to examine an alternative view of social psychology. This alternative will turn out not to be something entirely new; rather, it involves a development of several earlier ideas that were left by the wayside as the field rushed headlong into the technologies and thought forms of the twentieth-century.

The major theme of the several critiques argues against an emphasis on fundamental and universal intrapsychic cognitive and motivational properties that reside within the individual as part of human nature. An alternative would thereby have to include the following elements:

1. It should build on an interpersonal rather than an intrapsychic approach to understanding social behavior.

2. It should reflect more particularistic and contextual elements, understanding that what are now assumed to be properties of the individual are emergent properties of social interaction.

3. In this manner, sociohistorical effects and changes will be built in as constituent elements of the conceptual model of the field.

4. While its framework for understanding human social behavior should include the preceding, it should not be such as to cast aside existing knowledge; rather, it should offer a reconceptualization that casts what we now know in a different light and leads us to ask different questions and to seek different kinds of answers.

It is important to repeat here a message that I have noted elsewhere (Sampson, 1975, 1978). The critique of our discipline does not maintain that our discoveries to date (e.g., on equity theory or on androgyny, two examples on which I have previously written) are incorrect. Equity may indeed be a useful way to explain a great deal of exchange behavior today, even as androgyny seems to recommend itself as an adaptive sex-role ideal in contemporary society. But when these discoveries about human behavior are assumed to be something fundamental about the way the human psyche works—as our present paradigm suggests—we fail to understand the actual contextual bases of our discoveries about the human condition and to comprehend the value biases that may thereby creep into our work and our recommendations for policy and intervention, or alternatives that may be equally plausible under changed social conditions.

2. INTERPERSONAL FORMULATIONS OF PSYCHOLOGICAL PROCESSES

George Herbert Mead (1934), Lev Vygotsky (1978), C. Wright Mills (1975), and Harry Stack Sullivan (1953, 1954) are four analysts who stand somewhat apart from the dominant paradigm of contemporary psychological and social-psychological thinking. Each has offered a distinctly interpersonal theory of what are usually assumed to be basic intrapsychological processes. For Mead and Vygotsky, mind and self are interpersonal emergents of social interaction. For Mills, motivation is not an intrapsychic trigger to action but a public accounting practice or reason for action. For Sullivan, personality is a property of the interpersonal field and not a thing residing within the individual's psyche. In the hands of these theorists, psychology's most cherished intrapsychic qualities are given an alternate conceptualization.

2.1. G. H. Mead and L. S. Vygotsky on Mind

Mead's joining of mind, self, and society is no accident; it represents his view that society and social interaction form the field from which both thinking and identity emerge and to which both continually refer. There can be no thinking apart from a society (i.e., others) to which such thinking is addressed; there is no self apart from the others to whom that self refers.

To think is to engage in an inner conversation of gestures in which another is being addressed. People address themselves in thought through an audience whose perspective they adopt and from whose standpoint they view themselves. To hear ourselves involves adopting what we assume to be another's attitude as though that other had addressed us.

In saying that thinking is an inner conversation, Mead intended cognition to be akin to a dialogue in which we and some other are involved:

> in responding to ourselves we are . . . taking the attitude of another than the self that is directly acting, and into this reaction there naturally flows the memory images of the responses of those about us, the memory images of those responses of others which were in answer to like actions. Thus the child can think about his conduct as good or bad only as he reacts to his own acts in the remembered words of his parents. (p. 180)

Later, parents-as-audience expands to other people, including what Mead termed the "generalized other" of the community: an abstracted audience based on the many others with whom people have interacted and from whose perspectives they view (that is, address) themselves. Because we live in many different social environments and address many others, we thereby have many different minds and many different selves.

What is critical for our present concern is to observe that Mead's view of cognition gives it a distinctly interpersonal meaning. In the very act of thinking, including abstract thought, there is an audience that is being addressed and whose standpoint provides the thinker with the terms for understanding. Mead noted that even the abstruse thinker fills "out the bare spokesman of abstract thought . . . in seeking his audience" (p. 180). We will shortly return to this formulation of cognition.

Mead was not alone in offering an interpersonal theory of the mind and human thought. L. S. Vygotsky, a major figure in Russian psychology, independently offered a view of thinking that further develops the central thesis captured by Mead: interpersonal formulations precede and lay the foundations for the development of intrapersonal structures of the human mind.

A helpful way to understand Vygotsky's position is to work through one of the examples he introduced in his discussion of the process of *internalization:* "the internal reconstruction of an external operation" (1978, p. 56). Vygotsky described a little child who is reaching for a toy beyond his grasp. His fingers extend outward and wiggle as he tries to grasp the toy. His mother enters and comes to the child's aid; she views the child's extended hand and wiggling fingers as meaning that he wants a particular toy. For the mother, the child's uncoordinated and hesitant grasping means "pointing":

> The child's unsuccessful attempt engenders a reaction not from the object . . . but *from another person.* Consequently, the primary meaning of that unsuccessful grasping movement is established by others. (p. 56)

In time, the child transforms the vague movements of grasping into the more refined gesture of pointing:

> It becomes a true gesture only after it objectively manifests all the functions of pointing for others and is understood by others as a gesture. Its meaning and functions are created at first by an objective situation and then by people who surround the child. (p. 56)

As he outlined the process by which mind, as an internal representational system, comes into being, Vygotsky repeatedly emphasized the point that we have seen in Mead:

> An *interpersonal process is transformed into an intrapersonal one.* Every function in the child's cultural development appears twice: first, on the social level, and later, on the individual level; first, *between* people *(interpsychological)*, and then *inside* the child *(intrapsychological)*. This applies equally to voluntary attention, to logical memory, and to the formation of concepts. All the higher functions originate as actual relations between human individuals. (p. 57)

By linking the human mind to the social and the cultural world of interpersonal relations, Vygotsky, like Mead, provided us with a view of human thinking that necessarily is situated and subject to variations as a function of changing cultural and historical factors. Thought does not rise above or transcend its sociohistorical locus; it is invariably tied to it.

2.2. Mills on Motivation

In an impressive work, C. Wright Mills (1975) introduced a view of human motivation that removes motives from the individual as such and gives them a more social, interpersonal direction. More recently, in his critique of attribution theory, Buss (1978) adopted a similar perspective in differentiating between causes and reasons. But first Mills.

For Mills, motives were not conditions that exist prior to action and that serve as the springboards to action:

> Motives may be considered as typical vocabularies having ascertainable functions in delimited societal situations (p. 162). . . . Motives are accepted justifications for present, future, or past programs or acts. (p. 164)

Motives enter as reasons or public accounting practices that provide adequate grounds for action that has already been undertaken or that the individual intends to take.

Perhaps the best sense of Mills's analysis comes from examining one of his own examples:

> Individualistic, sexual, hedonistic and pecuniary vocabularies of motives are apparently now dominant in many sectors of twentieth-century urban America. Under such an ethos, verbalizations of alternative conduct in these terms is least likely to be challenged among dominant groups. In this milieu, individuals are skeptical of Rockefeller's avowed religious motives for his business conduct because such motives are not *now* terms of the vocabulary conventionally and predominantly accompanying situations of business enterprise. A medieval monk writes that he gave food to a poor but pretty woman because it is "for the glory of God and the eternal salvation of his soul." Why do we tend to question him and impute sexual motives? Because sex is an influential and widespread motive in our society and time. (pp. 166–167)

This passage describes the public and interpersonal quality of the motive concept. Insofar as sexual and economic reasons serve today as adequate grounds for behavior, they—and not religious fervor—serve as integral parts of our motive vocabulary. It is not a matter of our behavior's being mechanically triggered by sexual or economic drives, as the dominant psychological paradigm suggests. Rather, these have become socially acceptable formulations that render our behavior intelligible to others and to ourselves. What we take to be motives must be evaluated within a given sociohistorical context and its understanding of acceptable grounds for human action.

The implication of this view for the justice motive is clear. In Mills's hands, justice motivation is to be understood in terms of the vocabularies of justice that characterize a given society at a given point in its history. If we adopt Buss's terminology (1978) from the philosophy of the mind, we can speak of justice reasons: acceptable, public accounting practices or grounds for explaining past acts as "fair" and formulating future, intended acts that will be recognizable as "fair" or as "just." The intrapsychic quality of the motive is removed, and an interpersonal base and meaning are installed in its place.

2.3. Sullivan on Personality

Harry Stack Sullivan's definition of personality conveys an interpersonal rather than an individualistic–intrapsychic emphasis: *"personality is the relatively enduring pattern of recurrent interpersonal situations which characterize a human life"* (1953, pp. 110–111). In a manner reminiscent of both Mead and Vygotsky, Sullivan saw the individual as a portion of an interpersonal field and not as an entirely separate entity. Therefore, processes that we usually assume take place *within* the individual occur within the interpersonal field, *between* individuals. Sullivan's view of the self-system also parallels that of Mead: both saw the self as an emergent of social interaction (e.g., Sullivan, 1953, 1954). And, for Sullivan as for Mead and Vygotsky, language plays an important part in the process by which personality emerges and is sustained.

Because of his psychiatric focus, Sullivan, unlike the others, was especially concerned with distortions in self-system and personality. As fitting his interpersonal theory, Sullivan saw mental disorders as involving problems of interpersonal life and association. Indeed, Sullivan's concept of parataxic distortions gives a rather Meadian flavor to the entire issue of transference. The parataxically distorting perceiver is addressing an audience in thought and action that includes "the other people in the room," in addition to those who are actually present:

> The great complexity of the psychiatric interview is brought about by the interviewee's substituting for the psychiatrist a person or persons strikingly different in most significant respects from the psychiatrist. The interviewee addresses his behavior toward this fictitious person who is temporarily in the ascendance over the reality of the psychiatrist, and he interprets the psychiatrist's remarks and behavior on the basis of this same fictitious person. . . . Such phenomena are the basis for the really astonishing misunderstandings and misconceptions which characterize all human relations. (1954, p. 26)

What may appear to be a crazy pattern of perception and experience is seen to be an address to someone not actually present. It is interpersonal, however, in that it involves addressing another and viewing the self from that other's perspective.

3. THE ADDRESS FRAME

This brief overview of the key ideas of Mead, Vygotsky, Mills, and Sullivan introduces the outlines of an alternative, interpersonal approach for social psychology. Thinking, motivation, and personality are conceptualized as interpersonal processes and not as intrapsychic events. Using these ideas as its basis, I would like to introduce the concept of *address frame* (or frame of address) as an important new formulation for social-psychological inquiry.

I define an address frame as the standpoint that people assume or adopt in "imaginatively" completing the responses to their own gestures. An address frame, therefore, is *the whom* that people address in their external or internal conversation of gestures, *the other* whose responses render persons' own gestures meaningful. For example, Jane adopts what she imaginatively construes to be Sandra's response to her own (that is, Jane's) gesture and thereby addresses herself as she would be addressed by Sandra. She sees and hears herself as an object from Sandra's standpoint.

Address frames fulfill Mead's basic formulation of thought and meaning and are likewise consistent with Vygotsky's interpersonal view of mind. Mead spoke of a three-phase process involving (a) P's gesture, (b) the completed social act that it indicates, and (c) O's response to the gesture. The initial phase of the social act involves a gesture that P makes or intends to make. This gesture is said to indicate some resultant or later phase of the social act that the gesture has initiated. Finally, there is the adjustive response of the other person (his reply) to the gesture and what it has indicated.

By adopting O's standpoint (i.e., taking O's role), P can imaginatively complete O's response to the gesture. In this way, the response called out in P is similar to the response called out in O, because by adopting O's standpoint, P can imaginatively construe O's response to the gesture. I refer to this as an *address frame* because thinking involves another whose standpoint is taken (addressed) in order to render gestures meaningful.

We formulate our thoughts within the context of an address frame involving the implied responses of another to our gestures. The very act of thinking requires the imposition of some other whose responses we enact, if only in our imagination. That is what Mead meant when he noted that even those who engage in abstract thought refer the elements of their thinking to some audience whose implied response renders their thoughts intelligible. It is what Vygotsky meant in viewing human thought as having interpersonal origins. This is what Mills meant in arguing that motive vocabularies are socially accepted grounds for explaining and justifying action; in other words, the motive term locates an action within an address frame given in terms of the particular social group within which such frames operate. Likewise, it is what Sullivan meant in his conception of personality as an interpersonal process and distortions as experiencing, in terms of an address frame, "the others in the room."

Mead's three-phase process of social interaction can be conducted *externally*, in which case thinking is addressed to those actually present; or it can be conducted *internally*, in which case there is an implicit conversation of gestures in which the other is addressed imaginatively. This process (gesture–resultant–response) permits people to indicate to themselves what they indicate to others; they can call out within themselves, in response to their own gestures, the response that they assume will be called out in others.

For example, as I make the preliminary gesture of raising my arm with my fist tightly clenched, you, seeing this gesture, can project my intent as being harmful and so withdraw in a defensive posture as your adjustive response. In that I can imaginatively complete your response to my gesture (withdrawal), I can thereby render the meaning of the gesture to be menacing rather than playful. I can put myself in your shoes and view myself making my gesture and eliciting a response. In this manner, I can see myself as an object reflected through your perspective. I can also think about what meaning my behavior will have as a function of the particular others whose responses to my gestures I assume.

To say that thinking occurs within a particular address frame, therefore, is to note, along with Mead, Vygotsky, Mills, and Sullivan, that major psychological processes refer to some audience or other whose framework we adopt in conversing with ourselves and in grounding our action. Mind is an interpersonal process that is internally represented. While there are many implications that flow from this formulation, for our present purposes the most critical involves the way in which we have now brought context, including sociohistorical setting, directly into our understanding of social-psychological processes. Both in the development of thinking, motivation, and personality and in their ongoing existence, others are invariably involved.

3.1. Literal versus Indexical Meaning

Because our existing formulations of social-psychological processes delete or ignore the particular context in a search for fundamental universals of psychological functioning, they inappropriately treat meaning in a *literal* rather than an *indexical* manner (Wilson, 1970). To be literal is to assume that the meaning of a gesture has a transcontextual, objective quality. To treat a gesture as indexical, on the other hand, is to note, as Garfinkel (1967) suggested, that gestures "do not have a sense that remains identical through the changing occasions of their use" (p. 40).

Both Gergen (1978) and Harré (1977) made this same point. Gergen noted, for example, how "The pointing of a finger . . . may signify aggression . . . may be used to indicate an altruistic giving of information, a positive or negative attitude, egocentrism, or high achievement" (p. 1351). Recall Vygotsky's example of pointing, in which the meaning is engendered in the child through the interpersonal process of interacting with the mother. Insofar as the child continues to address his acts as though he were addressing his mother, he will evoke a similar meaning for the gesture.

Harré (1977) described a handshake, noting that

> I may not know what you are doing when you take hold of my hand. I might be very puzzled as to what the gesture means, not knowing in which structured sequence of actions it occurs. It might be part of a betting ritual, or an opening move in a karate encounter, or part of various other action sequences. (pp. 284–285)

Intelligibility, therefore, emerges only contextually (i.e., indexically) and does not have a literal, transcontextual quality.

The concept of address frame requires that meaning be established in terms of the particular other whose standpoint is assumed; thought and meaning are thereby situated within a social context. The meaning or sense of a gesture is not rendered independently of the social process within which it emerges and to which it refers. While the meaning of a given gesture may appear to achieve a transcontextual objectivity, this appears only insofar as members of a common culture have learned to adopt similar address frames in locating their gestures, for example, the addressing the abstracted, generalized other.

4. JUSTICE MOTIVATION AS ADDRESS FRAME

There are several implications that follow once we conceptualize what heretofore has been an intrapsychic motive force or need (e.g., justice motivation) as an interpersonally based address frame. We will consider four ideas: (a) the meaning of justice as address frame; (b) the role of sociohistorical factors; (c) the negotiated quality of justice; and (d) the changing intelligibility of the justice address frame.

4.1. The Meaning of Justice Motivation as Address Frame

Our first task is to specify what justice motivation means within the context of an address frame. An address frame grounds action in the social process; it makes people's acts intelligible to others and to themselves. In this view, justice becomes part of a conversation taking place between people. Justice motivation is removed from inside the mind to become an interpersonal process, because, in the view of an address frame, mind itself is interpersonal; that is, the process of thinking involves addressing oneself within the frame or standpoint of others. By locating justice within social interaction, we are better equipped to understand how specific situational and general sociohistorical events influence the ways in which justice functions in guiding social behavior.

In address-frame terms, justice is a way in which peoples' actions become intelligible to and may be evaluated by themselves and others; it does not refer to an internal state of need. Intelligibility requires locating an intended or completed act within a particular address frame and seeing the act from the standpoint of the others with whom people are interacting or who represent their social community. For example, if Bill adopts the address frame of his social group (if he views his intended acts from their standpoint), then he and they can judge his acts as being warranted ways to constitute "justice" and "fairness." This is what most of us routinely do.

To say that the act of equitably allocating resources by giving more to those with the greatest inputs is a just act requires that we employ an address frame that constitutes that meaning. In this view, there is nothing intrinsically or objectively or fundamentally or literally "fair" about such an act. The sense of fairness emerges by virtue of its evoking a response from certain others that indicates this meaning. By addressing ourselves to those others, we can adopt their standpoint and view our intended act of equitable allocation as being one that will mean "just" and "fair."

4.1.1. Some Relevant Research

It should be apparent that when different frames are addressed, different meanings for the same act will emerge. To act equitably when addressing certain audiences, therefore, would not render the act "just" and "fair"; perhaps it would appear to be exploitative, competitive, or status-seeking. It is in light of this and because actors address themselves to others in guiding their actions that just meanings emerge as part of the social process.

A recent study on the "naive psychology of motives" reported by Maki, Thorngate, and McClintock (1979) is relevant in the context we have been developing. Maki et al. had subjects make predictions about the motivations of what were actually preprogrammed game players. Their data indicated that motives of individualism and competition were more readily understood than motives of altruism and cooperation. Further,

> To the extent that an actor repeatedly behaves in an atypical fashion (i.e., displays a preference for negative over positive outcomes to self) . . . observers . . . have difficulty identifying the actor's choice rule. (p. 211)

These findings support a view of motives as socially situated ways that people employ to render actions intelligible. It was difficult for subjects to comprehend a selfless player in a competitive game. It is in cognizance of this social siting of motives and because thinking involves adopting an address frame that people are able to gear their actions into intelligible forms.

Although the authors did not explore the intelligibility that did emerge when actors were selfless and altruistic, we can speculate on several possibilities. If your acts lose points for you and give them to me, you are not addressing me as an opponent. Your overgenerosity might mean that you find me "needy" or perhaps even "weak" and "incompetent." In this view, our acts communicate the qualities we assume that our audience has. Members of that audience fit the meaning of our behavior into a framework that is also informative to them about the kind of person we think them to be.

A variation on this theme was recently suggested by Morgan and Sawyer (1979). They noted, for example, that for one nonfriend or stranger to permit another to receive a greater allocation in an exchange would be to communicate an admission of inferior status on the part of the former. This led the authors to suggest that in order to avoid this admission in the absence of good evidence to support it, a person would behave competitively in order to restore equality.

This formulation is consistent with the address-frame concept. Actors address themselves to a nonfriend-other; from that standpoint, they would appear to be of inferior status unless they behave competitively. They do not seek equality in order to convey a meaning of solidarity; this might occur were they addressing themselves to a friend. Rather, they seek equality in order to create the meaning of noninferior standing, given that they are addressing themselves to a nonfriend.

Research reported by Clark and Mills (1979) further affirms the usefulness of the address-frame approach to the social psychology of justice. They distinguished between exchange relationships (as in business dealings) and communal relationships (as in families and friendships). They found that attraction between people varies as a function of whether an act is addressed within the proper frame (this is my interpretation of their findings).

To frame an act within an exchange relationship, for example, when people believe that communal bonds exist, is to invite lesser attraction than to frame the same act within its more proper (i.e., expected) address frame. Specifically, Clark and Mills reported that when two people are in a communal relationship (e.g., expect to be in a friendship rather than a more impersonal, businesslike relationship), and one asks for a favor in return for a favor given, attraction is actually reduced: "treating a communal relationship in terms of exchange compromises the relationship" (p. 23).

The authors commented on the implications of this finding for the applicability of equity theory to liking relationships. Their findings indicate that to treat a liking relationship as though it were a business exchange by insisting on equity as the defining rule of justice is to challenge the communal nature of the relationship. These data further affirm the point I have been making: the context within which an action occurs defines the meaning of that action, in this case, the sense of justice that occurs. To repeat, because people engage an address frame in thinking and formulating both their intended and their completed actions, they introduce contextual factors (i.e., the standpoint of the other) and are thereby able to develop their ongoing course of action in light of the knowledge about its meaning to those with whom they are interacting.

From the perspective of justice as address frame, we can see how addressing oneself to another as business partner when that other expects to be taken as a friend can lead the act to be defined as inappropriate. Therefore, people typically address their acts in ways that take the other's standpoint into consideration. In order to do this, they think in terms of what their acts would appear to be from the other's point of view, for example, "What meaning would my request for a return favor have when the other believes us to be friends?"

4.2. The Role of Sociohistorical Factors

The concept of address frame translates sociohistorical factors into a distinctly social-psychological perspective and gives them a central role in our understanding of human social behavior. People in different historical eras have learned to formulate their thought and understanding in different interpersonal contexts. Furthermore, in their everyday activities, people continue to address themselves to significant others as figures embedded within particular sociohistorical contexts. Events that produce social change, therefore, establish a different context within which psychological processes develop and to which they continually refer.

By rooting thought, motivation, and personality in society and in history, we have a basis for expecting that what we now discover about human psychology (e.g., justice motivation) is not inevitable nor even a necessarily universal view; rather, it represents a formulation thoroughly embedded in the contemporary social context.

4.3. The Negotiated Quality of Justice

The conceptualization of justice motivation in terms of an address frame also suggests that issues of justice have a kind of negotiated quality. Because people do not invariably address the same others in their thinking, they may fail to sustain a shared conception of social reality; all social interaction has a potentially negotiated quality.

Imagine a student who wants to take an exam one day late because of an important family affair on the exam day, and a professor who claims that it would be unfair to give this one student special privilege. Both parties try to ground their behavior in terms of a justice analysis, attempting to get the other to accept their formulation.

When we treat justice motivation intrapersonally as a fundamental psychological force that operates in a similar manner in all people, we tend to overemphasize consensus over conflict and negotiation. However, if we view justice motivation interpersonally, in terms of a concept such as address frame, we introduce the process of people interacting together and attempting to negotiate some agreement or compromise over what will be accepted as just and fair.

The negotiated quality of justice motivation is especially relevant in the context of social change. It seems reasonable to suggest that changing social forms introduce competing formulations of social reality. As a result, justice does not involve simply fitting human behavior into a known and pregiven formula; it involves a complex process whereby people address their views to those with competing views and attempt to accomplish some reasonable joint formulation of social reality. This brings us to the next point.

4.4. The Changing Intelligiblity of the Justice Address Frame

Even to know when a situation involves justice requires something akin to a social psychology of address. I am not aware of work to date that has examined this matter. The standard research model has the investigator implicitly adopt an ongoing cultural framework that already assumes when justice is or is not involved. People's responses to these predefined situations are then studied. From my point of view, the determination of when an issue involves justice or fairness and when some other matter is involved is itself a matter of substantial importance. It is likely that what is defined as an issue of justice has changed historically and culturally and that thus, there is not even a universal understanding about when a particular question can intelligibly call upon justice motivation.

For example, youngsters today cry foul and insist that they are being treated unfairly when they cannot stay up to watch the late show, since "All the other kids' parents let them stay up." But does this mean that the issue is really—that is, profoundly and fundamentally—one of justice, or that the justice frame exists today as proper grounds for these family issues even if it did not throughout human history? It is likely that in earlier periods, children did not even contemplate comparing their fate in their own home with the fate of other children in another home. This type of social comparison represents a more contemporary formulation that reflects the intrusion of the public world and public values into the once private domain of the family and the household.

There is some basis for arguing in this way. Mannheim's (1971) account of romantic as contrasted with natural-law thought in the area of justice makes this matter clear. In the romantic formulation, fairness was not at issue when someone in Household or Village B was treated as a criminal for an action that in Household or Village A would not be considered criminal. It was understood that each household or village was its own organic whole with its own rules and regulations governing the conduct of its members, and that comparisons across such universes were not appropriate. In other words, the issue was not one of justice or fairness. In natural-law formulations, however, in which universal principles that applied to all persons everywhere were sought, it was clearly a matter of injustice if a crime in Setting B were not also a crime in Setting A.

Most of our present social-psychological views of justice ignore the sociohistorical context by taking the contemporary form for granted; thus, they have not, to my knowledge, even attempted to study the circumstances under which justice is introduced as a relevant issue. This endeavor would appear to be a crucial matter, however, if our goal is to understand the relationship between social change and justice; changed social arrangements should facilitate the reasonableness of certain formulations of justice motivation while casting others aside.

5. JUSTICE AND SOCIAL CHANGE

The foundations have now been laid for viewing justice motivation in the context of changing social forms and arrangements. Sociohistorical factors establish the terms within which different formulations of justice motivation render social behavior intelligible and serve as guides to behavior. To understand justice, we must explore some of the important sociohistorical currents and movements that affect the conceptions of justice motivation that develop. I would like to offer a brief historical perspective on justice motivation that will carry us to the present and suggest a somewhat bleak view of the future of justice.

5.1. Particularism versus Universalism

Two competing themes form the basis of this overview: particularism and universalism. A particularistic form of social organization, and the resulting particularism of social thought, selects certain, usually intrinsic, qualities of the individual and treats them as warranting special consideration. A universalistic formulation deemphasizes the particular and the concrete in the pursuit of certain abstract and broadly universal standards that can be applied to all people, that is, standards that are not intrinsic to the person but that are "detachable" and thereby applicable to anyone. Particularism emphasizes the embedded and contextual qualities of human life. Universalism seeks to disembed the person from the context and to establish transsituational principles that are blind to special contextual features; objective standards are sought to make judgments regarding deserving and justice. These standards are said to be objective only insofar as they reflect the currency of the total human realm and are not restricted to any particular group, time, or location.

In much of human history, particularism has been associated with special privilege based on kin and family bonds, while universalism has been associated with privilege as a function of certain equally available and readily calculable qualities that everyone, regardless of heritage, could potentially possess. Deserving special treatment is still a part of universalistic formulations; the emphasis, however, is on establishing criteria to warrant this privilege that are devoid of the particularism of primary bonds and that stress calculable (i.e., measurable and quantifiable), impersonal standards, especially of individual achievement.

A particularistic view would argue, for example, that because Jane is Dr. Benson's daughter, she should have special access to medical school. A universalistic view would argue that while certain people should clearly be admitted over others to medical school, the criteria for selection should minimize particularistic (i.e., kin-based) factors and emphasize universalistic standards, such as grades or test scores. The latter not only reflect individual achievement rather than family ties but also involve a universal yardstick that can be applied equally to all. Grades and test scores are usually considered more objective than the unquantifiable features that particularism suggests. Basically, grades are something that can be detached from the particular individual and measured in a standard form; one's family ties cannot be similarly detached nor universally standardized.

5.2. A Capsule History

I believe that an appreciation of human history in the Western world can be obtained by noting the emphasis that an era gives to particularism or universalism. These emphases define the sociohistorical terms within which different formulations of justice motivation appear and render behavior intelligible to members of the society. To appeal to universalistic standards when particularistic forms are being addressed by others, for example, would not serve as adequate grounds for guiding one's intended acts or justifying one's completed acts. The reverse would likewise be correct, that is, appealing to particularism in an era of universalism. Let me first sketch in the historical shifts in the particularistic–universalistic emphasis before further developing certain points.

5.2.1. Primary Particularism

In the years before the sixteenth century, particularism described the ethos of the Western world. People were not treated as individuals but as members of a village, a tribe, a group, a family, and so forth. Each collectivity had its own organic quality and was governed according to principles unique to its own manner of being. Background and heritage were the important warrants for receiving special privilege; primary group ties of family and kin were the accepted grounds of justice.

5.2.2. The Enlightenment and Individualism

The sixteenth, seventeenth, and eighteenth centuries saw the emergence of the individual as the key element of society and the introduction of universalistic formulations to all facets of human organization and activity:

> In religion, this . . . took the form of Protestantism; in economics the form
> of *laissez-faire*; in social ethics the form of the greatest good for the greatest
> number; in progressive politics the form of universal suffrage and repre-
> sentative, parliamentary government. (Ratner, 1973, p. 28)

This was the era of individualism: the individual was disembedded from the context of kin and family and became the atom of society. Each individual was to be treated as roughly equivalent to and subjected to the same universal principles as were applied to every other individual. Readily calculable standards of individual achievement became the centerpiece whereby universalism could be accomplished. What you did, how well you did it, and how much energy and effort you expended— all became the measurable, universalizing guideposts to override who you were as determinants of privilege and worth. Later, standardized tests became available to provide further universalistic ways to calculate individual deserving.

5.2.3. An Incomplete Task

The universalistic pursuit of equal treatment for all, however, was incomplete. It did not include blacks and other ethnic or religious minorities, nor did it include women. These wore particularistic badges that were generally considered sufficient to warrant less than equal treatment, regardless of whatever other universal standards were generally applied to "white males." Furthermore, the range of universalism covered the public world of business and economics and the supporting ideology-creating and sustaining institutions of religion, politics, science, and education. The private domain of the home and the family continued to be particularistically governed. Justice in the public domain differed from justice in the private and intimate domain of the home. The former was considered within the purview of universalistic standards; the latter, within the particularism that made each home unique and noncomparable with any other.

5.2.4. Extending Universalism

The present era involves an effort to complete the incomplete work of the earlier period of universalism. Universalistic formulations are being brought to those groups that had originally been excluded, for example, blacks and women. In addition, the particularism of the private and familial domain is slowly yielding to this same press of universalism. Universalistic doctrines have found their way into more and more parts of the social world as part of the publicization of all society and social life. As these efforts continue, tensions among competing concepts of justice become increasingly apparent. There is no full consensus concerning what is and what is not just. Where once universalism fought to win the moment over the particularism of traditional society, we now see signs that particularism is reemerging to counter the universalism of the modern world.

5.2.5. A New Particularism

The effort to extend universalism has produced what appears to be a new kind of particularism. If the earlier era emphasized particularism based on primary-group bonds, this new era is beginning to stress particularism based on more secondary bonds, that is, the less personal and nonfamilial ties of human association including gender (e.g., women's rights), race (e.g., civil rights), sexual preference (e.g., gay rights), and consumer status (e.g., consumer rights). People are cast together into the same particularistic category and warrant special treatment because of their secondary qualitites. Paradoxically, the extension of universalism is giving rise to new forms of particularism and introducing further dissensus about the proper conception of justice.

5.2.6. A Future Scenario

My conjecture about these trends offers a somewhat bleak portrait of the future. Within a dominant sociohistorical ethos of atomistic individualism, I anticipate that the extension of secondary-group particularism may eventually lead to a personal particularism. We might look forward to a time in which what is just and fair is defined primarily in terms of personal self-interest, the narcissistic era's view of justice motivation. And this will occur simultaneously with a declining resource base and a demand for greater mutual cooperation and a pulling together among strangers.

Having briefly sketched in the outlines of the historical picture, let me now develop several parts of this outline. This should help to provide a better sense of the way in which justice motivation changes with sociohistorical currents and movements and thereby can be appreciated only within its proper setting.

6. TENSION AND DISSENSUS

In spite of the apparent dominance of universalism in the contemporary world, including psychology's favorite theories about the ideals of justice motivation (e.g., the use of highly abstract principles that are disembedded from any sociohistorical context), there remains a vital undercurrent of particularism. Social commentators have noted, for example, the persistence of particularistic bonds involving language, ethnicity, and nationality that have continued to join people in more traditional ways in spite of the press of modernism (e.g., Chirot, 1977; Gordon, 1978; Williams, 1978).

How can particularistically bonded people address themselves to a universalistic formulation of justice? What intelligibility derives from the universalistic equations for human justice among people who continue to define themselves as members of very particular segments of society and who believe that they and their people deserve certain special rights and privileges? The persistence of traditional loyalties in an impersonal and universalistic setting helps fuel tensions and disagreements over competing conceptions of justice motivation: there is less consensus and more effort to negotiate settlements than most of our existing social-psychological theories of justice motivation would lead us to expect or understand. Let me develop this point with an example.

6.1. The *Bakke* Case

The recent uproar over the case of the white medical-school applicant who claimed to have been passed over in favor of a black applicant with lesser qualifications has brought to light the emerging tensions between different formulations of social justice; the case highlights a particularistic (affirmative-action) and a universalistic (reverse-discrimination) view. The case did not resolve the complexities of race relations; its argumentation, however, reveals the several dimensions of the issue.

According to the universalistic conception, all people openly compete to achieve desired goals, for example, entry to medical school; no particularistic standards should be applied in determining whether or not a person reaches the goal. To employ a particularistic category such as race or gender appears to violate the universalistic demand for equal treatment for all. Basically, it is regarded as being unfair and as involving a kind of reverse discrimination. In its deliberations, the U.S. Supreme Court revealed the several competing and as yet unsettled elements of the universalistic–particularistic formulation.

The court recognizes a principle of equality under the law. This principle says that justice is blind and so must disregard any particularistic standards, but especially those of race. Furthermore, the court searches for

> principles sufficiently absolute to give them roots throughout the community and continuity over significant periods of time, and to lift them above the level of the pragmatic political judgments of a particular time and place. (Powell, 1978, p. 5)

These two elements—(a) justice is blind to particularism based on collectively ascribed but not individually achieved characteristics and (b) justice must be based on universalistic and absolute principles that transcend their sociohistorical context—describe more than the guidelines employed by the court. They represent what I have presented as the Enlightenment or natural-law formulations that have dominated all areas of our contemporary world. To note this, of course, is not to say that this view is incorrect; it is simply to locate such thought in its sociohistorical context.

Justice Brennen, however, noted that

> we cannot . . . let color blindness become myopia which masks the reality that many "created equal" have been treated within our lifetimes as inferior both by the law and by their fellow citizens. (1978, p. 7)

He continued:

> Properly construed, therefore . . . a state government may adopt race-conscious programs if the purpose of such programs is to remove the disparate racial impact its actions might otherwise have and if there is reason to believe that the disparate impact is itself the product of past discriminations, whether its own or that of society at large. (p. 10)

It is not my purpose to examine the complex legal details of this case. Rather, I wish to use it to illustrate the nature of the tensions between universalistic and particularistic formulations of justice, especially with respect to race and perhaps gender. I say "perhaps gender" in that Justice Powell, for example, commented in his statement on the *Bakke* case that "the perception of racial classifications as inherently odious stems from a lengthy and tragic history that gender-based classifications do not share" (p. 5). It appears that the politics of justice motivation is not yet ready to perceive gender discrimination in the same light as race discrimination. My guess is that in time, gender and undoubtedly other particularistic factors will be added to the new understanding of what is just and fair.

The basic dilemma about justice highlighted in the *Bakke* case involves the failure of equality, as a universalistically applied principle, to establish actual equality for groups that, because of their particularistic characteristics, had been systematically discriminated against. Some people speak of *reverse discrimination*, in which an attempt is made to impose a nonparticularistic principle across the board, and any particularistic selection is defined as giving unfair advantage to one group over another. This approach is seen as reverse discrimination in that blacks are given advantages over whites or, perhaps, women an advantage over men. The concept of *affirmative action* does not wash for many people who understand justice universalistically.

As the U.S. Supreme Court noted, the reintroduction of particularism is necessary in order to undo the effects of universal equality that was never universally applied. A general principle of equality fails to obtain equality when those who were initially disadvantaged are treated equally with all others; that is, as though the initial disadvantage were itself blind. Equity also fails; it ignores disadvantages that those participating from different starting points bring.

In other words, both equity and equality pose problems for certain contemporary conceptions of justice—equality, because it ignores early disadvantage that is only exacerbated by treating all equally today; equity, because it fails to take into account that not all begin the race from the same starting point, for example, that whites, men, and those born to money have an advantage well beyond their individual efforts and achievements.

The *Bakke* case and its argumentation inform us of the tensions now present between two different justice motives. The one is based on a universalistic conception, born in the Enlightenment and seeking universal standards for all, which, in actual practice, were never applied to all. The other is based on a particularistic conception and its competing outlook; special claims are made for special interests—in the present instance, for those who were initially disadvantaged.

In highlighting the tensions between competing formulations of justice, the *Bakke* case returns us to the view that justice motivation must be seen contextually and not as some fundamental trigger or force driving individual psychology. In many respects, the U.S. Supreme Court's deliberations and their decisions attempt to settle the negotiated meaning that the present sociohistorical moment will grant to the concept of justice and fairness. The court's *Bakke* decision, however, has not yet firmly settled just what these qualities of justice will be.

7. JUSTICE MOTIVATION AND THE FUTURE

The Enlightenment ushered in a period of universalism that has not yet been completed. It seems reasonable to suggest that the era before us will see attempts to extend universalism more widely than ever before in human history. The sociohistorical context will require that justice come to mean universalism for all. Paradoxically, as we have seen with the *Bakke* case, this extension will demand that particularistic qualities be emphasized in order for the range of universalism to be extended.

One of the primary features of the near future will be its emphasis on particularistic formulations of justice among groups competing in a context that has heretofore denied them that equality. Practically speaking, this will mean that what some now think of as reverse discrimination will be in competition with a formulation in which particularistic standards and preferences are held to be just.

Once particularistic standards reemerge as critical elements in justice decisions, the claims of specific interest groups are likely to become a deafening roar. I cannot foretell how this conflict will be met or negotiated other than to note that the crowd for the inside track will be very large and competition will be substantial.

If this scenario is played out, tribal marks based on primary bonds are likely to give way to marks based on temporary, secondary-group ties. These new particularistic markings will play a critical role in defining the nature of justice and deserving. The gray flannel suit was once the mark of homogenized corporate man; the new look will be in the form of proudly proclaimed particularistic identity badges, such as "men," "women," "blacks," "gays," "ethnics," and consumers. The old forms of particularism based on kinship will yield to these new forms. To make their demands for the equalities denied to them, groups will bind themselves in terms of previously unchartered interests: consumers, taxpayers, the elderly, the young, mental patients, prisoners, welfare recipients, and so on. This means that being old, for example, will be a particularistic warrant for special treatment, even as the other badges will serve similar purposes.

7.1. Beyond Intrapsychology

What these special qualities will be or how they will become negotiated warrants for justice is not readily captured by our existing intrapsychic formulations of the justice motive, for example, equity theory (Walster & Walster, 1975) or even the just-world view (Lerner & Miller, 1978). Our present formulations assume some "basic nature to man." In equity theory, people are said to be concerned with the proportionality of investments and outcomes; the just-world theory speaks of a need to believe that people get what they deserve. Once we assume a basic human nature, we ignore the sociohistorical context in which that nature originates and in which it operates to render behavior intelligible. We are ill prepared to deal with changing sociohistorical circumstances in which a different nature can emerge and engender the cognitive or motivational terms to guide and explain behavior. Likewise, we are ill prepared to comprehend the competing versions of justice that emerge within a society or the kinds of negotiated settlements that carry the message of justice, if only for a historical moment.

Both equity theory and the just-world approach refer primarily to a segment of contemporary society. They both originated in the Enlightenment and the subsequent era of individualism. Equity requires a society in which universalism is held as an ideal and in which each person is compared with others in universalistic terms. Equity would not make the same sense in a more traditional society or in a domain in which the integrity of particularistic groupings is maintained, that is, where non-comparables are involved as members of unitary wholes and not as individualized parts.

The just-world view likewise requires a society in which individual action and accomplishment have a reasonable degree of correspondence. Again, this is a description of the era of the Enlightenment and the emphasis on individual action as a means to individual accomplishment. More traditional social forms would not be as congruent with a just-world view as the contemporary forms seem to be. It would not render behavior intelligible in the same way (i.e., as just) in a society in which action and accomplishment were not coordinate, one in which one's identity was a greater key to one's future than what one did.

Only by locating justice within its sociohistorical context, as the address-frame view accomplishes, can we hope to understand better how social change will affect the nature of justice behavior. If the analysis I suggested as describing history works out, then the years ahead will see a growing tension between universalistic and particularistic formulations of justice. In such a setting, equity, the just-world need, and other intrapsychic formulations are likely to line up on the side of universalism. Leaving aside the value issues that might thereby arise (e.g., affirming the existing shape of society), it is clear that such views will not be adequate to deal with the complexities of justice in a changing sociohistorical and conflict-ridden world. These formulations address themselves to a particular social form of which they are generally oblivious, given their pursuit of a basic human nature that transcends society or history.

8. SCARCITY AND JUSTICE

The historical scenario I have suggested says that the extension of universalism will itself demand particularistic emphasis. This emphasis will tend to increase the distinctions among people even while it binds smaller, temporary groupings together. The organic quality of these groupings will be minimal as compared with the unity of the traditional kin groups; thus, the bonds that unite will not significantly reduce the atomistic quality of the era of individualism.

Particularism will provide a temporary bond for special-interest groups seeking their justice. The tribal quality will diminish as the public-interest quality of such temporary alliances enlarges. The isolation and segmentation of the individual will become more widespread, and a scarcity of caring is likely.

Recall that justice motives offer sanctioned reasons for behavior in situations in which fairness is said to be an issue. The scarcity of caring will define as "just" actions designed to maximize self-gain. Several social critics have commented on the era of narcissism into which we are now moving (e.g., Lasch, 1976; Sampson, 1977; Sobo, 1975). If this era has a central theme, it is the emphasis on self-gain and personal interest. Is it so farfetched to imagine a day in which justice will be defined primarily in self-interested terms, with the boundaries of particularism going no further than one's own immediate person? After all, if it is considered fair to grant special privilege to temporary groupings based on collective self-interest, then surely there is no greater sense of particularism than one's own immediate self.

The obvious problem, however, is that while narcissism may appear to be clearly oriented to personal benefit and gain, in a world of scarce resources a scarcity of caring can lead only to loss of benefits for all. In such a setting, the temptation to yield to those of great power will be substantial. The autocracy of brute force or the autocracy of charisma can readily rise up and seem to offer rest from the conflicts and uncertainties that such a world will bring. Furthermore, it is unlikely that this rush for particularism and self-interest will truly serve to extend full rights to those who were initially abandoned in the Enlightenment's limited formulation of human justice.

9. CONCLUSION

This brief venture into history and the future is merely speculation and should be regarded as such. The overriding point, however, is that our conceptions of what is just and fair are not motives that serve as springs to action from within but are socially sanctioned, public procedures that serve as the grounds for behavior. Justice motivation thereby has a sociohistorical dimension that purely intrapsychic formulations cannot handle.

When P does X in the name of being fair, what *fair* means and whether it suffices as an acceptable reason for that action is a function of the prevailing form of social organization and the cognitive and motivational systems that emerge within that form. Whereas once it was fair for one village to have a law that punished a person for something that was not thought a crime in a nearby village, in the modern era universalism became the essential definer of what was fair and just: all villages should be governed by the same guiding principles of justice.

It now appears that universalism will demand a new kind of particularism for its realization: the process of extending universalism to once denied groups will require reintroducing particularism into our understanding of justice. This process will set up a search for particularistic standards that will define people as members of certain groups deserving special treatment. New forms of justice will emerge to compete with universalistic views; these forms will accept particularistic criteria as deserving special consideration, as long as the particularistic group can establish a negotiated identity of "once having been denied." Deserving will be linked to group membership and not invariably to individual accomplishment.

I have suggested that as this new particularism gains its foothold, it will serve to segment and separate more than it will unite: the new groups will be temporary and piecemeal; they will not permeate all aspects of human life as family and kin bonds once did. Thus, we will see an even greater atomization of humanity and the emergence of a scarcity of human concern and caring. As a result, what is just and fair may well be defined in terms of personal self-interest.

None of the preceding is inevitable. It is merely one possible version of human history. It emphasizes a direction that, in my view, now seems to be emerging in the context of an already atomistically individuated society. But, as C. Wright Mills (1959), among others, noted, people have the capacity for making their own history. It is not written; it remains for us to write.

REFERENCES

Brennen, W. J., Jr. Text of the opinion by Justice Brennen. *The Chronicle of Higher Education*, July 3, 1978.

Buss, A. R. Causes and reasons in attribution theory: A conceptual critique. *Journal of Personality and Social Psychology*, 1978, 36, 1311–1321.

Cartwright, D. Contemporary social psychology in historical perspective. *Social Psychology Quarterly*, 1979, 42, 82–93.

Chirot, D. *Social change in the twentieth century*. New York: Harcourt-Brace-Jovanovich, 1977.

Clark, M. S., & Mills, J. Interpersonal attraction in exchange and communal relationships. *Journal of Personality and Social Psychology*, 1979, 37, 12–24.

Garfinkel, H. *Studies in ethnomethodology*. Englewood Cliffs, N.J.: Prentice-Hall, 1967.

Gergen, K. J. Social psychology as history. *Journal of Personality and Social Psychology*, 1973, 26, 309–320.

Gergen, K. J. Toward generative theory. *Journal of Personality and Social Psychology*, 1978, 36, 1344–1360.

Gibbs, J. C. The meaning of ecologically oriented inquiry in contemporary psychology. *American Psychologist*, 1979, 34, 127–140.

Gordon, M. M. *Human nature, class and ethnicity*. New York: Oxford University Press, 1978.

Harré, R. The ethogenic approach: Theory and practice. In L. Berkowitz (Ed.), *Advances in experimental social psychology*, Vol. 10. New York: Academic Press, 1977, pp. 284–314.

Kelman, H. C. Human use of human subjects: The problem of deception in social psychological experiments. *Psychological Bulletin*, 1967, 67, 1–11.

Kelman, H. C. The rights of the subject in social research: An analysis in terms of relative power and legitimacy. *American Psychologist*, 1972, 27, 989–1016.

Lasch, C. The narcissistic society. *New York Review of Books*, 1976, 23, 5–13.

Lerner, M. J., & Miller, D. T. Just world research and the attribution process: Looking back and ahead. *Psychological Bulletin*, 1978, 85, 1030–1051.

Levine, N. On the metaphysics of social psychology: A critical view. *Human Relations*, 1976, 29, 385–400.

Maki, J. E., Thorngate, W. B., & McClintock, C. G. Prediction and perception of social motives. *Journal of Personality and Social Psychology*, 1979, 37, 203–220.

Mannheim, K. *From Karl Mannheim* (K. H. Wolff, Ed.). New York: Oxford University Press, 1971.

Mead, G. H. *The social psychology of George Herbert Mead* (A. Strauss, Ed.). Chicago: University of Chicago Press, 1934.

Mills, C. W. *The sociological imagination*. London: Oxford University Press, 1959.

Mills, C. W. Situated actions and vocabularies of motive. In D. Brissett & C. Edgley (Eds.), *Life as theater*. Chicago: Aldine, 1975, pp. 162–170.

Morgan, W. R., & Sawyer, J. Equality, equity, and procedural justice in social exchange. *Social Psychology Quarterly*, 1979, 42, 71–75.

Powell, L. F., Jr. Text of the opinion by Justice Powell. *The Chronicle of Higher Education*, July 3, 1978.

Ratner, J. Introduction to John Dewey's philosophy. In R. Handy & E. C. Harwood (Eds.), *Useful procedures of inquiry*. Great Barrington, Mass.: Behavioral Research Council, 1973.

Riegel, K. F. From traits and equilibrium toward developmental dialectics. In W. J. Arnold (Ed.), *Nebraska Symposium on Motivation*. Lincoln: University of Nebraska Press, 1976.

Rosnow, R. L. The prophetic vision of Giambattista Vico: Implications for the state of social psychological theory. *Journal of Personality and Social Psychology*, 1978, 36, 1322–1331.

Sampson, E. E. On justice as equality. *Journal of Social Issues*, 1975, 31, 45–64.

Sampson, E. E. Psychology and the American ideal. *Journal of Personality and Social Psychology*, 1977, 35, 767–782.

Sampson, E. E. Scientific paradigms and social values: Wanted—A scientific revolution. *Journal of Personality and Social Psychology*, 1978, 36, 1332–1343.

Smith, M. B. Perspectives on selfhood. *American Psychologist*, 1978, 33, 1053–1063.

Sobo, S. Narcissism and social disorder. *Yale Review*, 1975, 64, 527–543.

Sullivan, H. S. *The interpersonal theory of psychiatry*. New York: Norton, 1953.

Sullivan, H. S. *The psychiatric interview*. New York: Norton, 1954.

Vygotsky, L. S. *Mind in society*. Cambridge, Mass.: Harvard University Press, 1978.

Walster, E., & Walster, G. W. Equity and social justice. *Journal of Social Issues*, 1975, 31, 21–43.

Williams, R. M., Jr. Competing models of multiethnic and multiracial societies: An appraisal of possibilities. In J. M. Yinger & S. J. Cutler (Eds.), *Major social issues*. New York: Free Press, 1978.

Wilson, T. P. Conceptions of interaction and forms of sociological explanation. *American Sociological Review*, 1970, 35, 697–709.

7

Retributive Justice

ROBERT HOGAN and NICHOLAS P. EMLER

> From what sources things arise, into them
> also is their destruction, as is ordained; for
> they give satisfaction and reparation to one
> another for their injustice according to the
> ordering of time.
>
> —Anaximander

The reader might pause and reflect for a moment on the foregoing sentence. It is the only surviving fragment from the first known work of our civilization, Anaximander's cosmology. The sentence hints at the role of retribution as an organizing force in nature and therefore in human affairs. For the Greeks, retribution was the iron law of the universe; it was equally the silent assumption on which Greek morality and Greek science was based. It is curious that a notion so fundamental at the beginning of a civilization is so infrequently remarked on in its later history. Piaget (1932/1965) captured the modern attitude when he observed that young children think of justice in terms of retribution. Older, more mature, and therefore more cognitively advanced children think of justice in terms of the equal distribution of rewards. The moral is clear: young children and persons with similarly limited intellectual perspectives endorse retributive justice. But as they grow, mature, and

ROBERT HOGAN • Department of Psychology, Johns Hopkins University, Baltimore, Maryland 21218. NICHOLAS P. EMLER • Department of Psychology, University of Dundee, Dundee, Scotland.

progress, they transcend their earlier limitations and take on a more enlightened viewpoint. The purpose of this paper is to suggest that despite this modern attitude, retribution remains a critical perspective from which to understand how the concept of justice functions at the level of the individual.

The paper is organized in six sections. The first section describes some definitions of justice, principally to get our vocabulary straight. The second section points out some reasons that the concept of distributive justice is an inappropriate vantage point from which to analyze how individuals deal with the concept of justice. In the third section, the concept of retribution is introduced, primarily from a sociological perspective. The fourth section describes some connections between retribution as defined here and equity theory as it is now understood. The fifth section analyzes the morality of retribution. Finally, the last section of the paper is concerned with the connections among power, status, and retribution.

1. JUSTICE

One searches more-or-less in vain for a definition of *justice* in the psychological literature. One reason for forgoing a definition may be that if all those philosophers have failed, how can we successfully define the term? Although it may not be possible to produce a satisfactory definition of *justice*, we can qualify the term in certain ways and thereby render it more manageable. Kelsen (1957) provided a very useful discussion of the concept, and three of his points are worth reproducing here. First of all, justice has to do with conflicts of interest; where there are no conflicting interests, there is no need for justice. Demands for justice occur, therefore, chiefly during times of discord and discontent; utopias, on the other hand, obviate the need for justice. Justice, consequently, is a regulatory principle: it provides a means for regulating existing social activities rather than providing a basis for those activities in the first place. It is necessary, therefore, to consider the activities that justice regulates; if they raise no conflicting claims, then justice is not a problem.

Kelsen's second point is that justice has nothing to do with individual happiness. Justice is a function of a given social order; but no social order can guarantee individual happiness. Rather, as Kelsen remarked, "by happiness we must understand the satisfaction of certain needs, recognized by a social authority, the lawgiver, as needs worthy of being satisfied, such as the need to be fed, clothed, housed, and the like" (p. 3). This idea is closely analogous to Freud's view that individual happiness is incompatible with social living; social living elevates the quality of life in rational and objective ways, but this elevation is often bought at the cost of individual subjective happiness.

Kelsen's third point is that man seems to need absolute justification for his actions, and that the concept of justice is often used to provide this justification; it frequently serves as a God term for philosophers. This approach, said Kelsen, reflects the delusion that reason can provide knowledge of certain fundamental principles from which absolute values can be deduced. For example, the central problem of Plato's philosophy is justice, and he solved the problem with his doctrine of ideas. The most important idea is that of the absolute good, and it is this that gives rise to the notion of justice. The absolute good is itself never justified as a first principle. Consequently, it plays the same role in Plato's philosophy as does the idea of God in any other religion: justice becomes an ineffable, revealed truth, and yet at the same time the bedrock of a moral social order. Kelsen demonstrated that the same analysis holds for Aristotle, Kant, Marx, and other writers who used justice as the central justificatory notion in their system; that is, on close analysis, justice cannot be established rationally as the ultimate principle underlying society. The fundamental role of justice must be taken on faith. It has no doubt occurred to the reader that Rawls (1971) and Kohlberg (1976) have both used justice in this quasi-religious manner, although their discussions are couched in appropriately rational terms.

What are the consequences of denying the validity of an overarching, sempiternal concept of justice? First, the concept will no longer serve as the principal legitimizing concept in social theory. It also raises the possibility that there are no abstract features of justice common to all human situations where there is a conflict of interest. Rather than talking about universal features of justice, one might do better by trying to identify the specific problem of justice or injustice at stake in each situation. This second approach reduces the concept of justice to H. L. A. Hart's (1961) maxim simply to treat like cases alike and different cases differently. Our final point, then, is that one cannot define justice in the abstract because there are no universal features of justice to be abstracted from the various areas of human conflict. Justice or injustice cannot be defined apart from the context of action where it might be relevant. Indeed, we would like to suggest that justice or injustice, far from being a universal justificatory principle à la Plato or Kant, exists principally in the eyes of the beholder, that people vary considerably in their disposition to recognize justice–injustice in specific human affairs (Hogan & Dickstein, 1972), and that they differ considerably with regard to what constitutes justice–injustice in each situation.

2. DISTRIBUTIVE JUSTICE

The argument of this section can be summarized in terms of two claims: (a) psychological discussions of justice focus largely on distributive justice; and (b) distributive justice is an inappropriate analytical focus for understanding how the concept of justice functions at the level of the individual.

As noted above, Piaget (1932/1965) regarded distributive justice as the most advanced level that children can attain in the evolution of their moral reasoning. Piaget noted the interest that children have in the punishment of transgressors. But subsequent research inspired by his book has been devoted almost entirely to children's views of positive justice and how resources, prizes, and payments should be distributed (e.g., Ugurel-Semin, 1952; Loughren, 1976; Damon, 1975). Similarly, equity theory (Berkowitz & Walster, 1976) turns chiefly on the notion that people are concerned with getting what they deserve from social encounters, that is, that rewards are distributed equitably and that they get their share. Lerner, Miller, and Holmes (1976) presented data and arguments showing that it is sometimes necessary to qualify this central claim of equity theory, but even they took distributive justice as the first principle in social relations. In general, then, current psychological literature contains a wealth of information regarding how people distribute resources or payments; see, for example, the extensive bibliographies recently compiled by Adams and Freedman (1976) and by Hook and Cook (1979). There is no comparable literature on punitive justice.

Kohlberg (1969, 1971, 1976) has attempted to develop a comprehensive account of the development of justice concepts, but this account, too, is only marginally concerned with punitive justice. Punishment is an issue in Kohlberg's system that arises in the context of other, more basic considerations, such as upholding the law and legitimate authority. His theory is explicitly concerned with the nature of people's criteria for resolving conflicts, but these are conflicts of moral imperatives, not personal interests. Ultimately, therefore, this theory is also about distributive justice—the just distribution of rights and duties. It is not addressed to and neither can it deal with the question of what to do when others break or ignore the rules.

Lerner's extensive treatment of the justice concept (Lerner, 1975, 1977; Lerner et al., 1976) also turns out in practice to be about people's concern with getting the benefits that they and others are entitled to. Despite the emphasis that the concept has received, several considerations suggest that distributive justice is a limited perspective from which to understand the psychology of justice. First, the heavy emphasis on distributive justice (e.g., in equity theory) suggests that human relations are largely about how pools of goods and resources are to be distributed.

But as Wordsworth remarked, "Getting and spending we lay waste our powers"; that is, the pursuit of worldly goods is an irreversable expense of spirit. This moral criticism aside, it is obvious that other noneconomic perspectives on human affairs are possible. So, for example, one can plausibly argue with Freud that people are largely concerned with scratching the coital itch, or with Jung that people mostly seek selfhood and a sense of personal meaning, or with Hogan that people chiefly seek status and the opportunity for differential reproductive success that status affords.

A second problem with distributive justice is that it reflects a power holder's view of the world. How to distribute resources defensibly is a problem that concerns deans, parents, political leaders, and other persons who must keep peace among the flock that they tend. They are concerned with keeping the system running. The members of the flock are concerned with keeping themselves running—they rarely consider how to allocate the available resources fairly. They are primarily interested in getting their fair share, and this is quite a different issue. Insofar as they consider distribution, they usually do so in the context of evaluating the powerholder's performance in his job. Consequently, to ask children to act in the role of a disinterested resource allocator, as many researchers have done, seems to be an irrelevant exercise.

A related issue is that discussions of distributive justice generally assume that people are in a position to know all the relevant inputs and outcomes. We would guess that this condition rarely holds in practice, and that power holders contrive to prevent it for sound practical reasons. Homans (1976) made this point in his comment that it may be better for the morale of professors, for example, if only their dean knows their individual salaries. Practically, then, most empirical treatments of distributive justice place subjects in positions that they would not normally occupy and in possession of information that they would rarely have.

A third problem with distributive justice is that it assumes that the persons among whom the resources are to be distributed are strangers. (Actually, many of the problems that concern contemporary social psychologists—impression formation, nonverbal communication, attribution processes—are not problems for people who are acquainted with one another, prompting Emler [Hogan & Emler, 1978] to remark that contemporary social psychology is largely the social psychology of strangers.) As Lerner et al. (1976) pointed out, when people identify with one another (i.e., when they are acquainted or associated as opposed to being strangers), they tend to regulate the distribution of rewards according to norms of need or parity rather than equity. We would go further than this and argue that in the context of personal associations that have a past and future and that are embedded in a social network of mutual acquaintances, the just distribution of benefits plays a small part in each person's calculations (i.e., people are more concerned about past insults to themselves, or about avoiding appearing to give insult to others, or about saving face, or about protecting their reputation among their acquaintances, or about the advantages of doing X a favor on this occasion, the desirability of not antagonizing Y, etc.).

A fourth problem with distributive justice as an analytical perspective is that it is relevant only under very particular socioeconomic circumstances. There is, after all, a prior question of where the rewards to be distributed come from. David Hume, in his *Enquiry Concerning the Principles of Morals*, observed that in a world of universal plenty, there would be no problem with the allocation of resources and that distributive justice would be a pointless consideration. Thus, when gasoline supplies are plentiful, how much you get depends only on your ability to pay for it. Similarly, under conditions of extreme scarcity, personal sacrifices are necessary and it would be pointless to insist on the niceties of reciprocal compensation. The same point is made by Lerner *et al.* (1976) and by Homans (1976): competition tends to nullify the claims of distributive justice.

Finally, many of the things that people want from and exchange with one another are not tangible, quantifiable, material resources. They are, as Talcott Parsons (1951) noted, "particularistic" resources; they have no meaning except in relation to the particular individual who gives them. These include such social and psychological resources as esteem, respect, affection, acceptance, attention, and sympathy. In these matters, it is irrelevant to talk of distributive justice. Insofar as there is a justice notion operating in such everyday social exchanges, we would guess that the only one with any generality is likely to be reciprocity: "If you don't continue to be pleasant to me, keep me informed about our mutual acquaintances, sympathize with my problems, take my side in arguments, I won't continue to do the same for you."

The foregoing remarks suggest that distributive justice is a concept with a markedly restricted "range of convenience." It is relevant only under rather delimited socioeconomic conditions, and it is useful chiefly for interpreting the actions of power holders who must distribute scarce material resources among groups of individuals. For the non-power-holder interacting with acquaintances, questions of distributive justice seldom arise.

3. RETRIBUTIVE JUSTICE

Justice, "rendering each his due" (Matson, 1978), always contains a positive and a negative side, as reflected in the terms *distributive justice* and *retributive justice*. Psychologists have focused almost exclusively on the positive side, on allocating and exchanging benefits on a just basis. Despite this contemporary emphasis, we believe that the process of retribution is older, more primitive, more universal, and socially more significant. The two concepts are related in that they both rest on the notion of reciprocity, but distributive justice is a thin, unstable, and narrowly restricted derivative of retributive justice. Our view is based on four considerations.

First, although people may be rewarded for conformity to social rules (for example, with status as Hollander, 1958, claims), such rewards are notoriously unreliable and are not part of most people's moral expectations. One is supposed to be virtuous and obey the rules because it is right, not because one expects to be rewarded for so doing. On the other hand, people normally do expect punishment to follow rule violation and lapses from virtue. Consequently, retributive justice is more salient than distributive justice in most people's ordinary social expectations. Second, most societies have institutionalized and labor-intensive mechanisms for dealing with violations of social prohibitions. No such comparably complex mechanisms exist for dealing with inequities in the distribution of rewards. Third, in the administration of justice, communities invariably punish culprits rather than compensate victims, because it is more efficient and economical to proceed in a retributive fashion. Finally, it may be assumed that most members of a community have the desire to break the rules from time to time but don't do so because of the social costs involved (cf. Box, 1972). Perhaps the reason that social control rests so much more heavily on punitive sanctions than on rewards is that all but the most improvident and incompetent have something to lose by deviant conduct.

As we have noted, retribution is not a common topic in social-psychological circles. The point of this section is to introduce the concept to the reader. The argument draws rather heavily on Hans Kelsen's (1943) discussion.

The first point to note about retribution is that it is not a "natural" concept like sympathy, jealousy, or aggression; that is, there is no reason to think that its sources are rooted in human biology. It is rather a "social" concept and, as such, presupposes the existence of a social order; retribution also entails some level of social or interpersonal conflict:

> The idea of retribution presupposes that the evil arousing the reaction has been unjustly inflicted. . . . [Retribution] implies the tendency to regard the evil sustained as a breach of norms, a violation of the social order which exists in the consciousness of the individuals between whom retribution is exercised. (Kelsen, 1943, p. 50)

Retribution is not merely a social concept; it is a legal concept as well. For example, "the institution of blood revenge which can be traced back to the beginnings of social development, indicates rather clearly that death is not only the oldest crime but also the oldest socially organized punishment" (Kelsen, 1943, p. 54). Consistent with his position as a legal positivist, Kelsen believed that retribution is also a moral concept: "Inasmuch as retribution is possible only in a society, it consequently always represents in some degree a moral principle" (1943, p. 55).

A third point about retribution that may not have occurred to some readers is that the process is a cultural universal, found in all known societies and going as far back as historical records will allow us to trace it. Colson (1974) argued that among human groups as diverse as American Indians and African tribesmen that are not subject to a central government, freedom of action must still be controlled, and the principal mechanism of control is retribution or fear of retribution. Fürer-Haimendorf's (1967) account of social controls among the Dafla hill tribes of northeast India shows clearly that prior to the institution of formal, community-managed grievance procedures, people tended to take justice into their own hands: they retaliated against those who offended them. The reason is quite clear. The man who did not revenge the wrongs done him left himself and his family open to further abuse. The same lesson can be derived from DuBois's (1961) analysis of precolonial New Guinea society. And the indefatigable Kelsen (1943) exhaustively documented the universality of retribution as an organizing force in human affairs. From this mass of ethnographic data, it appears that retribution ranks with religion, the family, language, incest taboos, and sex roles in terms of its importance in human affairs.

The process of retribution can be identified in all societies and human groups. This is not to say, however, that it always has the same form. Although the function remains similar across cultures, two markedly different forms can be identified. In the first case, when a member of a group feels that he or she has been wronged, that person takes action directly. This form of retribution is most common in those societies remote from the influence of centralized structures of governance. Examples of institutionalized direct retribution can still be found among some Indian tribes in Amazonia, in southern Asia, in certain isolated tribes in the Philippines, and in Micronesia. It is an unofficial form of governance in perhaps all rural areas of the world: mountain villages in Appalachia, Sicily, Cambodia, Chile, and New Guinea. The process of direct retribution, although permitting immediate gratification of a powerful impulse, can lead to the destruction of a family or a community through incessant internecine warfare.

As we move from preliterate and precolonial society to ancient Greece (Homer) and the Middle East (the Old Testament), retribution becomes less direct and more impersonal. Vengeance becomes a devine prerogative and obligation: for Homer, it was an inexorable law of nature; in the Old Testament, vengeance belongs to the Lord. As Western culture evolved, retribution increasingly became the responsibility of the worldly representatives of the divine lawgiver: kings, monarchs, and other chiefs of state who were often given careful guidance by the priesthood.

The bureaucratization of retribution produces two immediate consequences. First, it significantly reduces the level of group or fratricidal warfare and family feuds, making civil life more pacific, predictable, and lawlike, and thereby contributing to the sum total of human happiness. Second, it frustrates a deep human need (for revenge) because the machinery of the bureaucracy is not very efficient, and this inefficiency increases the level of unhappiness among many victims of injustice. Hence, vigalante movements continuously and spontaneously erupt, and not merely in the backward parts of the American rural South— highly sophisticated Jewish vigilantes still pursue Nazi war criminals 40 years after the original atrocities were perpetrated. As in all other forms of social change, there are distinct benefits and costs associated with the bureaucratization of revenge.

A further point is that formal legal codes cover only a small fraction of the possible offenses against persons, and typically, they are concerned with those most likely to aggravate community relations beyond tolerable limits. But beyond these formally actionable offenses, there is a vast array of everyday insults, slights, petty slurs, taunts, humiliations, and tricks over which the victim can only take action himself. The point is that insofar as philosophers and social scientists have considered retribution at all, it has been in the context of formal legal processes. Retribution is also and perhaps more frequently an issue in the minutiae of our everyday informal dealings with one another. Even in the most trivial details of their encounters, people cannot ignore the claims of retributive justice. For example, Goffman's (1971) examination of "remedial interchanges" testifies to the care people give to avoiding the appearance of offending and thus inviting retribution. They are constantly taking remedial action to demonstrate that, despite possible inferences to the contrary, they are maintaining the rules of proper social conduct.

Political theorists in the Machiavellian or *realpolitik* tradition argue for self-interest as the fundamental regulatory principle in human affairs. Self-interest, however, has two disadvantages as a sovereign regulatory concept. First, it is an intrapsychic rather than an interpersonal concept; that is, the dynamics are in an actor's head rather than between actors, where these issues would be negotiated. Second, these writers, at least, are not persuaded that people always perceive the course of action that reflects their best interest or that they choose it when they do perceive it. Lévi-Strauss (1968) argued that the exchange relationship—the social obligations that are incurred by the recipient of a gift or a favor—is perhaps the most important single structural phenomenon regulating the interaction between people. But Lévi-Strauss's notion ignores the problem of ingratitude, as exemplified by the well-known propensity of children to ignore the wishes, desires, and hopes of their parents. It also ignores the fact that in some cases, being the recipient of gifts may only inspire animosity and resentment toward the donor. The possibility of retaliation and revenge against persons who violate shared social norms (including the obligations incurred after receiving a gift or favor) seems to be a more realistic organizing principle in human affairs. The universality of revenge as a motive in world literature—Achilles, Hamlet, Samson, William Tell, Medea, Ahab—testifies to its power as an organizing force in human affairs.

4. RETRIBUTION AND EQUITY THEORY

Earlier, we outlined some reasons for regarding distributive justice as an inappropriate perspective from which to analyze how individuals actually use the notion of justice. In this section, we would like to show how equity theory and "retribution theory" (if we may use such a term) entail rather different views of the world. This difference should indicate why the notion of retribution does not occur in equity theory, and why someone who takes the reality of retributional processes seriously has problems understanding equity theory. The two views reflect widely different assumptions about society and its governance that cannot be readily reconciled.

For equity theorists, justice is given by that social system in which people get what they deserve; their rewards are commensurate with their contributions, accomplishments, or achievements. Virtue is rewarded in a just society. As noted above, the equity theorist's conceptions of justice are closely tied to the notion of distributive justice, and moreover, special conditions must obtain in a society before distributive justice becomes applicable. Distributive justice has a clearly demarcated range of convenience.

Retribution theory, drawing on a central insight of psychoanalysis, defines justice as provided by that situation (social circumstances) wherein everyone suffers alike. Here we are not concerned about virtue, accomplishment, or rewards; we are concerned instead that everyone share equally in the burdens, privations, taxes, and renunciations that are the unavoidable concomitants of civilized living. Rewards for one's accomplishments are unreliable; the aggravations of social living are inevitable.

Lerner et al. (1976) linked deserving and justice; justice means getting what one deserves. Deservingness is a function of hard work, virtuous conduct, self-restraint, and other desirable behaviors. A child gives up the pleasure principle and

> adopts the alternative of incurring the costs of frustrated impulses and investing efforts on the *assumption* that an appropriately more desirable outcome will accrue to him in the future. . . . He learns and trusts that his world is a place where additional investments often entitle him to better outcomes, and that "earning" or "deserving" is an effective way of obtaining what he desires. As the child matures he places greater portions of his goal-seeking activities under the rules of deserving, so that in the normal course of events, he eventually finds that his life is committed and organized on the basis of "deserving" his outcomes. (p. 135)

Consider the sentence "Jones got his just deserts." For equity theory, this sentence conjurs up the image of a Horatio Alger success story. But the sentence is ambiguous and can be given at least one alternative interpretation. That is, when we say "Jones got what he deserved," we also evoke images of retribution as in "Nixon (or the Shah of Iran, or Idi Amin, or Macbeth, or Thomas Henchard) got what he deserved." Justice and deservingness are at least as appropriate to retribution theory as they are to equity theory. Indeed, for native speakers of English, the phrase "getting what one deserves" inevitably brings the principle embodied in Anaximander's fragment to mind.

According to Lerner et al. (1976), people carefully monitor external events to see if they get their "deserved outcomes," meaning just rewards. They are also concerned that others get their "deserved outcomes" because if they don't, then the persons themselves can't depend on rewards following their efforts. Thus, "any evidence that social reality is not preserving *justice for others* is threatening, and should motivate the individual to restore his belief that he does live in a just world." Thus, we monitor the external environment, staying alert for information that virtue continues to be rewarded, as evidence that our good deeds will similarly meet with fortune's approbation. Again, these information-processing goals are consistent with the program of distributive justice. But in other, less affluent times, it is easy to imagine different information-processing goals. Specifically, and recalling Freud's view of justice, one can postulate that people pay attention to social information for two reasons. First, to ensure that other people will not receive *more* than they deserve for their efforts—that other people's rewards do not exceed what the observer feels is appropriate. Second, from the perspective of retribution theory, people should monitor social information to make sure that other people get what they deserve for their transgressions. This is fundamental; unless this aspect of justice is maintained, the entire social order is threatened, not merely one individual's pursuit of happiness.

Equity theory and retribution theory also differ concerning the motivation for our actions when we see injustice. Equity theorists postulate that when we see that we are involved in relationships where equity is not preserved, where we or others are not receiving deserved outcomes, then we are "motivated" to restore our belief that we do, in fact, live in a just world. It is hard to tell exactly what the source of this motivation might be, but two possibilities come to mind. First, there seems to be a tendency in operation that is much like that assumed by balance theory (Heider, 1958): when we perceive injustice we feel some degree of cognitive strain and an aesthetic impulse to see order or balance restored. But the first axiom of equity theory (Walster, Berscheid, & Walster, 1973) states also that people are chiefly motivated by considerations of self-interest, and it is not in one's best interest to remain in relationships that are not equitable, where one contributes more than one gets out of the relationship. Lerner *et al.* (1976) also adopted this self-interest approach: when justice is not maintained for others, then one cannot depend on rewards' following on one's own efforts.

Calculations of self-interest strike us as too cool to explain the passion with which people react to perceptions of injustice (cf. Hogan & Dickstein, 1972). From the perspective of socioanalytic theory (Hogan, Johnson, & Emler, 1978), of which "retribution theory" is a corollary, the survival of a culture or a social group depends vitally on its members' maintaining the rules of the culture. This is particularly true when *culture* is considered from the perspective of evolutionary theory (cf. Campbell, 1975). The disposition to comply with authority, to respond to social expectations, and to invent metaphysical rationalizations for the rules of one's culture is seen as biologically preprogrammed. But it also makes sense, from an evolutionary perspective, that if one perceives that others are getting more than their share, or that malefactors are not appropriately punished, then one should react with moralistic aggresion. From this perspective, then, perceptions of injustice (defined as others' getting more than their fair share, or not getting their just deserts after misbehaving) are followed by aggressive reactions that may serve one's selfish best interests but are not prompted by considerations of self-interest. Rather, moralistic aggression is akin to a reflex following perceptions of injustice (as defined above).

Equity theory and retribution theory differ in yet another way that can be highlighted by the question, Who believes in a just world? Lerner *et al.* (1976) suggested that most people believe that the world is just in the sense that reward follows merit. Homans (1976, p. 240), speaking *ex cathedra*, remarked that some people never acquire this conviction. And, we suspect, among those who do, many soon lose it. From the present writers' perspective, Lerner's view seems somewhat naive, and we would argue in contrast that for most people, the relationship between merit and reward is highly tenuous. As a wise and seasoned associate of ours used to observe, "No good deed goes unpunished, and no bad deed goes unrewarded." Lerner himself hedged his generalization with qualifications. So, for example (p. 153), when resources are scarce and people must compete for them, winning is everything and deservingness is nothing. Moreover, when people identify with one another, the claims of distributive justice are set aside. Finally, and perhaps more ominously, Lerner pointed to a general law that the more undeserved a person's suffering, the more is that person derogated by observers. This suggests a curious lack of transitivity in the way that views of justice operate. One wonders how people evaluate those persons whose rewards are equally undeserved. Balance theory would predict that such persons would be praised, but that seems doubtful.

Finally, it is important to note, when comparing the claims of equity theory with those of retribution theory, that in those conditions where the requirements of equity break down, the requirements of retribution remain firm. Under conditions of affluence, it is still possible to imagine people committing malicious acts out of spite. In these conditions, we would still want malefactors to be punished and our injuries revenged. Under conditions of scarcity, where equity is also acknowledged to break down, people's malevolent impulses still must be constrained if there is to be any society. Hence retribution is still a relevant consideration. And friends, relatives, and acquaintances are also capable of mischief independent of their economic circumstances. The rules of equity do not hold within organic social groups, but the requirements of retribution certainly do.

Equity theory, the adopted child of experimental social psychology, implies that social life revolves around the questions of how to distribute resources. The theory was developed outside the study of natural communities, and the theory's proponents may not have noticed that in such communities grievances largely arise not over resources but over reputations. For example, Bailey (1971), in a series of studies of European village communities, pointed out the centrality of people's preoccupation with defending their community standing. The members of these communities expend enormous efforts to maintain their status and their reputations among their neighbors. Moreover, the desire to maintain one's status and reputation is generally more important than considerations of equity. Thus, Tajfel and others (Tajfel, Billig, Bundy, & Flament, 1971; Turner, 1975) have shown rather clearly that in the distribution of resources, options that promote the individual's status relative to other people are preferred to those that maximize equity. We believe that the members of any human group are in more-or-less constant competition for status, no matter how carefully the competition is disguised. If your neighbor competes with you, you cannot simply ignore him; you must respond or lose your standing in the group. One function of moral codes is to regulate this competition and prevent its becoming too destructive. In this view, retribution exists to punish those who take unfair advantage in this competition, that is, by cheating. This is yet another reason that retribution is more universal than equity: status is by definition a scarce resource, and it can be acquired only at another's expense.

5. THE MORALITY OF RETRIBUTION

Philosophers and social scientists have given a good deal of attention to the definition of morality and its justification. In comparison, relatively little effort has been directed at analyzing the functions of moralities—what they do in the context of everyday life—perhaps because this problem does not seem to offer the same intellectual challenge as the first two. One obvious function of moralities is that they provide the enabling conditions for social life; in the absence of a moral order, normal life is impossible.

Beyond providing a framework for normal life by establishing the ground rules for social congress, moralities have at least two other functions: they provide criteria or grounds for deciding what to do when specific duties conflict or when the requirements of specific rules are unclear; and they tell us what we can reasonably expect from others—that is, they provide us with criteria for evaluating the actions of others.

Knowledge of the function of moralities suggests two further, rather interesting conclusions. First, it is in principle possible to specify the conditions that determine whether two persons stand in a moral relationship to one another; moralities are sets of rules, and rules always have circumscribed applicability. Consequently, just because we coexist with others does not mean that a moral contract exists between us: we do not always stand in a moral relationship to other people; there are limits to our moral obligations. Second, knowledge of the moral rules is not very helpful in deciding how to respond to actions signifying that another person is ignoring the moral rules. Behavior in response to the transgressions of others lies under a kind of moral penumbra. Conventional morality provides little guidance concerning how to treat those who indicate by their actions that they do not believe they stand in a moral relationship to us.

In might be useful to consider for a moment the transgressor's perspective in rule violation. Mentally competent transgressors anticipate and attempt to avoid retribution or retaliation. If the violations are slight or unintentional, then one kind of action is taken to forestall retaliation. Goffman (1971) outlined the strategies typically used here, namely, explanations and apologies. If the transgressions are calculated, then a second kind of action is called for. In his essay "Cooling the Mark" (1952), Goffman described how deliberate offenders act to deflect the outrage of those they have duped. To succeed in the competition for status and still cheat, however, one must maneuver one's rival into a position from which he or she cannot readily retaliate. This is the object of many of the games described by Berne (1964).

People who break the rules (moral, contractual, legal) do more than violate legitimate expectations. There is a symbolic content to their actions as well; they are telling us, among other things, that they do not care to play the social game with us, that they have abrogated the norm of reciprocity. The question is, What are we to do in such circumstances? We would like to make a number of observations about responses to the transgressions of others without stating what should be done in any particular case.

First, if, as many have argued (Campbell, 1976; Hogan *et al.*, 1978), man evolved as a culture-using animal (with morality as a major part of the culture), then the biologically preprogrammed response to violations of the moral code may be aggression. All responses to others' transgressions, *no matter what form these responses take,* are to some extent motivated by moralistic aggression. These responses differ not in their aggressive intent but in the directness with which the aggression is expressed.

Second, the pacifist's response of turning the other cheek, of meekly accepting and then forgiving the transgressions of others, may actually be, as Nietzsche suggested, a form of *realpolitik* reflecting one's basic weakness and inability to retaliate. In this case, one attempts to make a virtue out of a necessity.

Third, overt responses to the moral violations of others are governed not by moral considerations but by social conventions and matters of taste—how one feels about retribution and revenge. Generally speaking, in the Anglo-American tradition of jurisprudence, we try to make the "punishment fit the crime." Such refinements in retributive justice seem not to prevail in the Middle East.

Fourth, one's stand on topics such as capital punishment is not a moral issue. Rather, one's stand reflects current opinion, political preference, and—following Nietzsche—the degree to which one has sublimated or disguised one's aggressive impulses. That is, there are no good moral grounds for preferring charity over retribution (or any other response to the transgressions of others). One's choice is governed by taste, custom, or—very infrequently—empirical evidence regarding the kinds of responses to rule violations that are most likely to suppress them in the future.

Evidence recently provided by Vidmar and Miller (1980) shows that people's desire for retribution is reliably related to dimensions of individual differences (e.g., authoritarianism), although there appear to be conditions under which everyone favors retribution. Piaget (1932/1965) reported an interesting sex difference in children's attitudes toward retribution. Boys were far more likely to favor giving the same or more blows for blows received than were girls, who typically favored giving fewer. This may be because, for boys, the ability to return physical aggression is more relevant to the competition for status than it is for girls.

Finally, the "natural" response to others' deliberate rule violation is, as we suggested above, moralistic aggression. This response flows naturally into retaliation and retribution. Retribution is not justified in any ultimate moral sense—but then neither is any other response, including turning the other cheek. But what are the practical and/or political consequences of not retaliating when someone deliberately transgresses against us? One consequence is that we are thereby deprived of some of our legitimate resources—money, property, honor, privileges. In addition, by not retaliating, we appear indecisive, irresolute, even cowardly. Bailey (1971) provided ethnographic examples of individuals who had dropped out of the struggle for status and, consequently, no longer retaliated when slighted. In every case, they were mocked as stupid and, in returning good for evil, were dismissed as *troppo buono*—not virtuous but foolish. As Fürer-Haimendorf (1967) observed of the Dalfa, if an injured party fails to seek redress, he acquires a reputation for being weak and indecisive and may therefore be cheated in other dealings. Thus, if we do not retaliate, we invite further transgressions; as a political scientist colleague points out, every act of appeasement leads to Munich. Consequently, retribution can never be given a final moral justification (again, neither can any other reaction to transgression). But retribution is often justified in a practical sense. Indeed, one might even argue that the urge to retaliation must be a part of one's nature if one is to survive.

6. POWER, RETRIBUTION, AND STATUS

There are interesting connections among power, status, justice, and retribution. Homans (1976) observed that

> the beneficiary of an inequity may do nothing to compensate the victim, and the victim nothing to force the compensation. For the former, his gain from injustice may be too great, and his fear of reprisal from the victim too low. For the latter, his loss by inequity may be too little in view of the danger to himself of trying to take reprisals. *More generally*, equity may cede to power. (p. 237)

Homans concluded that power is "the more primitive phenomenon that lies behind distributive justice" (p. 243).

With regard to distributive justice, Homans is probably correct. As noted in Section 2 above, distributive justice is the unique concern of power holders, especially those who wish to be regarded as just, fair-minded, and equitable. Not only are power holders concerned with distributive justice, but they must also have power at their disposal to enforce its claims. In short, distributive justice is linked to power in that it is a principal concern of fair-minded power holders, and power is necessary to ensure that the requirements of distributive justice are met.

The connections between retributive justice and power, however, are less obvious and more interesting. The first point to note is that although power is necessary to redress inequality, almost anyone can have revenge. Homans seems to feel that rational calculations of self-interest will stop the injured powerless from retaliating, but he is clearly wrong, as the activities of the North Vietnamese in the late 1960s reveal.

A second point worth noting concerning the connection between power and retribution is that the desire for revenge can be a powerful incentive for worldly success and the achievement of status. An apocryphal story regarding Joseph Kennedy is a case in point; he is reputed to have told his children, concerning the slights that they may have suffered at the hands of the Protestant Boston establishment, "Don't get mad, get even." Getting even in this case meant achieving a position whereby one could retaliate in kind, not by inflicting physical harm but by causing the others to suffer equal social humiliation and social disgrace. The desire for equity may perhaps serve as a motive for something, but the drive for revenge can lead very directly to the attainment of status and power.

A third point in this regard is that the higher one's status, the more carefully one must guard against giving gratuitous public offense. In *Totem and Taboo* (1950), Freud provides detailed documentation concerning how high-status persons in primitive societies must endure the most grievous insults from the other members of their group. Similarly, in this culture, notable public figures must endure with limitless good grace extraordinary invasions of privacy and violations of civil norms of conduct from the press corps and "the public." Otherwise, they risk severe retaliation. Freud's point is that success inevitably arouses jealousy and hostility in others. If they can't retaliate against the successful person directly, they may murder his or her children. Hence, once again, the more status one has, the more careful one must be not to arouse the resentment of the less fortunate—one dare not call them less deserving. This may explain the well-known relationship between status and conformity. Those with high status must be particularly careful in their conformity if they are not to give grounds for resentment and retribution.

Finally, it is important to note that there is a significant intransitivity between status and the ability to retaliate after suffering injustice. It is ignoble for superiors to seek personal retribution against inferiors. On the other hand, it is *de rigueur* for equals or inferiors to right the wrongs suffered at the hands of equals and superiors—as the legends of Robin Hood, William Tell, and others remind us.

REFERENCES

Adams, J. S., & Freedman, S. Equity theory revisited: Comments and annotated bibliography. In L. Berkowitz & E. Walster (Eds.), *Advances in experimental social psychology*, Vol. 9. New York: Academic Press, 1976.

Bailey, F. G. *Gifts and poison: The politics of reputation.* Oxford: Blackwell, 1971.

Berkowitz, L., & Walster, E. (Eds.) *Advances in experimental social psychology*, Vol. 9. New York: Academic Press, 1976.

Berne, E. *Games people play.* New York: Grove, 1964.

Box, S. *Deviance, reality and society.* New York: Holt, Rinehart and Winston, 1972.

Campbell, D. T. On the conflicts between biological and social evolution and between psychology and moral tradition. *American Psychologist*, 1976, *30*, 1103–1126.

Colson, E. *Tradition and contract.* Chicago: Aldine, 1974.

Damon, W. Early conceptions of positive justice as related to the development of logical operations. *Child Development*, 1975, *46*, 301–312.

DuBois, C. *The peoples of Alor.* New York: Harper & Row, 1961.

Freud, S. *Totem and taboo.* New York: Norton, 1950.

Fürer-Haimendorf, C. von *Morals and merit: A study of values and social controls in South Asian societies.* London: Weidenfeld and Nicholson, 1967.

Goffman, E. *Relations in public.* New York: Basic Books, 1971.

Goffman, E. On cooling the mark. *Psychiatry*, 1952, *15*, 451–463.

Hart, H. L. A. *The concept of law.* New York: Oxford University Press, 1961.

Heider, F. *The psychology of interpersonal relations.* New York: Wiley, 1958.

Hogan, R., & Dickstein, E. Moral judgments and perceptions of injustice. *Journal of Personality and Social Psychology*, 1972, 23, 409–413.

Hogan, R., & Emler, N. P. The biases in contemporary social psychology. *Social Research*, 1978, 45, 478–534.

Hogan, R., Johnson, J., & Emler, N. P. A socioanalytic theory of moral development. In W. Damon (Ed.), *Moral development*. San Francisco: Jossey-Bass, 1978.

Hollander, E. P. Conformity, status and ideosyncrasy credit. *Psychological Review*, 1958, 65, 117–127.

Homans, G. C. Commentary. In L. Berkowitz & E. Walster (Eds.), *Advances in experimental social psychology*, Vol. 9. New York: Academic Press, 1976.

Hook, J. G., & Cook, T. D. Equity theory and the cognitive ability of children. *Psychological Bulletin*, 1979, 86, 429–445.

Kelsen, H. J. *Society and nature*. Chicago: University of Chicago Press, 1943.

Kelsen, H. J. *What is justice?* Berkeley: University of California Press, 1957.

Kohlberg, L. Stage and sequence: The cognitive developmental approach to socialization. In D. Goslin (Ed.), *Handbook of socialization theory and research*. Chicago: Rand McNally, 1969.

Kohlberg, L. From is to ought: How to commit the naturalistic fallacy and get away with it in the study of moral development. In T. Mischel (Ed.), *Cognitive development and epistemology*. New York: Academic Press, 1971.

Kohlberg, L. Moral stages and moralization: The cognitive developmental approach. In T. Lickona (Ed.), *Moral development and behavior: Theory, research and social issues*. New York: Holt, Rinehart and Winston, 1976.

Lerner, M. J. The justice motive in social behavior: Introduction. *Journal of Social Issues*, 1975, 31(3), 1–19.

Lerner, M. J. The justice motive: Some hypotheses as to its origins and forms. *Journal of Personality*, 1977, 45, 1–52.

Lerner, M. J., Miller, D. T., & Holmes, J. G. Deserving and the emergence of forms of justice. In L. Berkowitz & E. Walster (eds.), *Advances in experimental social psychology*, Vol. 9. New York: Academic Press, 1976.

Lévi-Strauss, C. *The elementary structure of kinship*. Boston: Beacon Press, 1968.

Loughran, R. A pattern of development in moral judgements made by adolescents derived from Piaget's schema of its development in childhood. In T. Lickona (Ed.), *Moral development and behavior*. New York: Holt, Rinehart, & Winston, 1976.

Matson, W. What Rawls calls justice. *The Occasional Review*, 1978, 8/9(Autumn), 45–55.

Parsons, T. *The social system*. Glencoe, Ill.: Free Press, 1951.

Piaget, J. *The moral judgment of the child*. New York: Free Press, 1965. (Originally published 1932.)

Rawls, J. *A theory of justice*. Cambridge, Mass.: Harvard University Press, 1971.

Tajfel, H., Billig, M., Bundy, R. P., & Flament, C. Social categorization and intergroup behaviour. *European Journal of Social Psychology*, 1971, 1, 149–177.

Turner, J. C. Social comparison and social identity: Some prospects for intergroup behaviour. *European Journal of Social Psychology*, 1975, 5, 5–34.

Ugurel-Semin, R. Moral behavior and moral judgment of children. *Journal of Abnormal and Social Psychology*, 1952, 47, 463–474.

Vidmar, N., & Miller, D. T. Social psychological processes underlying attitudes toward legal punishment. *Law & Society Review*, 1980, 14, 565–602.

Walster, E., Berscheid, E., & Walster, G. W. New directions in equity research. *Journal of Personality and Social Psychology*, 1973, 25, 151–176.

8

The Social Psychology of Punishment Reactions

DALE T. MILLER and NEIL VIDMAR

1. INTRODUCTION

Social psychologists have devoted considerable attention to exploring the role that considerations of justice play in social behavior. People's concern with the issue of justice has been shown to affect strongly their interpersonal perceptions (Lerner & Miller, 1978; Walster, Walster, & Berschied, 1978) as well as many of their behavioral reactions (Lerner, 1977; Miller, 1977a,b). Despite general interest in reactions to injustice, social psychologists have devoted surprisingly little attention to punishment reactions, focusing instead primarily on reactions to victims of injustice.

The neglect of punishment reactions is puzzling since punishment is often regarded as constituting the cornerstone of the justice process

This chapter is an abridged and modified version of an article (Vidmar & Miller, 1980) appearing in *Law & Society Review*. The Law and Society Association holds the copyright to the original article.

DALE T. MILLER • Department of Psychology, University of British Columbia, Vancouver, British Columbia, Canada V6T-1W5. NEIL VIDMAR • Department of Psychology, University of Western Ontario, London, Canada N6A 5C2. This article and the chapter were written with the support of the National Science Foundation (Grant No. SOC 77-13266). Both authors also acknowledge support for their respective Canada Council grants. The second author also wishes to acknowledge support received from the Russell Sage Foundation and the Battelle Memorial Foundation during the formative stages of the research.

(Vidmar & Miller, 1980). Indeed, as Hogan notes in the present volume, the meting out of punishment often appears to people to be synonymous with the rendering of justice. In fact, when individuals are given a choice between punishing the perpetrator of an injustice *or* compensating the victim of an injustice, they generally choose to punish (Miller & McCann, 1979). This view, of course, also finds embodiment in the legal system, where the administration of justice generally necessitates only that a perpetrator of an offense be punished, not that the victim of the offense be compensated.

Justice considerations underlie two distinct rationales or motivations for punishment. The first of these is oriented toward the present or the future, and we will call this the *behavior control orientation;* the second is backward-looking, and we refer to this as the *retributive orientation.* The behavior control orientation is concerned solely with the elimination of some unjust behavior, either ongoing behavior or anticipated future behavior. It is the fear of continued acts of injustice that motivates this type of punishment. Retribution is concerned primarily with the elimination of a sense of injustice produced by some previous action. When people punish because of a desire for retribution, they are concerned not with deterrring future acts of injustice but with redressing a previously committed injustice. The future is relevant to the retributive orientation only in that the sense of injustice is likely to linger on unless punishment is administered. These two types correspond roughly to the traditional legal distinction between utilitarian and retributive motives. Despite the conceptual distinction that can be made between these two types of punishment, it should be clear that punitive action often involves elements of both types of motivation.

1.1. Definition of Punishment

The term *punishment* has been defined and conceptualized in confusing ways in the social sciences (see Lindesmith, 1968). Our working definition is straightforward and comports with most legal and philosophical conceptualizations, as well as with common dictionary definitions: punishment is a negative sanction intentionally applied to someone who is perceived to have violated a law, a rule, a norm, or an expectation. This definition is robust enough to incorporate the social-psychological dynamics involved in both legal punishment reactions and in those reactions evoked in most nonlegal settings. In accomplishing this flexibility, the definition also avoids the difficulties of specialized terminology that frequently arise in psychology and sociology. It also avoids the difficult issue of defining a legal act (see Abel, 1973) and the issue of "formal" (court-administered) versus "informal" punishment (see Lindesmith, 1968). Thus, punishment need not be limited to collective perceptions and official definitions, as is frequently the case in sociological writings (see Lindesmith, 1968).

Two important implications of this definition should be made explicit. First, punishment must follow and be a consequence of the perception that someone has violated a rule, norm, or law. Psychologists working with animals and humans have frequently used the term *punishment* to refer to aversive stimulation that conditions behavior, but for our present purposes, the shaping of the behavior of a rat in a Skinner box or the slapping of a toddler's hands to teach him to keep away from an electric outlet would *not* be considered punishment since no rule has been violated. Second, the term *sanction* should be broadly interpreted. It may take the form of a deprivation or an unpleasant experience, either of which may be physical, social, or psychological. Punishment, therefore, may include not only physical acts such as torture, confinement, a fine, or enforced restitution, but also attempts at status degradation, such as Eskimo song duels, ridicule, ostracism, or banishment from the social group.

Before continuing we should make it clear that punishment also serves functions that have little to do with justice considerations. First, as sociologists beginning with Durkheim have emphasized, punishment serves the function of maintaining in-group–out-group distinctions as well as that of strengthening the cohesion of the social group. Generally, these sociological functions are latent (Merton, 1957) and do not constitute intended consequences of the punishment process. In other instances, however, punishment can be consciously employed as an instrument of power to pursue social and political goals (e.g., Foucalt, 1977; Gusfield, 1949; Rusche & Kircheimer, 1939).

Punishment may also serve other, more psychological functions. A number of writers (e.g., Menninger, 1966; Weihofen, 1957) have observed that the punishment of others for their infractions often represents an unconscious impulse to punish the tendencies we deny or repress in ourselves. Finally, punishment has been viewed by some as providing an outlet for repressed sadism (Alexander & Staub, 1956).

Punishment, therefore, is not always solely or even primarily concerned with redressing injustices or preventing future ones. It may serve a variety of social and psychological functions as well. In the present chapter, however, we limit ourselves to a discussion of behavior control and retributive motives and their relationship to issues of justice. In discussing each of these types of punishment, we consider various classes of variables that affect them, including characteristics of the rule violation, the rule violator, and the reactor. In a final section, we discuss how changes in society are affecting the dominance and prevalence of these two types of punishment motives.

2. AN ANALYSIS OF THE COMPONENTS OF PUNISHMENT REACTIONS

As outlined in Table 8.1, in addition to distinguishing between two types of motives underlying punishment reactions (behavior control vs. retribution), our analysis also distinguishes between two types of targets to whom these reactions may be addressed: the offender or others in the social environment. Although punishment is almost always administered to the offender, the reactor may often be more concerned with its impact on a larger social group. The broader social group becomes the focal audience when the purpose of the punishing act is general social deterrence, increased in-group cohesion, or the reactor's need for enhanced status or for social consensus regarding the correctness of the rule.

The general punishment orientations we have introduced cannot be understood apart from other beliefs that the individual holds about punishment, such as the expectation that it will be effective in preserving obedience to a rule and the fear that the failure to punish may threaten

Table 8.1. A Classification of Purposes Underlying Punishment Reactions

Basic motives	Target	
	Offender	Others in the social environment
Behavior control	Deterrence, isolation, elimination, reeducation of offender; restitution to victim.	General deterrence or threat; prevention of vengeance by victim or others; upholding morale of conformers; disavowal of act.
retribution	Change in offender's belief system vis-à-vis victim or societal rule; reaffirmation of private self-image of victim or surrogates; status degradation and differentiation of offender; assertion of power over offender.	Vindication of rule; reestablishment of social consensus about rule; diffuse release of psychological tension through social comparison processes.

internalized beliefs about order, rightness, and justice. Such associated beliefs result from socialization experiences and may be accepted by most members of society as cultural "truisms." Individuals are additionally influenced by specific situational factors associated with perceptions of the rule violation and of the rule violator. Finally, individual differences among reactors influence the weight given to various punishment motives, as well as the perceptions of both rule violations and rule violators.

We begin our examination of these various aspects of the punishment process with a discussion of the conceptual issues and empirical findings relevant to behavior control.

3. BEHAVIOR CONTROL

Most criminal offenses pose a direct physical or material threat to someone. The victim, of course, wants the behavior stopped. But so do others who feel threatened, directly or indirectly. Thus, one major set of punishment motives is behavior control. The reaction may be directed toward the offender or toward others who might emulate the offender.

Punishment reactions directed toward the offender are designed to deter present or future violations. In addition to teaching the offender about the consequences of improper behavior, punishment may also seek to reeducate the offender about proper modes of behavior or to isolate or even eliminate the offender. Restitution may even be required in those cases where the offense involves redressable damages and where the situation can be returned to its state prior to the offense. Not only is the behavior control reaction in its pure form unconcerned about either the past or the moral implications of the offense, but even the offender's moral character is irrelevant, except as an index of corrigibility. If the reactor could be assured that the perpetrator's offense would never happen again, there would be no need to punish. The dispassionate attempt by Dr. Brodsky to condition Alex to be repulsed even by the simple thought of violence in *A Clockwork Orange*, epitomizes this perspective. In Brodsky's words, "We are not concerned with motive, with higher ethics. We are concerned only with cutting down crime" (Burgess, 1962, p. 126).

A crucial component in this type of reaction is the individual's belief that punishment is efficacious in deterring the offender's present or future behavior. This mediating variable helps explain what might otherwise appear to be paradoxical behavior in punishment reactions. For example, some studies of sentencing behavior (see Hogarth, 1971; Orland & Tyler, 1974) have shown that judges who tend to be more liberal, empathic, and nonretributive frequently give more severe sentences than conservative, nonempathic retributive judges—for minor or juvenile offenses, at least. The fact that liberal judges subscribe more strongly to the belief that "punishment corrects" appears to be the factor responsible for their harsher treatment in these instances.

Punishment of the offender may also be perceived as a "general deterrent" directed toward others who might contemplate similar offenses. A reactor may be motivated in this fashion to the degree that he believes that (a) other people need to be deterred; (b) the punishment of an offender can discourage other potential offenders; and (c) potential offenders will become aware of the sanctions applied against the offender. Although this "general deterrence" belief is usually associated with a belief in the efficacy of punishment in controlling the behavior of the offender himself, the two can be theoretically and empirically separated. For example, some white-collar offenders may be punished sufficiently by being caught: they lose their jobs, are socially disgraced, and are likely to be denied the opportunity to repeat their offense. Nevertheless, the reactor—whether a legal authority, the business organization, or the general public—may insist on additional punishment as a deterrent to other potential white-collar offenders. It is so difficult to detect offenders that some may believe it necessary, in order to achieve general deterrence, to impose highly publicized criminal sanctions on the rare offender who is caught. Similarly, punishment of "first-time" shoplifters as a deterrent to others is frequently advocated, not out of personal animosity toward the offender or because of the likelihood of recidivism, but simply to publicize the consequences and deter others.

Two additional sets of beliefs are implicated in behavioral control. First, punishment of the offender may be seen as serving to uphold the morale of those other potential offenders who have resisted the temptation to violate the rules but who might be upset at seeing an offender go unpunished: "when a defendant escapes who, people think, deserves punishment, they may lose faith in the social structure and may relax their own inhibitions" (Weihofen, 1957, p. 136). Second, punishment of the offender by a third party may be viewed as a means of preventing vengeance by the victim or by others identified with the victim. Punishment for the purpose of deflecting vengeance is probably more significant and common in tribal societies that contain groups capable of engaging in feuds, but it is also manifested in modern Western societies. The awareness of this possibility is frequently expressed in the sentiment that without formal legal justice, "people will take the law into their own hands."

These various behavior control motives, directed toward both the offender and a broader audience, are widely recognized phenomena and play a prominent role in legal and philosophical writing on punishment (e.g., see Andeneas, 1974; Ezorsky, 1972; Zimring & Hawkins, 1971). Interestingly, however, they have been given relatively little attention—and, in fact, are frequently ignored altogether—in much of the social-science literature concerning individual punishment reactions, probably because the concept of behavior control appears less rich and interesting, sociologically and psychologically, than the motive of retribution. Nevertheless, when behavior control motives have been studied, they have been found to be strongly and significantly related to punishment reactions. A number of nationwide polls on the death penalty, for example, have found the degree of support for capital punishment to be positively related to the belief that it is a deterrent to crime (Vidmar & Ellsworth, 1974). In particular, a 1973 Harris survey found that 76% of those persons who advocated capital punishment felt that it was a more effective punishment than a life sentence; in comparison, only 29% of those persons who opposed it thought it was more effective.

A study by Sarat and Vidmar (1976) provides even more direct evidence of the link between punitiveness and belief in the effectiveness of punishment. These researchers attempted to test Justice Thurgood Marshall's hypotheses, stated in *Furman v. Georgia* (1976), that (a) people are ill-informed about the limited deterrent value of the death penalty and (b) if they were accurately informed, the majority would oppose it. In an experimental survey with several conditions, adult respondents were presented with arguments and empirical findings concerning the utilitarian (behavior control), the humanitarian, or both aspects of capital punishment. The utilitarian information but not the humanitarian information reduced support for the death penalty. The perceived efficacy of punishment in deterring the offender would appear, then, to be a crucial factor in the operation of behavior control motives.

Thomas and his associates (Thomas, 1976; Thomas, Cage, & Foster, 1974; Thomas & Cage, 1974) have also investigated the role of behavior control motives in attitudes toward first-degree murder and lesser crimes of violence and have discovered an interesting fact. Their research suggests not only that belief in the efficacy of punishment results in increased willingness to punish, but also that fear of crime may increase the perceived efficacy of punishment. Clearly, beliefs in the perceived efficacy of punishment are complexly related to the punishment sequence.

3.1. Rule and Offense Characteristics

The most important factor affecting behavior control reactions is the reactor's perception of the personal threat caused by the offense. Rossi,

Waite, Bose, and Berk (1974), for instance, found that crimes of violence were generally judged as more serious than crimes involving property. But there are exceptions to this pattern; middle-class property owners, for instance, sometimes express more punitive attitudes toward crimes involving property than toward those involving physical harm. Since the middle class live and work in an environment relatively free of violent crimes, it is not surprising that they tend to be less concerned with the threat of physical harm than with the loss of material possessions.

The actual harm resulting from the offense may affect sanctioning responses as well, though its influence may be indirect or multifaceted. Obviously, an assault causing serious harm or a theft inflicting large losses is more threatening (see Rossi *et al.*, 1974). But there are also substantial research findings that suggest another reason that resultant harm might be related to punitiveness. The more severe the consequences, the more likely persons are to ascribe intention and responsibility to the offender (e.g., Shaver, 1970; Walster, 1966).

The extent to which the reactor believes that other persons might be tempted to commit the same offense if the offender is not punished should also influence behavior control responses. The more contagious the reactor believes the offense to be, the more he will favor punishment.

Perceived social consensus concerning the nature of the offense and its dangerousness is another factor influencing punitive reactions, particularly when the immediate threat is not readily apparent. For example, white-collar crimes or corporate misbehavior, such as price fixing, may evoke only strong sanction responses when the reactor perceives that others believe that the offense is serious and in need of control.

3.2. Rule Violator Characteristics

By definition, behavior control motives should not be affected by the offender's "moral" character. Nevertheless, offender characteristics do moderate judgments of culpability, predictability, and corrigibility—factors that bear on behavior control.

Generally, a person who has done something accidentally is seen as less responsible and is less likely to be punished, or is punished less severely, than a person who did something intentionally (see Rule & Nesdale, 1976). Intentional wrongdoing may be viewed as more predictive of subsequent wrongdoing by the offender. Conversely, offenses committed as a consequence of forces external to the person, (e.g., under duress) are seen as less predictive of subsequent similar behavior. For example, Kelman and Lawrence (1972) found that some people opposed the prosecution of Lieutenant Calley on the grounds that he and the other soldiers acted under momentary duress, perhaps reasoning not only that Calley would never have the opportunity to murder Vietnamese again but also that the punishment of Calley would have no deterrent value for other soldiers in similar circumstances. In a similar vein, research by Hamilton (1976, 1978) showed that when individuals were presented with a hypothetical Calley-type case, in which a soldier killed a prisoner under the orders of a superior, less responsibility was assigned to the defendant and less punishment was thought appropriate, the higher the status and power of the superior giving the orders.

There is the additional case, however, where there appears to be no plausible reason or motive for the rule violation. Wanton or capricious action, such as a thrill killing by a gang, is particularly disturbing to people. The desire to isolate or otherwise deter the offender may be stronger in such a case than it is where the intention is clear and understandable. The intensity and implications of the desire for a predictable and stable environment has been demonstrated in many studies (Lerner & Miller, 1978; Wortman, 1976).

The perceived intentionality of the offense, and other aspects of the offender's behavior, also bears on the reactor's perceptions of offender corrigibility. For example, a dispassionate or incontrite rule-breaker is more likely to be viewed as a potential recidivist than an offender who expresses remorse (see Sykes & Matza, 1957). Similarly, an offender who has a history of rule violations is more likely to be perceived as a potential recidivist than a first-time offender (Ebbessen & Konecni, 1975; Carroll & Payne, 1977).

The social, occupational, and educational status of the offender also affects the punishment response. Status attributes may affect *ascriptions* of responsibility for the offense, *perceptions* of corrigibility, and the perceived effectiveness of punishment, irrespective of whether or not these inferences are empirically valid. Research by Kipnis and Cosentino (1969) and others (Rothbart, 1968) in work contexts suggests that one of the reasons that work supervisors vary sanctions according to status characteristics is their belief that persons of different status are differentially responsive to sanctions.

A number of studies also show that the perceived similarity between the offender and the reactor may affect the punishment response: the greater the similarity, the lower the punishment recommended (e.g., Mitchell & Byrne, 1973). This relationship may exist because similarity tends to increase attraction, and as the evidence suggests, those who are attractive are less likely to be seen as immoral or capable of transgressions than are unattractive others (Heider, 1958). Future violations, therefore, may be expected less of similar than of dissimilar others.

3.3. Individual Differences

Any of the above factors may affect the punishment reactions of some individuals more than of others. Differences due to demographic variables such as age, sex, or social class, as well as personality dimensions, may be important. Unfortunately, the existing data bearing on individual differences in the context of behavior control motives are very sparse even if quite suggestive.

Before discussing the individual difference variables that have been examined in the context of behavior control reactions, it is important to recognize that these variables can affect the punishment process at a number of levels. Some individuals may recommend more severe punishment than others because they react more negatively to the characteristics of the rule violator, others because they accord the rule violator more responsibility for his actions, and others because they see the violation as more severe. Furthermore, individuals may differ in their perceptions of the effectiveness of punishment and, consequently, may vary in their punishment reactions. The mere fact that one group recommends more punishment than another is therefore often ambiguous. With no additional information, it is difficult to know at which level or levels of the punishment sequence the variable is relevant.

The findings reported by Hogarth (1971) and Wheeler, Bonacich, Cramer and Zola (1960) that liberal judges may sometimes punish more severely because they believe that "punishment corrects" has already been discussed. Kipnis and Cosentino (1969) studied corrective actions taken by supervisors in industrial or military settings and found variation by supervisor, by form of problem, and by other factors. From the present perspective, however, the important finding from this study is that the choice of corrective actions was directly related to the supervisors' personal beliefs about the efficacy of the various sanctions.

Authoritarianism, a general personality syndrome incorporating beliefs about authority, punitiveness, ethocentrism, and other factors (see Adorno et al., 1950), is another dimension found to be related to the endorsement of behavior control. Vidmar and Crinklaw (1974), for example, found that persons classified as high authoritarians believed more strongly in general deterrence as a justification for punishment than those classified as low authoritarians. These authors hypothesized that since other research has found strict punishment to be part of the childhood socialization experience of high authoritarians (Hart, 1957;

Levinson & Hoffman, 1955), these individuals may simply learn from an early age that punishment is effective in controlling behavior.

4. RETRIBUTION

The term *retribution* is used to refer to punishment reactions involved in the moral rather than the behavioral implications of the offense. Aside from the negative physical or material consequences that may result from a criminal act, an offense also has symbolic consequences for the individual reactor or his perceived social group. The victim (or someone who identifies closely with the victim) may view the offense as an affront to his values or status. Heider (1958) has discussed such consequences in the context of interpersonal relations. Heider noted that when one person intentionally harms another person (i.e., violates a rule) far more is involved than the physical, material, or social hurt itself, The offender is frequently perceived as demonstrating contempt for the person harmed, as asserting power over him, or as attempting to assert the superiority of the offender's belief or value system. For the victim-reactor, therefore, punishment helps to reestablish the psychological equilibrium by redressing the sense of inferiority that the act of harm has engendered. Indeed, failure to punish may lead to a greater sense of inferiority. As the Israeli court stated in sentencing Eichmann, "punishment is necessary to defend the honor or the authority of him who was hurt by the offense so that the failure to punish may not cause his degradation" (Arendt, 1976, p. 287). Punishment can also seek to extract an acknowledgment from the harm doer that his beliefs toward the victim-reactor are wrong. In short, punishment of the offender serves to maintain the self-image and the beliefs and values of the reactor.

Next, let us consider retribution from the perspective of the social group. Laws and rules derive from social groups. In addition to providing behavioral proscriptions and prescriptions, they help to define the boundaries and the social reality of the group, as many sociologists (e.g., Durkheim, 1964; Eriksen, 1966; Gusfield, 1963) and social psychologists (e.g., Mead, 1918; Thibaut & Kelley, 1958) have noted. Violation of these laws or rules constitutes a threat to the group. This phenomenon can be seen most clearly in so-called victimless crimes, where no physical or material harm results to anyone but where the reactor perceives a threat to the value system of the group.

As in the dyadic interpersonal context analyzed by Heider, punishment for violation of group rules asserts the group's power by lowering the status of the offender. Both phenomena may also contribute to changing the offender's belief matrix *vis-à-vis* the group's values, as well as the moral principles which they represent. The purpose of Room 101 in George Orwell's *1984* was not to control behavior but rather to foster love of the rules themselves as symbolized by Big Brother.

As is true of behavior control motives, however, the offense may be viewed as involving others besides the participants, and the punishment reaction may therefore be directed toward a wider audience. For instance, the offense may lead the victim-reactor to believe that his values and self-worth have been lowered in the eyes of others, especially if the offense is known or is likely to become known. Punishment lowers the offender's status relative to that of the victim and suggests to others that the offense should be attributed to the offender's bad character rather than the weakness of the victim.

The phenomenon of retribution is also similar to that of behavior control in that reactions directed toward the offender and those directed toward the wider audience have different requirements with regard to offender awareness. In the former case, the offender must both know that he is being punished and know why if the goal of retribution is to be achieved fully. In the latter case, the offender himself is merely a means through which group solidarity and consensus are achieved. The offender's reactions to the punishment, including acts of repentance, are relevant only to the extent that they are viewed as contributing to this end. Furthermore, in this latter case, a major contributor to the initiation of retributive action is the reactor's belief that the punishment will be publicized to the broader group or audience.

Punishment reactions based on retribution are embedded in psychological structure more deeply and completely than are behavior control reactions (see Hogan, this volume). Although retributive reactions may derive partly from perceiving a threat to group or individual values, they also arise from deeply held beliefs of "justice" or "oughtness": the reactor feels that it would not be "right" or "just" for the offender to escape with impunity. The offender has violated a moral rule that transcends the specific victim and even the social group. The response to a crime like murder involves far more than just the perception that the act has challenged the group's values; it confronts essential belief systems—an impersonal objective order has been disturbed. The affective reaction in these instances is strong, and a compelling need to see the moral order set right is aroused (Heider, 1958; Lerner, Miller, & Holmes, 1976; Lerner, 1977). And, as Durkheim (1964) observed, punishment is viewed as "the mystical procedure that will effect this restitution" (p. 212).

Let us now turn to a consideration of how characteristics of the offense, the offender, and the reactor affect the phenomenon of retribution.

4.1. Rule and Offense Characteristics

The more important the rule is to the belief or value system of the individual reactor or the reactor's society, the more likely the punishment reaction and the stronger it will be (see Heider, 1958). Often the moral disapprobation associated with a rule violation parallels the physical or material harm caused by the act (as in murder), but this need not be the case. Violations of certain rules—those against homosexuality, for example—though posing no physical or material danger, may threaten important values held by the individual reactor or his group. Moreover, the cultural context may define certain rules as important or unimportant. For example, though stealing cattle may be considered *malum in se* in our culture, among the Sards the act is rather *malum prohibitum* (see Nader, 1975) and does not evoke punitive responses.

Even when the rhetoric generated in defense of a rule emphasizes the behavior control function of the rule, there often exists an underlying threat to values or beliefs. Evidence relevant to this hypothesis is found in studies of attitudes toward capital punishment (see Harris, 1968; Vidmar, 1974; Vidmar & Ellsworth, 1974; Sarat & Vidmar, 1976). When questioned about the basis of their support, most people who favor the death penalty initially invoke deterrence and other utilitarian reasons, but ultimately, many of them indicate that even if they were convinced that capital punishment does not deter, they would favor it anyway. There is reason to believe that a similar process operates in noncapital crimes.

The greater the perceived degree of social consensus about a rule or law, the more likely the individual is to respond punitively. Consensus is an indication of the validity of the violated rule. In many instances, consensual support is related to the moral significance of the rule, with the violation of the most important and most widely supported rules (e.g., those against violence) evoking the strongest punishment reactions. The importance of consensual support may be seen in the response to marijuana. As public attitudes have changed from strong moral opposition to mild opposition mixed with attitudes favoring legalization, the magnitude and actual imposition of sanctions have diminished. In another example, Cook's (1973) study of the sentencing behavior of judges in draft evasion cases indicates that those judges who perceived there to be a public consensus against evasion were more likely to sentence severely.

A final factor that may influence a retributive response is the amount of harm caused by the offense. In general, the greater the harm, the greater the punishment reaction. In discussing issues relevant to behavior control, we noted that this effect may be due, in part, to the fact that adults tend to ascribe greater intention and responsibility to the perpetrator when outcomes are more severe (Shaver, 1970,). We deal more fully with the relationship between perceived intentions and punishment reactions in the next section. The important point to be made here, however, is that the seriousness of the offense outcome seems to mediate punishment responses in more ways than perceived intentionality. Kalven and Zeisel (1966) noted that the jury's reaction to a criminal defendant is sometimes directly related to the amount of harm suffered by the victim. In one of several murder cases, where the evidence of intention was approximately the same, the victim, left for dead, recovered, crawled a great distance, and suffered a great deal before finally expiring. Though the murderer clearly did not intend this outcome, the jurors were more punitive in this case.

Shaw and Reitan (1969) examined responsibility ascription and sanctioning responses among samples of lawyers, policemen, military personnel, and ministers. Although both dependent variables were directly affected by outcome intensity, sanctioning was influenced more strongly than responsibility ascription. Recently, Vidmar (1977; 1978) has attempted to separate experimentally the effects of outcome and intention on punitiveness. Subjects were presented with examples of murder cases where the offender's intent to kill the victim was clear and unequivocal but where the severity of outcome varied. In one condition, for example, the bullet was deflected and caused only minor harm; in another, it caused moderate harm to the victim; in the third condition, death resulted. Subjects rated the offender's intention to kill equally in all conditions, but the magnitude of recommended punishment varied directly with the magnitude of harm. In one of the experiments (Vidmar, 1977), where subjects were exposed to all three possible outcomes, their punishment reactions showed the same relationship to outcome severity. Most interestingly, when they were later asked why their responses had varied, they did not articulate severity as the reason but continued to contend that offenders should be punished only according to intentions. Perhaps these harm-proportional punishment reactions reflect cultural learning that is not customarily verbalized or even recognized (Nisbett & Wilson, 1977). The amount of harm may also increase punitiveness by affecting the perceived threat to the rule itself. Additional research will be required to answer this intriguing question.

4.2. Rule Violator Characteristics

Four main classes of offender variables affect punishment reactions: ascriptions of intentionality and responsibility, the violator's subsequent reactions to the crime, the violator's prior behavior, and the relationship of the violator to the reactor.

4.2.1. Responsibility Ascription

Legal scholars and social philosphers have discussed at great length the role that responsibility ascription plays in reactions to crimes or transgressions (e.g., Hart, 1968). Psychologists have also been intrigued by the question of responsibility, though with the exception of a few studies by Shaw and his associates (e.g., Shaw & Reitan, 1969), not much effort has been devoted to the connection between responsibility and punishment. Building on the conceptual work of Piaget (1948) and Heider (1958), Shaw and Sulzer (1964) distinguished five levels of responsibility: responsibility through association, simple commission, foreseeability, intentionality, and justification, the last level being a case where responsibility is reduced if the actions producing the outcome are justified. Although adults in Western culture generally do not acknowledge responsibility by association, the other levels—commission, foreseeability, intentionality, and justification—correspond roughly to the legal concepts of strict liability, negligence, willfulness, or premeditation, and the various forms of mitigating excuses, respectively. Recently, Hamilton (1977) has argued that the perpetrator's social role may affect the level of responsibility assigned to him. For instance, a person in a high-status or power role is more likely to be assigned responsibility on the basis of liability or negligence than a low-status person. Unfortunately, it is beyond the scope of this paper to examine all the antecedent causes of responsibility ascription, and we limit our discussion here to a consideration of the ways in which attributions of responsibility mediate punishment reactions.

The perceived intentions of the offender are a prime mediator of punishment reactions. Generally, a person who is perceived as having done something accidentally is seen as less "responsible" and is punished less severely than someone who is perceived as having done a similar act intentionally. When the reactor's motivation is behavior control, we speculated that this relationship may exist because an intentional act is thought to reveal a higher probability of recidivism. But intentional wrongdoing also produces more punitiveness than accidental or negligent wrongdoing because it indicates a direct disregard of the rule and is therefore more threatening to the integrity of the rule (see Heider, 1958).

Aside from the issue of whether the offender has broken a rule intentionally or not, the perceived motivation of the actor is also crucially important. There are better and worse reasons for intentionally breaking a rule. If an offender does so for personal gain, for example, punitive reactions are likely to be harsher than if the actor was seeking to benefit others, such as his family, or to avoid some negative fate (Savitsy & Babel 1976). Similarly, transgression to avoid loss is less disturbing than is transgression for the purpose of gain (Kelly, 1971).

Wanton violation of a rule tends to evoke the most punitive reaction because it is the most severe affront to the rule and the values it reflects. An early empirical study by Sharp and Otto (1910b) illustrates this point nicely. Those authors presented a number of cases of criminal rule violations to young males. A substantial number of their subjects initially rejected retribution as a principle of punishment. As the crimes increased in severity and wantonness, however, more and more of the respondents abandoned their scruples about retributive punishment and came to recommend suffering for the wrongdoer, even though the examples were contrived so that no deterrent or other end would be served by the punishment.

As noted earlier, responsibility ascription is not independent of outcome severity. In a number of social-psychological studies, it has been shown that outcome and assignment of responsibility are highly related: the more severe the harm or damage experienced by a victim, the greater the responsibility attributed to the perpetrator of the action (see Lerner & Miller, 1978; Vidmar & Crinklaw, 1974; Wortman, 1976, for reviews). This relationship apparently stems from people's belief that there is order and stability in their environment and that bad things— at least, very bad things—happen only because someone or something was responsible for them (see Lerner & Miller, 1978). Since maintaining this belief in the "justness" of the world is so important to people, the more severe the harm or injustice, the more strongly motivated they will be to explain it by assigning responsibility for it.

But severe outcomes may not result in more punitive reactions only because they elicit stronger attributions of responsibility. In some instances, the strength of the punitive reaction may actually precede and determine the responsibility attribution. This latter relationship may emerge for two reasons: (a) the greater the magnitude of harm, the stronger the feeling that someone should be punished (Drabeck & Quarantelli, 1964); and (b) people's sense of justice requires that punishment be proportionate to offender responsibility. In other words, people may often be motivated to perceive culpability in order to justify their desire to punish. As Lasswell and Donnelly (1959) observed, "When defendant X is declared to be responsible, a characteristic is imputed to X that makes him an eligible target for negative sanctioning measures" (p. 872).

Attributed responsibility, therefore, is a two-way sword; it is both a cause and an effect on reactions to injustice or harm. People have stronger reactions to rule violators the more they feel the offenders are responsible for their actions, and they are more likely to ascribe responsibility to rule violators the more distressed they are by the outcome of the rule violation.

Another interesting aspect of responsibility assignment is revealed in contexts where there is more than one potential offender. In such instances, it is frequently the case that the target of blame assignment shifts as the severity of the outcome increases. Generally, this shift is guided by the need or tendency to match causes and effects: the larger the effect, the larger the presumed cause. As a consequence of this tendency, certain individuals are more likely to be selected for blame and punishment than others. Veltfort and Lee's (1943) study of responsibility assignment and desire for punishment following the Coconut Grove Nightclub fire in which 488 people were killed is instructive in this regard. As their analysis indicates, not just anyone was accepted as being responsible for the disaster. The 16-year-old boy who lit the match, for example, was not viewed as a sufficient cause to explain the catastrophe. Other, "larger" causes had to be identified. Ultimately, a county grand jury indicted 10 men, including the principal owner of the club; the Boston Building Commissioner; and a fire department inspector. Similar reactions are evident in the plethora of conspiracy theories surrounding the Kennedy assassination; many people simply cannot accept the possibility that such an insignificant figure as Lee Harvey Oswald could, by himself, have single-handedly changed the course of history.

4.2.2. Subsequent Violator Behavior

Aside from the responsibility attributed to the perpetrator, his subsequent behavior is also important. A dispassionate or incontrite rulebreaker generally incurs greater hostility and more punishment. Contrition tempers the punishment response for two interrelated reasons. First, the more contrition displayed, the more confident the reactor may be that the violator will not violate the rule again, a phenomenon already discussed with respect to behavior control motives. Second, and perhaps more importantly, contrition acknowledges the validity of the violated rule. It is the threat to the validity of the rule that is the most serious one for the social group and that is most responsible for the intensity of reaction to the incontrite rule violator. The fact that contrition appears frequently to be of paramount importance to reactors is evidence that concern with the broader group can transcend concern for the individual offender. Sometimes, the act of full contrition may serve to affirm the validity of the rule so completely that punitive reactions against the offender totally dissipate and may even be replaced by positive responses. As Alexander and Staub (1956) observed, people respond much more warmly to a "penitent sinner" than to a "righteous man."

4.2.3. Prior Violator Behavior

A history of previous violations by the offender also affects punishment reactions. Viewed in the context of a present offense, prior offenses not only suggest increased likelihood of recidivism but also indicate that the offender does not share the reactor's commitment to the rules. What is more, the new offense impugns any earlier expressions of remorse.

The opposite of this last proposition also holds. Rule violations by individuals who have demonstrated their commitment to the group and its values may evoke weaker punishment reactions than those by individuals who have not shown such commitment (Hollander, 1958). If the individual has accumulated social credits for his past adherence to the rules, he will not appear to threaten the structure of his group or society, both because he will seem unlikely to violate rules in the future and also because, even were he to do so, his earlier compliance demonstrated a commitment to the basic values of the group.

4.2.4. The Relationship between Violator and Reactor

The relationship between the violator and the reactor is also an important mediator of punishment responses. As a general principle, the more attractive the rule violator is to the reactor, the weaker the punishment response. As mentioned earlier, one reason may be that attractive others are less likely to be seen as immoral or capable of transgressions (Heider, 1958). Responsibility ascriptions, therefore, may be attenuated in the case of an attractive violator.

The similarity that the reactor sees between himself and the violator bears a complex relationship to his punishment reaction. On the one hand, because similar others are perceived as more attractive (Byrne,

1971), punishment reactions may be weaker, and there is evidence to support this proposition (see Mitchell & Byrne, 1973; Brooks, Doob, & Kirshenbaum, 1975). On the other hand, the violation of a rule by a similar other has a number of disturbing aspects. First, it can reflect on the reactor as well. If one risks being tarred with the same brush, one may be particularly concerned that rule violations by members of one's own group do not happen. This concern is another instance where punishment is administered more for its effect on a broader audience than for its effect on the rule violator himself. Second, offender-focused punishment reactions might also be more intense against a "similar" rule violator if the reactor assumes that similar others should "know better" or have a higher level of morality.

Finally, if the reactor takes the rule violation of a similar other to reflect a rejection of rules that define or embody the social group, then the punitive reaction may be harshest of all. Renegades, heretics, and apostates evoke particularly strong condemnation and punishment because they present the strongest possible challenge to the value system of the group (see Coser, 1956).

4.3. Individual Differences

Punishment reactions based on retributive motives, like those motivated by the goal of behavior control, are affected by individual differences. Some of these directly reflect the person's concern with retribution, and others are a result of more complex interactions with the type of rule or the offender. Such individual differences not only are interesting in their own right but also help to validate and illuminate some of the propositions already set forth.

4.3.1. General Differences

Cross-cultural psychological studies of the socialization of aggression suggest some of the ways that individual differences in punishment reaction may arise within a culture (see Devos & Hippler, 1969; Whiting & Whiting, 1960). Different socialization practices result in differences in the degree to which peoples internalize moral precepts and consider punishment appropriate for wrongdoing. For example, Barry, Child, and Bacon (1959) found cultural differences in the degree to which people internalize feelings of responsibility, conformity, and obedience. Inkeles and Levinson (1969) have noted how differences in the internalization of norms and rules is directly tied to the application of negative sanctions for wrongdoing.

Shaw and his associates (e.g., Shaw, 1967; Shaw & Schneider, 1969; Schneider & Shaw, 1970) reported differences in the sanctioning responses of whites, blacks, and Puerto Ricans, although their findings did not clarify the source of these differences. Segerstedt's (1949) study of Swedes documented differences in punishment reactions between urban and rural populations, among social classes, among religions, and between males and females. Segerstedt's data further suggest that some of these differences may be explicable in terms of the degree of commitment to moral values.

Other research has attempted to assess more directly the influence of differences in retributive motives on punishment reactions. As mentioned earlier, Sharp and Otto (1910a,b) found substantial differences in the extent to which people approve of retribution. More recently, research on attitudes toward capital punishment has also shown individual differences in the degree of retributiveness (e.g., Hamilton, 1976, 1978; Harris, 1968; Sarat & Vidmar, 1976; Vidmar, 1974; Vidmar & Dittenhoffer, 1979). For example, a number of surveys indicate that approximately half of those who favor capital punishment state that they would favor it even if they were convinced that it had no deterrent or other instrumental effects. The studies by Sarat and Vidmar (1976) and Vidmar and Dittenhoffer (1979) extend our knowledge on this issue by indicating that though nonretributive people tend to change their positions on the death penalty after being exposed to evidence challenging its deterrent effect, retributive proponents remain firm supporters. Retributiveness as the basis for endorsing the death penalty is correlated with more punitive attitudes toward wrongdoers in noncapital offenses (see Hamilton, 1976, 1978; Vidmar, 1974; Vidmar & Ellsworth, 1974).

The most comprehensive research on the significance of individual differences is that examining the relationship between the personality variable of authoritarianism (cf. Adorno et al., 1950) and retributive punishment reactions. Many studies have shown that high authoritarians are more punitive than low authoritarians (e.g., Boehm, 1968; Jurow, 1971; Mitchell & Byrne, 1973; Roberts & Jessor, 1958; Sherwood, 1966; Vidmar, 1974). Although Vidmar and Crinklaw (1974) have provided evidence that some of this greater punitiveness may arise from behavior control motives (high authoritarians are more inclined to believe that punishment is effective in deterring crime), there is both conceptual and empirical evidence to suggest that much of this punitiveness stems from retributive motives. Moral punitiveness is part of the syndrome of authoritarianism: such people tend to adhere rigidly to conventional values, to view deviations from them in moral terms, and to favor severe punishment. Research focusing on sanctioning behavior, moreover, shows that they tend to endorse punishment as an end in itself (Sherwood, 1966; Vidmar, 1974; Vidmar & Crinklaw, 1974).

4.3.2. Interaction of Individual Differences with the Form of Rule Violation

Although the study of individual differences suggests that groups of individuals may differ generally in the degree to which their punishment motives are based on retribution, these differences are clearly dependent on the type of rule that is violated and the severity of the outcome. Recall that the Sharp and Otto studies (1910a,b) indicated that persons who were relatively nonretributive toward criminal offenders responded retributively if the offense was heinous enough. More importantly, however, some research has shown a reversal in the retributiveness of high and low authoritarians, depending on the kind of rule that is violated. Vidmar (1977, 1978), for example, constructed accounts of criminal offenses in which the rule violated would be regarded differently depending on the subject's authoritarianism. An attempted killing was followed by three possible outcomes: the victim was killed, suffered major injury, or incurred minor injury. It was predicted that in a "conventional" murder attempt, high authoritarians would be more punitive than low authoritarians and that only among the former would the severity of the punishment recommendations increase with the severity of the outcome. Another condition, however, involved a "crime of obedience" like My Lai (see Hamilton, 1976, 1978; Kelman & Lawrence, 1972; Suedfeld & Epstein, 1973), in which a soldier attempted to kill an unarmed prisoner of war. It was predicted that because low authoritarians tend to evaluate such crimes more severely than do high authoritarians (see Kelman & Lawrence, 1972; Suedfeld & Epstein, 1973), they would impose harsher punishments and increase the severity of these sanctions as the severity of the outcome increased. Both predictions were confirmed. Thus, the study of individual personality differences supports the previously stated proposition that the more important the rule and the more serious the outcome of the offense, the more severe will be the retributive punishment reaction. Moreover, these findings argue against the hypothesis that retributive responses are limited to those who fall at the "conservative" end of a personality continuum (cf. Garcia & Griffith, 1978). Persons with "liberal" social beliefs, who might ordinarily reject retribution as justification for criminal punishment, may well respond more retributively than those with "conservative" social beliefs in the case of an offender, like Adolf Eichmann or Dr. Joseph Mengele, who has, in their view, perpetrated an especially heinous crime.

4.3.3. Interaction of Individual Differences with Characteristics of the Offender

The authoritarianism of the reactor also interacts with characteristics of the offender. At least in common criminal cases, high authoritarians tend to respond more strongly to personal characteristics of the offender, such as status, moral character, or similarity (Berg & Vidmar, 1975; Boehm, 1968; Centers, Shomer, & Rodrigues, 1970; Jurow, 1971). The research of Centers *et al.* (1970) suggests that this may be because high authoritarians tend to view offenders as personally responsible for their actions, whereas low authoritarians tend to stress environmental factors more. Other researchers have explained the results in terms of differing susceptibility to affective feelings, or conditions of status and authority (Mitchell & Byrne, 1973; Berg & Vidmar, 1975). All of these factors are probably relevant, and in addition, high authoritarians may be more likely to ascribe responsibility to an offender who is of low status or who is dissimilar.

Notwithstanding these generalizations, there are exceptions where low authoritarians are more punitive. Mitchell (1973), for example, devised a case of a tavern brawl and a subsequent shooting in which an off-duty policeman and a draftsman were the participants. In one condition, the draftsman shot the policeman; in the other, the roles were reversed. Mitchell reasoned that high authoritarians would defer to the policeman because of his status, whereas low authoritarians would hold him to a higher standard of conduct for the same reason. It was found, as predicted, that though high authoritarians were more punitive when the draftsman was the defendant, the pattern was reversed when the policeman was the defendant.

5. A FINAL PERSPECTIVE

Our present analysis has employed two main dimensions for understanding punishment reactions: retribution and behavior control. We have also attempted to indicate how these interact with characteristics of the offense and how individual differences among reactors relate to each variable. Unfortunately, in focusing on the importance of behavior control and retributive motives, we were not able to discuss the role that other justice-related motives, especially equity considerations, play in punishment reactions (e.g., Austin, Walster, & Utne, 1976; Brickman, 1977). Some very interesting insights into some of the possible differences between equity and retributive motives can be found in Hogan's chapter in the present volume.

The present chapter also avoided discussion of the basic conflict between treatment or rehabilitation of the offender and punishment motives, though to some extent, treatment motives are involved in the concept of behavior control. Finally, our analysis has centered on initial impulses or reactions without exploring how these may diminish over time or be modified when the reactor learns of the offender's actual fate. As was the case with equity motives, these are topics reserved for future exploration.

Most relevant to the theme of the present volume is the question of how changes in society's physical, economic, and social structure might affect the relative prevalence and strength of retributive versus behavior control motives. This is a complex question, and only speculative responses can be offered at this point. One such speculation is that as the crime rate increases, the concern with deterrence will also increase (Furstenburg, 1971; Thomas, 1976). In other words, the threat of material and physical harm may increase concern with behavior control and thereby increase punitiveness.

Economic and social threat might also induce changes in the punishment process. Since the celebrated study of Hovland and Sears (1940), in which it was reported that the price of cotton in the southern United States (an index of economic prosperity) was negatively correlated with the lynchings of blacks, psychologists have acknowledged that economic factors can affect punishment responses. This particular finding and other similar ones are generally discussed in terms of the frustration–aggression hypothesis, but the archival work of Sales (1972, 1973) suggests that changes in punitiveness under economic stress may actually reflect increased preoccupation with retribution. More specifically, Sales's work suggests that economic hardship produces increases in a number of authoritarian tendencies, including retributiveness, moralistic thinking and in-group–out-group differentiation. The speculation that macrosocioeconomic changes can induce microattitudinal changes is an interesting one and merits further research.

On the basis of our analysis, it appears that there may also be other types of social changes that affect the punishment process. For example, if social cohesion diminishes in a society and people identify less with group norms and values, retributive motivation may diminish—at least, in removed, "disinterested" cases. Concern with deterrence and behavior control, however, may increase under these same conditions (Durkheim, 1911, 1949; Schwartz & Miller, 1964).

Whatever social changes occur, it is clear that punishment will continue to be a complex social and psychological phenomenon and one of much importance to society. Though it has been badly neglected in contemporary psychological literature, it must be a cornerstone of any complete social-psychological theory of justice. Hopefully, the present analysis, as well as that by Hogan, will help to draw the conceptual and empirical attention to this phenomenon that it deserves.

REFERENCES

Abel, R. L. A comparative theory of dispute institutions in society. *Law & Society Review*, 1973, *8*, 217–347.

Adorno, T. W., Frenkel-Brunswick, E., Levinson, D. J., & Sanford, R. N. *The authoritarian personality*. New York: Harper, 1950.

Alexander, F., & Staub, H. *The criminal, the judge and the public*. Glencoe, Ill.: Free Press, 1956.

Andeneas, J. *Punishment and deterrence*. Ann Arbor: University of Michigan Press, 1974.

Arendt, H. *Eichmann in Jerusalem*. New York: Penguin Books, 1976.

Austin, W., Walster, E., & Utne, M. K. Equity and the law: The effect of a harmdoer's "suffering in the act" on liking and assigned punishment. In L. Berkowitz & E. Walster (Eds.), *Advances in experimental social psychology*, Vol. 9. New York: Academic Press, 1976.

Barry, H. H., Child, I., & Bacon, M. K. Relations of child training to subsistence economy. *American Anthropolgist*, 1959, *61*, 51–63.

Berg, K., & Vidmar, N. Authoritarianism and recall of evidence about criminal behavior. *Journal of Research in Personality*, 1975, *9*, 145–157.

Boehm, V. Mr. prejudice, Miss sympathy, and the authoritarian personality: An application of psychological measuring techniques to the problem of jury bias. *Wisconsin Law Review*, 1968, 734–738.

Brickman, P. Crime and punishment in sports and society. *Journal of Social Issues*, 1977, *33*, 140–164.

Brooks, W. N., Doob, A. N., & Kirshenbaum, H. M. Character of the victim in the trial of a case of rape. Unpublished manuscript, University of Toronto, 1975.

Burgess, A. *A clock-work orange*. New York: Ballantine, 1962.

Byrne, D. *The attraction paradigm*. New York: Academic Press, 1971.

Carroll, J., & Payne, J. Crime seriousness, recidivism risk, and causal attributions in judgments of prison terms by students and experts. *Journal of Applied Psychology*, 1977, *62*, 595–602.

Centers, R., Shomer, R., Rodrigues, & Rodrigues F. A field experiment in interpersonal persuasion using authoritative influence. *Journal of Personality*, 1970, *38*, 392–403.

Cook, B. Sentencing behavior of federal judges: Draft cases, 1972. *Cincinnati Law Review*, 1973, *42*, 597–633.

Coser, L. *The function of social conflict*. New York: Free Press, 1956.

Cunningham, J., & Kelley, H. H. Causal attributions for interpersonal events of varing magnitude. *Journal of Personality*, 1975, *43*, 74–93.

Devos, G. A., & Hippler, A. E. Cultural psychology: Comparative studies of human behavior. In G. Lindzey & E. Aronson (Eds.), *Handbook of social psychology*, Vol. 4. Reading, Mass.: Addison-Wesley, 1969.

Drabeck, T. E., & Quarantelli, E. L. Scapegoats, villains and disasters. *Trans-Action*, 1964, 12–17.

Durkheim, E. *The division of labor in society* (G. Simpson, trans.) Toronto: Collier-Macmillan Candada, Ltd., 1964.

Ebbessen, E. B., & Konecni, V. J. Decision making and information integration in the courts: The setting of bail. *Journal of Personality and Social Psychology*, 1975, *32*, 805–821.

Erikson, K. T. *Wayward Puritans: A study in the sociology of deviance*. New York: Wiley, 1966.

Ezorsky, G. (Ed.). *Philosophical perspectives on punishment*. Albany: State University of New York Press, 1972.

Foucalt, Michel. *Discipline and punish*. New York: Pantheon Books, 1977.

Funstenberg, F. Public reaction to crime in the streets. *The American Scholar*, 1971, *40*, 601–622.

Garcia, L. T., & Griffith, W. Authoritarianism–situation interactions in the determination of punitiveness: Engaging authoritarian ideology. *Journal of Research in Personality*, 1978, *12*, 469–478.

Gusfield, J. R. *Symbolic crusade: Status politics and the American temperance movement*. Urbana: University of Illinois Press, 1963.

Hamilton, V. L. Individual differences in ascriptions of responsibility, guilt, and appropriate punishment. In G. Bermant, C. Nemeth, & N. Vidmar (Eds.), *Psychology and the law*. Lexington, Mass.: Lexington Books, 1976.

Hamilton, V. L. Who is responsible? Toward a social psychology of responsibility attribution. *Social Psychology*, 1977, *41*, 316–323.

Hamilton, V. L. Obedience and responsibility: A jury simulation. *Journal of Personality and Social Psychology*, 1978, *36*, 126–146.

Harris, L. Changing public attitudes toward crime and corrections. *Federal Probation*, 1968, *32*, 9–16.

Hart, H. L. A. *Punishment and responsibility*. Oxford: Oxford University Press, 1968.

Hart, I. Maternal child-rearing practices and authoritarian ideology. *Journal of Abnormal and Social Psychology*, 1957, *55*, 232–237.

Heider, F. *The psychology of interpersonal relations*. New York: Wiley, 1958.

Hogarth, J. *Sentencing is a human process*. Toronto: University of Toronto Press, 1971.

Hollander, E. P. Conformity, status, and idiosyncrasy credit. *Psychological Review*, 1958, *65*, 117–127.

Hovland, C. & Sears, R. Minor studies in agression: VI. Correlation of lynchings with economic indices. *Journal of Psychology*, 1940, *9*, 301–310.

Inkeles, A., & Levinson, D. J. National character: The study of modal personality and sociocultural systems. In G. Lindzey & E. Arsonson (Eds.), *Handbook of social psychology*, Vol. 4. Reading, Mass.: Addison-Wesley, 1969.

Jurow, G. L. New data on the effect of a "death-qualified" jury on the guilt determination process. *Harvard Law Review*, 1971, *84*, 567–611.

Kalven, H., & Zeisel, H. *The American jury*. Chicago: University of Chicago Press, 1966.

Kelley, H. H. *Attribution in social interaction*. New York: General Learning Press, 1971.

Kelman, H. C., & Lawrence, L. H. Assignment of responsibility in the case of Lt. Calley: Preliminary report on a national survey. *Journal of Social Issues*, 1972, *28*(1), 177–212.

Kipnis, D., & Cosentino, J. Use of leadership powers in industry. *Journal of Applied Psychology*, 1969, *53*, 460–466.

Lasswell, H., & Donnelly, R. The continuing debate over responsibility: An introduction to isolating the condemnation sanction. *Yale Law Journal*, 1959, *68*, 869–899.

Lerner, M. J. The justice motive: Some hypotheses as to its origins and forms. *Journal of Personality*, 1977, *45*, 1–52.

Lerner, M. J., & Miller, D. T. Just world research and the attribution process: Looking back and ahead. *Psychological Bulletin*, 1978, *85*, 1030–1051.

Lerner, M. J., Miller, D. T., & Holmes, J. G. Deserving and the emergence of forms of justice. In *Advances in experimental social psychology*, Vol. 9. New York: Academic Press, 1976.

Leventhal, G. S. The distribution of rewards and resources in groups and organizations. In L. Berkowitz & E. Walster (Eds.), *Advances in experimental social psychology*, Vol. 9. New York: Academic Press, 1976. (a)

Leventhal, G. S. Fairness in social relationships. In J. Thibaut, J. T. Spence, & R. Carson (Eds.), *Contemporary topics in social psychology*. Morristown, N.J.: General Learning Press, 1976. (b)

Levinson, D. J., & Hoffman, P. E. Traditional family ideology and its relation to personality. *Journal of Personality*, 1955, 23, 251–273.

Lindesmith, A. R. Punishment. In D. Sills (Ed.), *International encyclopedia of the social sciences*. New York: Macmillan & Free Press, 1968.

Makela, K. Public sense of justice and judicial practice. *Acta Sociologica*, 1966, 10(1), 42–67.

Mead, G. H. The psychology of punitive justice. *American Journal of Sociology*, 1918, 23, 577–602.

Menninger, K. *The crime of punishment*. New York: Viking, 1966.

Merton, R. K. *Social theory and social structure*. Glencoe, Ill.: Free Press, 1957.

Miller, D. T. Altruism and threat to a belief in a just world. *Journal of Experimental Social Psychology*, 1977, 13, 113–126. (a)

Miller, D. T. Personal deserving versus justice for others: An exploration of the justice motive. *Journal of Experimental Social Psychology*, 1977, 13, 1–13. (b)

Miller, D. T. The effects of perpetrator-relevant characteristics on reactions to victims of injustices. Presented at the Symposium on Equity, Retribution, and Other Factors in Legal Justice Reactions, Annual Meeting of the Eastern Psychological Association, Washington, 1978.

Miller, D. T., & McCann, C. D. Children's reactions to the perpetrators and victims of injustices. *Child Development*, 1979, 50, 861–868.

Miller, D. T., & Ross, M. Self-serving biases in the attribution of causality: Fact or fiction? *Psychological Bulletin*, 1975, 82, 213–225.

Mitchell, H. Authoritarian punitiveness in simulated juror decision-making: The good guys don't always wear white hats. Presented at the Annual Meeting of the Midwestern Psychological Association, Chicago, 1973.

Mitchell, H., & Byrne, D. The defendant's dilemma: Effects of juror's attitudes and authoritarianism on judicial decisions. *Journal of Personality and Social Psychology*, 1973, 25, 123–129.

Nader, L. Forums for justice: A cross-cultural perspective. *Journal of Social Issues*, 1975, 31(3), 151–170.

Nisbett, R. E., & Wilson, T. Telling more than we know: Verbal reports on mental processes. *Psychological Review*, 1977, 84, 231–259.

Orland, L., & Tyler, H. R. *Justice in sentencing*. Mineola, N.Y.: Foundation Press, 1974.

Pepitone, A. Social psychological perspectives on crime and punishment. *Journal of Social Issues*, 1975, 31, 197–216.

Piaget, J. *The moral judgment of the child*. New York: Macmillan, 1948.

Ranulf, S. *Moral indignation and middle class psychology*. New York: Schocken Books, 1964.

Roberts, A. & Jessor, R. Authoritarianism, punitiveness and perceived social status. *Journal of Social Psychology*, 1958, 56, 311–316.

Rossi, P., Waite, E., Bose, E., & Berk, R. The seriousness of crimes: Normative structure and individual differences. *American Sociological Review*, 1974, 39, 224–237.

Rothbart, M. Effects of motivation, equity, and compliance on the use of reward and punishment. *Journal of Personality and Social Psychology*, 1968, 9, 353–362.

Rule, B. G., & Nesdale, A. R. Moral judgment of aggressive behavior. In R. Geen & E. O'Neal (Eds.), *Perspectives on aggression*. New York: Academic Press, 1976.

Rusche, G., & Kircheimer, O. *Punishment and social structure.* New York: Columbia University Press, 1939.

Sales, S. M. Economic threat as a determinant of conversion rates in authoritarian and nonauthoritarian churches. *Journal of Personality and Social Psychology,* 1972, *23,* 420–428.

Sales, S. M. Threat as a factor in authoritarianism: An analysis of archival data. *Journal of Personality and Social Psychology,* 1973, *28,* 44–57.

Sarat, A., & Vidmar, N. Public opinion, the death penalty, and the eighth amendment: Testing the Marshall hypothesis. *Wisconsin Law Review,* 1976, *1,* 171–206.

Savitsky, J., & Babel, J. Cheating, intention and punishment from an equity theory perspective. *Journal of Research in Personality,* 1976, *10,* 128–136.

Scheler, M. *Resentiment.* New York: Free Press, 1961.

Schneider, F., & Shaw, M. Sanctioning behavior in negro and in white populations. *Journal of Social Psychology,* 1970, *81,* 63–72.

Schulhofer, S. J. Harm and punishment: A critique of emphasis on the results of conduct in the criminal law. *University of Pennsylvania Law Review,* 1974, *122,* 1497–1607.

Schwartz, R. Legal evolution and the Durkheim hypothesis: A reply to Professor Baxi. *Law & Society Review,* 1974, *8,* 653–668.

Schwartz, R., & Miller, J. Legal evolution and societal complexity. *American Journal of Sociology,* 1964, *70,* 159–171.

Segerstedt, T. Research into the general sense of justice. *Theoria,* 1949, *15,* 323–338.

Sharp, F., & Otto, M. Retribution and deterrence in the moral judgments of common sense. *International Journal of Ethics,* 1910, *20,* 438–453.

Sharp, F., & Otto, M. A study of the popular attitude toward retributive punishment. *International Journal of Ethics,* 1910, *20,* 341–357.

Shaver, K. G. Defensive attribution: Effects of severity and relevance on the responsibility assigned for an accident. *Journal of Personality and Social Psychology,* 1970, *14,* 101–113.

Shaw, M. Some cultural differences in sanctioning behavior. *Psychonomic Science,* 1967, *8,* 45–46.

Shaw, M., & Reitan, S. Attribution of responsibility as a basis for sanctioning behavior. *British Journal of Social and Clinical Psychology,* 1969, *8,* 217–226.

Shaw, M. & Schneider, F. Intellectual competence as a variable in attribution of responsibility and assignment of sanctions. *Journal of Social Psychology,* 1969, *78,* 31–36.

Shaw, M., & Sulzer, J. An empirical test of Heider's levels in attribution of responsibility. *Journal of Abnormal and Social Psychology,* 1964, *69,* 39–46.

Sherwood, J. Authoritarianism, moral realism and President Kennedy's death. *British Journal of Social and Clinical Psychology,* 1966, *5,* 264–269.

Spitzer, S. Punishment and social organization: A study of Durkhein's theory of penal evolution. *Law & Society Review,* 1975, *9,* 613–638.

Suedfeld, P., & Epstein, Y. Attitudes, values and ascription of responsibility: The Calley case. *Journal of Social Issues,* 1973, *29,* 63–71.

Sykes, G., & Matza, D. Techniques of neutralization: A theory of delinquency. *American Sociological Review,* 1957, *22,* 664–670.

Thibaut, J. W., & Kelley, H. H. *The social psychology of groups,* New York: Wiley, 1958.

Thomas, C. Perceptions of crime, punishment and legal sanctions. Paper presented at the American Society of Criminology, 1976.

Thomas, C., & Cage, R. Correlates of public attitudes toward legal sanctions. Paper presented at the meeting of the Western Sociological and Anthropological Association, 1974, Banff, Canada.

Thomas, C., Cage, R., & Foster, S. Public opinion on criminal law and legal sanctions: An examination of two conceptual models. Paper presented at the American Society of Criminology, Chicago, 1974.

Veltfort, H. R., & Lee, G. E. The Coconut Grove fire: A study in scapegoating. *Journal of Abnormal and Social Psychology*, 1943, 138–154.

Vidmar, N. Retributive and utilitarian motives and other correlates of Canadian attitudes toward the death penalty. *Canadian Psychologist*, 1974, *15*, 337–356.

Vidmar, N. Effects of degree of harm and retributive motives on punishment reactions. Paper presented at the Annual Meeting of the Canadian Psychological Association, Vancouver, B.C., June 1977.

Vidmar, N. Outcome, offense type, and retribution as factors in punishment reactions. Paper presented at the Annual Meeting of the Eastern Psychological Association, Washington, D.C., April 1978.

Vidmar, N., & Crinklaw, L. Attributing responsibility for an accident: A methodological and conceptual critique. *Canadian Journal of Behavioral Science*, 1974, *6*, 112–130.

Vidmar, N., & Dittenhoffer, A. Public opinion and the death penalty: The effects of knowledge on attitudes. Unpublished manuscript, 1979.

Vidmar, N., & Ellsworth, P. Public opinion and the death penalty. *Stanford Law Review*, 1974, *26*, 1245–1270.

Vidmar, N., & Miller, D. T. Social psychological processes underlying attitudes toward legal punishment. *Law & Society Review*, 1980, *14*(3), 565–602.

Walster, E. Assignment of responsibility for an accident. *Journal of Personality and Social Psychology*, 1966, *3*, 73–79.

Walster, E., Walster, G. W., & Berscheid, E. *Equity, theory and research*. Boston: Allyn & Bacon, 1978.

Weihofen, H. *The urge to punish*. London: Gollancz, 1957.

Westermarck, E. A. *Ethical relativity*. New York: Harcourt Brace, 1932.

Wheeler, S., Bonacich, E., Cramer, R., & Zola, I. Agents of delinquency control: A comparative analysis. In S. Wheeler (Ed.), *Controlling delinquents*. New York: Wiley, 1960.

Whiting, J. W., & Child, I. *Child training and personality*. New Haven, Conn.: Yale University Press, 1953.

Wortman, C. B. Causal attributions and personal control. In J. H. Harvey, W. J. Ickes, & R. F. Kidd (Eds.), *New directions in attribution research*. Hillsdale, N.J.: Erlbaum, 1976.

Zimring, F., & Hawkins, G. The legal threat as an instrument of social change. *Journal of Social Issues*, 1971, *27*, 33–48.

9

Microjustice and Macrojustice

PHILIP BRICKMAN, ROBERT FOLGER, ERICA GOODE, and YAACOV SCHUL

1. INTRODUCTION

If every individual in a society has been fairly rewarded, does this guarantee that rewards have been fairly distributed in the society as a whole? Surprisingly, it does not. It is our contention that people use different criteria to assess microjustice (the fairness of rewards to individual recipients) and macrojustice (the aggregate fairness of reward in a society).

Recent concern with race and sex differences has begun to sensitize social psychologists to the possibility that treating individuals fairly may produce what seems to be an unfair distribution of rewards among groups. This view has been most clearly formulated in the area of test bias (Cole, 1973; Thorndike, 1971). Suppose we have two groups, and that all members of the first group have a 60% chance of making good doctors, while all members of the second group have a 30% chance of making good doctors. If we admit people to medical school strictly on the basis of individual merit, we will wind up with a society in which 100% of the doctors come from the first group. Macrofairness, however, would lead us to feel that some percentage of the doctors in society

PHILIP BRICKMAN • Department of Psychology and Research Center for Group Dynamics, University of Michigan, Ann Arbor, Michigan 48106. ROBERT FOLGER • Department of Psychology, Southern Methodist University, Dallas, Texas 75275. ERICA GOODE • Department of Psychology, University of California at Santa Cruz, Santa Cruz, California 95064. YAACOV SCHUL • Department of Psychology and Research Center for Group Dynamics, University of Michigan, Ann Arbor, Michigan 48106.

should come from the second group. Some members of that group would make good doctors; some members of society might benefit from having doctors who come from that group; and, most fundamentally, the distribution of medical training (or any valued resource) in a way that entirely excludes potentially eligible persons from certain groups violates our sense that all groups are being treated fairly. This is especially the case, to be sure, when we feel that any differences in the average chances for success of members of the different groups have been socially created by prior differences in the way members of each group have been treated. If people are advantaged or handicapped because of their group membership and not simply because of their individual qualities, it is not fair to judge them solely on the basis of their individual qualities.

Macrojustice is not, however, simply a question of treating groups as well as individuals fairly. Principles of subgroup or between-group fairness can be the familiar principles of microjustice (e.g., need and merit) applied to groups or to social units other than individuals. There is, however, another set of qualitatively different principles that apply to the perception of the overall distribution of rewards in a system, whether that system is composed of individuals or larger social units. Even if all groups, as well as all individuals, receive what their aggregate merits appear to deserve, the distribution of rewards in the society as a whole can be judged unfair. As Brickman and Stearns (1978) have demonstrated, observers instructed to provide rewards or incentives for contributions distribute resources among hypothetical workers in a way that vastly enhances, rather than diminishes, the differences between the more able and the less able groups. Perhaps people are simply unaware that providing greater rewards for those capable of superior performance tends to produce an increasing degree of concentration of resources in the hands of the more talented, or in the hands of those who are believed to be more talented. But people are not, we suspect, happy with this result when they are confronted with it. Brickman (1975) found that members of four-person groups differentiated among overall distributions of rewards that were produced by uniform application of the same principle, namely, merit or reward in proportion to performance. In particular, they rejected a negatively skewed distribution (in which one member did worse than all the rest) as least fair, despite the fact that they were on the whole most satisfied with their own positions in this distribution. It would seem, thus, that people have preferences for the overall shape of the outcome distribution *per se,* and not only for the microlevel principles by which such outcome distributions may be produced.

A similar conclusion is suggested by the studies of Brickman and Bryan (1975, 1976). Once again, these studies explored people's feelings about the distribution of resources in four-person groups. The first study demonstrated that subjects (fifth-graders) are more likely to endorse the surreptitious transfer of resources when it is done by an altruistic or disinterested party than when it is done by a party who profits from the transfer. The second study demonstrated that subjects are more likely to endorse the secret transfer of resources by a third party when this transfer rectifies a previous injustice in the assignment of rewards than when it creates an injustice. But both studies also showed that, independently of how the final distribution of rewards was created, subjects preferred certain distributions (more equal ones) to other distributions (less equal ones).

The first evidence for a conception of macrojustice distinct from principles of individual reward, then, is indirect. It comes from the observation that people resist taking the sum of outcomes earned or deserved by each individual as defining all that they mean by the term *justice*. This might be called the demonstration from insufficiency. It occurs, we suggest, because an overall distribution is also directly apprehended as an integrated whole, a gestalt, with properties as a whole that make the whole seem either good or bad (cf. Thurow, 1971). These properties can conflict with the judgment that emerges from people's aggregation of the microjustice of a myriad of individual cases.

It is, of course, the potential for conflict that makes the distinction between microjustice and macrojustice interesting and important. It may be impossible in all cases to satisfy the demands of microjustice and macrojustice simultaneously, just as it is impossible in all cases to find an overall preference for society that represents, in a satisfactory manner, the sum of all members' individual preferences (Arrow, 1951). Anyone who has ever tried to assign grades in a classroom intuitively knows how difficult it is to satisfy simultaneously the claims of micro- and macrojustice. When the most elaborate and precise assessments of each individual's performance are assembled into a whole, the resulting distribution invariably seems flawed in a number of ways. There are too many high grades or too many low ones. Some distinctions appear to be too sharp and others not sharp enough. Experienced teachers may save their assignment of "class participation points" until after the final distribution is constructed so that these bonus points can be used to help things come out right.

The strongest assertion of conflict between micro- and macrojustice is that such conflict is not only possible but inevitable. From this perspective, microjustice serves as a social trap for macrojustice in the same way that people's pursuit of their own individual well-being serves as a social trap for the collective well-being of society (Platt, 1973). In pursuing the short-run goal of trying to make each individual situation as fair as possible, people overlook the very considerations that must be taken into account to make the overall situation fair in the long run. A series of microdecisions, all rational or fair in themselves, leads to an overall conclusion that is neither rational nor fair. Our distinction between microjustice and macrojustice has in turn, we believe, an important implication for understanding social traps. The adoption of a macrojustice perspective is the key to preventing people from pursuing their own selfish interests in situations where this pursuit is ultimately self-destructive (pollution, arms races, and overfishing are common examples).

Other authors have been aware of the conflict between individual deserving and a more global sense of social justice. Lerner, Miller, and Holmes (1976) have shown that people's concern in protecting the sense that individuals get what they deserve leads them to repudiate potential claims by victims in a way that violates a wider sense of justice. Runciman (1966) has demonstrated that important differences in political orientations reflect the fact that people draw a distinction between whether or not they personally have been treated unfairly (personal deprivation) and whether or not an important reference group of theirs has been treated unfairly (fraternal deprivation). Cohen (1979) has suggested that the conflict between individual deserving and distributive justice can be understood only if we carry out studies with groups larger than dyads, in which we can distinguish between people's feelings about any single comparison and their feelings about the overall shape of a reward distribution. Reflections on need and equality as principles of justice (Deutsch, 1975; Leventhal, 1976; Sampson, 1975) have also, implicitly, been concerned with the distinction between individual deserving and social justice, or between micro- and macrojustice. No previous work, however, has provided a clear analytical framework for understanding the exact nature of this difference. The ambiguity in previous research can be traced in part, we believe, to the lack of such a framework. This paper is an effort to provide one.

2.   ELEMENTS OF MICROJUSTICE AND MACROJUSTICE

Principles of microjustice are individualizing. Any principle can be recognized as a principle of microjustice if it requires the determination or assessment of the attributes of individuals. Any principle of the form "To each according to their X" is a principle of microjustice. Need ("To each according to their need") and merit ("To each according to their contribution"; cf. Adams, 1965) are two common principles of microjustice. Principles of microjustice imply that the overall distribution of rewards or resources in a group is determined by an empirical assessment of the distribution of certain desirable features in the group. These principles place no *a priori* restrictions on the shape of this outcome distribution. If the principle for distributing a given resource is merit and merit is equally distributed, the resource will be equally distributed. If merit is normally distributed, the resource will be normally distributed. And, as Mark (1980) put it, if merit is distributed in the shape of the Manhattan skyline, the resource will be distributed in the shape of the Manhattan skyline. The same is true of the distribution of resources according to need.

Principles of macrojustice are deindividuating. They do not call for a determination of the attributes of individuals and a fit between the individual's attributes and his or her rewards. Principles of macrojustice simply specify, *a priori*, constraints on the allowable form or pattern for an overall distribution of resources. Such principles not only fail to require that the rewards of individuals match the characteristics of individuals but often make it impossible for this to be the case. Since our culture, and social psychology, has paid more attention to principles of microjustice than to principles of macrojustice, it is not surprising that principles of macrojustice are less familiar and less well understood. They have, nonetheless, made a periodic appearance, as Mark (1980) illustrated in the following illuminating discussion:

> Examples [of macrojustice principles] would be: "The resource allotment of the best-off shall not exceed three times the resource allotment of the least well-off;" "No one's resource allotment shall fall below one half the median resource allotment" (Fuchs, 1967); "Resources shall be distributed normally." Notice that these rules are not of the form "To each according to . . ." Thus, knowing a system-level [macro] rule does not tell us how much each individual should receive, even if we know the characteristics of each individual. System-level rules do tell us, however, such things as whether there should be a guaranteed minimum income, or a ceiling on income, or a limit to income inequalities. Rawls' (1971) second principle is a system-level distributional rule: "Social and economic inequalities are to be arranged so that they are . . . to the greatest benefit of the least advantaged" (p. 83). This rule is less specific than other society-level rules we have mentioned. It specifies that inequalities are permitted only when they increase the allotment of those at the bottom of the resource distribution. It does not, however, specify who receives how much within whatever inequality is allowable . . ."
>
> [It should now be] clear that unless the empirically-based distribution prescribed by a micro level rule happens to fall within the *a priori* limits prescribed by a society-level rule, micro level justice and system level justice will be in conflict. For example, someone might perceive it fair that income is distributed "in proportion to one's merit (micro level rule), and they might also perceive it fair that no one's income fall below a certain level, say one-half of the median (system-level rule). Unless merit is distributed such that no one has less than one-half the median amount of merit, these two rules would be in conflict. (pp. 6, 8)

The primary difference between microjustice and macrojustice arises from their different focus of concern. Microjustice is concerned with the qualifying attributes of individuals. Macrojustice is concerned with the appropriate order of society or the goodness of a society. These two different concerns are, interestingly enough, the respective concerns of the two major source figures in the history of Western moral philosophy, Plato and Aristotle. Macrojustice is the major concern of Plato's *Republic*. Microjustice is the more prominent focus of Aristotle's *Nichomachean Ethics*. The imbalance of research in social psychology is symbolized by Aristotle's being mentioned seven times in Walster, Walster, and Berscheid's (1978) book on equity, while Plato is mentioned only once, and then without reference to the essential concerns of the *Republic*. As Mark (1980) pointed out, social-psychological theories of justice have essentially restricted themselves to examining only three microjustice principles: equity, need, and equality (which has been incorrectly, though understandably, conceptualized as a microlevel principle; see below). Minimum, average, maximum, or range principles are ignored. Mark went on to note that most social-psychological research has involved observing direct exchanges (or people's feelings about direct exchanges) with a small number of others, usually one. Walster *et al.* (1978) restricted their theorizing to "actual relationships," which they defined (following Thibaut & Kelley, 1959) as requiring direct interaction between individuals. Such microjustice theorizing is unlikely to be sufficient for understanding what happens when there are struggles, among large populations who do not interact directly, over resources that cannot be easily assigned to any of them on an individual basis.

We may specify the difference between micro- and macroprinciples of justice in another way. Microprinciples are principles of correspondence. They specify that an appropriate correspondence should be established between an individual's standing on a meaningful qualifying or "input" dimension (e.g., need, merit) and his or her standing on a reward or "outcome" dimension (resources, reward). Mathematically, they specify that $Y = f(X)$, or that some quantity, Y, is to be distributed as a function of people's standing on some dimension, X. Macroprinciples, on the other hand, are self-referential principles. They specify a reflexive adjustment involving only the outcome dimension. Mathematically, macroprinciples specify that $Y = f(Y)$, or that people's standing on some dimension, Y, is to be adjusted as a function of the amount of Y they already have. Some self-referential or self-adjusting operations are characteristic of all living systems and all functional units (Hofstadter, 1979). It should not be surprising, therefore, that some version of the kind of principles that we are calling macrojustice principles should be characteristic of the operation of any social unit that makes claims to embody a system of justice.

The major defining properties of micro- and macrojustice are summarized in Table 9.1.

The distinction between micro- and macrojustice can help clear up an enduring confusion about the status of need and equality as principles of justice. Equality is a macrojustice principle because it places constraints on, and indeed completely specifies, the nature of the total distribution of resources. In this sense, it is merely the limiting case of a number of range and minimum principles. Unlike other macrojustice principles, however, equality also specifies exactly what each individual person should get. It can thus be thought of as a microjustice principle as well, and it should not be surprising that researchers unfamiliar with the idea of macrojustice should think of it only as a microjustice principle. Equality would be a microjustice principle if it rested on the assessment of each individual in the situation, a comparison of their particular needs or merits, and a decision that they were all, in fact, equally needy or equally deserving. But instances in which the principle of equality is invoked rarely involve such measurement and comparison. If they did, error of measurement alone would be sufficient to ensure that people would not be found equal even if they were in fact equal. Asserting that people are equally deserving without assessing their particular attributes is specifying a desirable property of a distribution rather than an empirical property of a set of individuals.

The ambiguity in the application of a needs rule has been pointed out by Eckhoff (1974). Need is generally thought of as an individualizing or microjustice principle, in which it is necessary to know each individual's state before a determination can be made of what resources they should receive. However, as Eckhoff (1974) pointed out, a principle of need can also be invoked with the implicit aim of leveling out a distribution, or of putting those who are disadvantaged on a particular dimension (and hence determined to be in need) on a more equal footing with those who are advantaged on that dimension. In this sense, need is actually a stalking horse for the macrolevel principle of equality, or,

Table 9.1. Properties of Microjustice and Macrojustice

	Microjustice	Macrojustice
Focus of concern	Relationships between individuals (Aristotle)	Structure of social order (Plato)
Metaprinciple	Assessment	Citizenship
General nature of principles	Individualizing Correspondent	Deindividuating Self-referential
Example of principles	Need Merit	Average Minimum

at least, the macrolevel principle of some maximum allowable range. Need can be recognized as a disguised version of a macrojustice principle like equality whenever it takes on a self-referential form, or whenever people are held to need Y if they simply have less of Y than others do. The fact that need can often be used to justify what is actually a search for equality, and vice versa, has obscured the extent to which need and equality, like any pair of micro- and macrolevel principles, can conflict. A distribution in which the parties actually received resources in strict proportion to their need for these resources would almost certainly be a highly unequal distribution. In many domains, the handicapped, for example, need much more than nonhandicapped individuals, if they are to function. Marxism is no less vulnerable to this possible contradiction than liberalism. The principle of "from each according to his ability, to each according to his need" may or may not coincide with a general state of equality.

Just as need can serve as a stalking horse for the macrojustice principle of equality, merit can serve as a stalking horse for a macrojustice principle of inequality. People and parties who would not be able to defend inequality as good in itself can in effect defend inequality by defending merit as a principle of microjustice, especially if the existing distribution of wealth and power in society is seen as corresponding to this existing distribution of merit. It may be a major triumph for egalitarian interests simply to shift the domain of discourse from micro- to macroconcerns, where the question of equality versus inequality can be debated directly rather than by the proxy issues of need and merit. Furthermore, once equality is understood as a macroprinciple, it becomes apparent that there are egalitarian principles (e.g., minimum or range) that can be endorsed without abandoning the vital concerns of microjustice (e.g., some sense of personal control over outcomes).

The content of salient micro- and macroprinciples will, of course, differ in different cultures. In a peasant society, the prime criterion for seeing an individual as worthy may be the performance of filial duties rather than effort and accomplishment at the tasks valued by Western society. But, we would argue, some version of microjustice principles and some version of macrojustice principles are present in every group and every culture. The main difference is how important the one set of principles is relative to the other. The main predictor of relative importance, in turn, is the degree of collective consciousness or group identification in the culture. If microjustice principles appear weakly developed in many peasant or tribal societies, it is precisely because people in such societies are judged primarily as members of social networks rather than as single individuals.

From this perspective, it is especially interesting to note that Rawls

(1971), in his famous book on justice, predicted the emergence of what we would call macrojustice principles under exactly the circumstances that force people to contemplate society as a whole rather than the merits of individuals. Rawls defined a hypothetical "original position" in which the members of a society know the possible distributions of resources in the society but are behind a "veil of ignorance" when it comes to where they or any other individuals stand in that society. In this position, he argued, rational individuals prefer an egalitarian order as one that minimizes their maximum possible loss or suffering. Brickman (1977a) has found some evidence that subjects in an experimental analogue to the original position behave much as Rawls predicted. They may, of course, do so for reasons quite different than the philosophical argument offered by Rawls. They may do so because the original position high- lights their sense of empathy with potentially disadvantaged others or because it stimulates a primitive, functional jealousy against potentially advantaged and powerful others. The important point, however, is that people have preferences for the shape of an overall outcome distribution, that these preferences emerge most clearly when attention is focused on the group as a whole, and that these preferences appear to be strongly egalitarian.

3. MAJORITY RULE AND MINORITY RIGHTS

The model underlying the general set of microjustice principles can be called an *assessment model*. It attends to, assesses, and respects the particular attributes of individuals. The model underlying the general set of macrojustice principles can be called a *citizenship model*. It assigns people a standing—or constrains their possible standing—as a function of their group membership. Viewed in this way, it is not surprising that the distinction between micro- and macrojustice can be correlated with a number of other distinctions among types of justice. For example, it can be argued that microjustice refers more easily to the economic do- main and macrojustice to the political domain. Microjustice principles may appear to apply more easily to informal or decentralized groups, while macrojustice requires some form of centralized authority. Macro- justice, because it requires some consciousness of a collective whole, may appear more easily associated with cooperation, group spirit, and the production of public goods, and microjustice may be more associated with competition, selfishness, and private goods.

We believe that macrojustice is correlated, empirically, with some of these dimensions but that it cannot be reduced, analytically, to any of them. With regard to cooperation and competition, for example, it must be remembered that macrojustice principles are not necessarily egalitarian. A macrojustice principle may specify that there be a large gap or large difference between two classes as well as that there be a small gap or a small difference. Caste societies, aristocracies, and hierarchies are just as well described by macrojustice principles as are democracies or classless societies. Thus, macrojustice principles can be used to specify arrangements that will foster antagonistic as well as harmonious relationships, although they may obviate some of the more painful forms of competitive individual comparison (cf. Brickman & Bulman, 1977). Conversely, the microjustice principles of need or merit (especially where merit is defined as working for the good of the group) can be vital to the functioning of a cooperative enterprise (cf. Kanter, 1973).

If these various dimensions are partially correlated and partially distinct, it is, of course, possible to make any one of them the primary dimension and the others secondary. We favor micro- and macrojustice as the most useful way to begin dividing up the domain of justice for two reasons. First, the distinction between micro- and macrojustice can be most clearly specified in analytical terms. Second, the nature of this distinction appears most highly general, or most likely to be useful in coordinating or clarifying different distinctions in areas far removed from our initial realm of inquiry.

There are, however, two other distinctions that seem to us to be sufficiently relevant to micro- and macrojustice as to be worth mentioning at this point. The first is the distinction between distributive and procedural justice (cf. Thibaut & Walker, 1975; Folger, 1977; Folger, Rosenfield, Grove, & Corkran, 1979). Distributive justice is concerned with outcomes. Procedural justice is concerned with procedures. There is a sense in which microjustice is more like a principle of procedural justice and macrojustice is a more pure form of distributive justice. If we follow a procedural principle like informed consent, we say that any outcome distribution is fair, provided it has the informed consent of all concerned. Microjustice principles say that any outcome distribution is fair, provided it has been determined by appropriate assessment of individuals' relevant attributes (cf. Nozick, 1975). Macrojustice principles, on the other hand, say that only certain outcome distributions are fair, and they say nothing about the procedures by which these distributions will be created. The distinction between equality of opportunity and equality of outcome is partly a distinction between procedural and outcome justice. Equality or inequality of opportunity constrains only the procedures to be followed in allowing individuals to pursue outcomes and is indifferent to the range of outcomes produced. Equality of opportunity is a presumption of the microjustice principle of equity, a necessary precondition for subsequent differences to be ascribed to effort or ability. Equality of outcome, on the other hand, specifies that all individuals will arrive at the same state and is indifferent as to the means or procedures used to get them there. It is, of course, a principle of macrojustice.

Another carrier of the distinction between micro- and macrojustice, at least in Western culture, is the distinction between individual deserving, whether on the basis of need or merit, and majority rule or democracy. The authors of this paper are not in complete agreement among themselves on the value of this parallel (or on the above discussion of concern with procedures versus outcomes). Nonetheless, we believe that the overemphasis on microjustice in previous research can be explained in part by the difficulty of identifying a schema for macrojustice that is as vivid, coherent, and available as the equity schema for microjustice. The only comparably vivid schema for macrojustice—democracy—comes from the political domain and perhaps has been overlooked by social psychologists for that very reason. It seems to be worthwhile to call the attention of future researchers to this schema, even if the fit with other parts of the present analysis is not perfect.

Democracy means that the majority of a group decides what is in the collective interest of that group. It often does so by voting, but there are other possible procedures. Democracy is a principle that defines what distributions of opinion toward an adopted course of action will be considered fair and what distributions will be considered unfair. It does not, in itself, provide safeguards for the rights of individuals or minorities against a tyranny of the majority. This has been the classic basis for criticisms of democracy. The addition of a principle of minority rights to a majority-rule system places a constraint on the majority's ability to adjust the general distribution of goods and services in society to its liking. Minority rights are generally protected by the microjustice principle of individual deserving, that all people are entitled to what they need or what they have earned. The rights of the disadvantaged can also in some measure be protected by the macroprinciple that all people are entitled to some minimum. But this principle does not protect the entitlement of the more advantaged, who have often been in the forefront of the struggles for individual rights and have also been the prime beneficiaries of their existence. Just as principles of macrojustice place limits on what is under individual control (to the extent that individual attributes are under individual control to begin with), principles of microjustice place limits on what is under collective control. On the other hand, while macrojustice principles like democracy may threaten microjustice principles of individual deserving, microjustice principles like equity or need threaten democracy every time they dictate a distribution of resources that interferes with the ability of all members of society to exercise their democratic rights (cf. Okun, 1975).

The analysis of this paper is primarily focused on questions of distributive justice; this paper does not speak directly to the analysis of legal systems or moral codes. Nonetheless, it is worth noting that retributive justice or the law in general may involve both micro- and macroprinciples, and that, here as elsewhere, the conflict between these principles is highly illuminating. The controversy over flat versus indeterminate sentencing is to some extent a conflict over macro- versus microprinciples of justice. Flat sentencing requires that all individuals who have committed a particular category of crime be given the same sentence, regardless of possible individual differences in the circumstances of the crime or the background of the defendant. Indeterminate sentencing allows these individual differences to be taken into account. The controversy over capital punishment is also to some extent a controversy over micro- and macroprinciples of justice. Advocates of capital punishment begin with the premise that it is a punishment richly deserved by certain criminals. (Advocates of capital punishment also base their case on the presumed deterrent effects of an occasional execution, but this is not, properly speaking, a principle of justice; cf. Brickman, 1977b. To be an effective deterrent, a punishment should be out of proportion to the crime and can as readily be applied to the innocent as to the guilty. This may be why slaves and sailors were so often flogged for minor offenses.) Opponents of capital punishment may argue that its appropriateness in certain cases cannot outweigh the fact that it will inevitably, by the very frailty of human institutions, be meted out by mistake to an occasional individual who does not deserve it. More commonly, however, opponents of capital punishment have rested their case on certain macrojustice arguments, that the penalty falls disproportionately on certain subgroups of society or that it represents a degree of severity in punishment that is incompatible with humane beliefs about the permissible distribution of rewards and punishments in society.

It should be clear from the above discussion why our distinction between micro- and macrojustice does not articulate precisely with the familiar distinction between rewards and rights. Macrojustice is typically concerned with specifying the outcomes people expect by right, but microjustice, too, can establish the basis on which people feel entitled to certain things (cf. Goldman, 1979). Moreover, whether a reward is something to be earned or something that is a person's by right is itself a matter than can change over time.

4. LOTTERIES: MACROJUSTICE WITHOUT MICROJUSTICE

Lotteries are an elegant device for studying the operation of macrojustice. By specifying how many winners there will be, and how much each winner will receive, we can completely determine what the final distribution of outcomes will be. At the same time, by assigning these outcomes entirely on the basis of chance, we can ignore all features of the individual participants and can rule out the application of any possible principles of microjustice. Thus, people's feelings about lotteries under different circumstances are a relatively pure measure of their feelings about the macrojustice of these situations.

According to Jaynes (1976), primitive humans had no conception of chance. They saw in divination, lots, and oracles a connection that we do not see to the individual circumstances of the people whose fate was being decided. Indeed, we attach great importance to the events that follow the assignment of outcomes by lot, under certain circumstances, precisely *because* they have no connection to the prior properties of the individuals involved. This is, of course, the principle of experimental design that allows us to infer that randomly assigned events were the causes of any subsequent differences between groups that were or were not exposed to them.

We may expect that lotteries will evolve as the basis for assigning scarce resources (prizes) or unwelcome obligations (military service) when no acceptable principle of microjustice appears to be available for the assignment of these outcomes. At the same time, however, we may expect that lotteries will be transitional devices that operate only during periods in which one microjustice principle of allocation has disappeared while another has not yet emerged. This appears to describe the situation of the draft lottery during a war that became too large to be served by a standing professional army and too unpopular to be sustained by patriotic volunteers. Lotteries may persist for longer periods of time in relatively minor domains, such as legalized gambling, in which participation is voluntary and the resources involved are marginal to the society as a whole.

Lotteries will always be vulnerable institutions for two reasons. First, people are ambivalent and uncomfortable about any process of allocating truly scarce and vital resources. Second, a lottery cannot, by its nature, give individuals an explanation of why they, personally, have been selected or rejected, or of what variations in skill and effort are responsible for their selection. People's belief that their outcomes are responsive to variations in their behavior is an important component of their ability to cope with the world (Seligman, 1975) and an important source of their motivation to believe that the world is a just place (Lerner, 1975). Lotteries should be more acceptable to the extent that they can defer to this belief, or to the extent that they can incorporate principles of microjustice along with principles of macrojustice. We set out to test this proposition in the following experiment.

In general, a lottery respects the interests of microjustice to the extent that it (a) guarantees that the winners will be people who deserve their outcomes; (b) gives people some sense of control over their participation and over, if not their chances of winning, their use of their outcomes; and (c) deals with quantities for which there appears to be no valid or noncorrupt microjustice allocation process. To make a preliminary test of some of these ideas, the senior author and Mary Kay Lauritzen asked 134 undergraduates their opinion about a number of hypothetical lotteries. Of these subjects, 78 were male and 56 were female; 46 came from Northwestern University and 88 from the University of Illinois at Chicago Circle. Each item consisted of a description of two lotteries that differed in only one critical feature. The subjects were asked to rate how fair they thought each lottery would be on a scale ranging from 0 for "not at all" to 6 for "very much." The subjects also filled out a number of background questions and Rubin and Peplau's (1973) Just World scale. The results are reported in Table 9.2.

The idea that lotteries limited to those who deserve to win are considered fairer than lotteries not so limited was tested by the first three items in Table 9.2. The prescreened-for-merit lottery compared a college admissions lottery for all applicants with an admissions lottery for only those applicants whose records conformed to a certain standard. The prescreened-for-need item compared a public-housing lottery for all applicants to a housing lottery for only those families whose financial position fell below a specified poverty level. Not surprisingly, the lotteries were rated as much fairer when those eligible for the drawing had been prescreened for need or merit. Prescreening guarantees that the winners will be deserving of the resources they receive, though it does not guarantee that they will be the only deserving winners or the most deserving of all possible winners. It is interesting to note that quotas, like lotteries, are viewed more favorably when they are seen as drawing only on candidates who are deserving. Jemmott and Tebbets (1980) found that newspaper articles supporting affirmative action in the *Bakke* case stressed that special treatment of minority candidates did not mean admitting or hiring the unqualified.

The third item relevant to our screening hypothesis is perhaps most interesting. It indicates that people want the screening to be done before the lottery rather than after, or early rather than late. In one alternative in the third item, a company used a lottery to select for processing 10%

Table 9.2. Fairness of Lotteries

Alternatives		$F(1,32)$ of difference
Prescreened for merit 3.81	Not prescreened 2.12	41.89[a]
Prescreened for need 4.13	Not prescreened 2.35	42.85[a]
Prelottery screening 3.56	Postlottery screening 2.45	17.61[a]
Choice of participation 3.71	No choice of participation 2.63	26.06[a]
Individual outcome 4.13	Shared outcome 2.72	43.39[a]
Know 50% odds 3.98	Guess 50% odds 3.25	12.32[a]
Choose own number 3.85	Computer chooses number 3.58	1.76
Valid alternative criteria 2.54	No valid alternative criteria 3.77	19.05[a]
Corrupt alternative 3.20	Correctable alternative 2.08	20.95[a]
Screening alternative 3.40	Auction alternative 3.34	0.07

[a] $p < .001$.

of a large initial pool of job applicants. It then chose a few qualified people to be hired. In the second alternative, the company first selected the 10% of the initial pool of applicants who were qualified, then used a lottery to decide which of the people were to be hired. As Row 3 of Table 9.2 indicates, people clearly preferred a lottery that considered only qualified candidates to a lottery that selected a list of candidates whose qualifications were then to be ascertained.

The idea that lotteries are considered fairer to the extent that people feel they have some control over their participation or their outcomes was tested in the next four items. People have meaningful control over a lottery if they have a choice of whether or not to participate in it, that is, if they are prescreened for willingness to participate. This criterion alone differentiates the popular revenue lotteries from the unpopular draft and admissions lotteries, and it has been found to make subjects more favorable toward the principle of random assignment in a scientific experiment (Wortman, Hendricks, & Hillis, 1976). We compared a situation in which college students could choose between participating in a lottery for fellowships or having their applications screened by a committee with a situation in which there was no alternative to the lottery.

As can be seen in Row 4 of Table 9.2, the optional lottery was rated as fairer than the mandatory lottery. People also have a kind of meaningful control over a lottery if the outcome is one that they control by themselves as opposed to when the outcome is one that must be shared with other people of possibly different views. We compared the fairness ratings of a housing lottery for four desirable single rooms with the ratings of a lottery for four places in a desirable four-person suite. The lottery for the single rooms was seen as fairer than the lottery for places in the suite.

It is interesting to note that American Airlines has made use of the optional participation principle in handling the difficult problem of bumping passengers with confirmed reservations from overfull flights. Airlines overbook flights in order to make up for last-minute cancellations or no-shows that would leave empty seats on popular flights. If there are fewer cancellations than expected, someone has to be bumped. The question is, who? In such cases, the airline now plans to ask passengers if they would voluntarily give up their seats in return for an immediate cash payment of between $25 and $100 and a confirmed seat on a later flight. About 3–6% of passengers will do so. Passengers to be bumped are then selected by lot from among those willing to accept the compensated delay ("American Smooths Out the Bumps," 1978).

Lotteries may also give people an illusion of control (cf. Langer, 1975), a mistaken belief that the lottery outcomes themselves, as opposed to events preceding or following the lottery, are under their control. Subjects may have the belief that they are more in control of a lottery—and are presumably more likely to get what they deserve—if they know the exact odds of the lottery than if they are given mathematically equivalent but variable odds (cf. Tversky & Kahneman, 1974). We compared a situation in which people were told that their chances of winning tickets to a concert could be as little as 10% or as much as 90% or anything in between, with a situation in which they were told that their chances of winning tickets were exactly 50%. The best guess of a person's chances of winning is identical (50%) in both cases. Nonetheless, as Row 6 of Table 9.2 shows, people rated the lottery as significantly more fair if exact odds were given than if variable odds were given. In a related vein, both Langer (1975) and Wortman (1975) found that subjects feel more confident of their ability to control a random outcome when they are allowed to choose their own lottery ticket or to draw their own prize than when their ticket or prize is drawn for them. We compared an undergraduate housing lottery in which each student picked a number to determine his own priority with a lottery in which a computer generated a random number to determine each student's priority. This variation had no significant effect on respondents' ratings of the fairness of the two lotteries.

The idea that a lottery is considered more fair when the alternative is unfair was tested in the last three items. The first described a training program in the new field of agrodynamics, or underwater agriculture, and indicated that past academic and athletic achievements either would or would not predict success in this new enterprise. As can be seen, the use of a lottery to admit people to the program was seen as fairer when there were no valid alternative criteria. The next item described college basketball recruiting as a situation in which the official rules were broken more often than not, and it indicated that realistic enforcement of these rules either was or was not likely. Subjects were more favorable to the idea of assigning high school athletes to colleges by lottery when recruiting practices appeared irredeemably corrupt than when they appeared correctable. Finally, an item compared the perceived fairness of a lottery to assign tickets to a popular show when the alternative was auctioning them off to the highest bidders versus having a committee screen ticket applications and decide who would get tickets. We had thought that the lottery would be seen as fairer when the alternative was an auction, since the auction would favor the wealthy. As can be seen in Table 9.2, the lottery was seen as equally fair in the two cases. The fact that the tickets represented a luxury rather than a necessity may have allowed the subjects to see an auction as an appropriate means of allocating them.

Despite the fact that our sample of respondents was fairly heterogeneous, their lottery ratings were unaffected by age, sex, race, religion, parental education, place raised, family size, marital status, political orientation, employment, belief in a just world, or perception of self as above average or advantaged.

5. THE SALIENCE OF MICROELEMENTS AND MACROELEMENTS IN SOCIAL POLICY DECISIONS

We did not, in the lottery study, test people's preferences for different possible macrojustice principles or for macrojustice principles versus microjustice principles. Lotteries are a very special case, albeit an instructive one, in which alternative principles are not ordinarily seen as equally available. As a first step toward assessing the more general pragmatic importance of micro- and macrojustice, we need to determine the relative weight that people attach to different principles in a variety of social policy domains.

To do this, we carried out a simple study in which we asked people to consider, in turn, six different problems of distributive justice: how scholarship money should be distributed to medical school applicants, how subsidies for rehabilitation should be distributed to accident victims, how gasoline should be allocated to commuters (during an oil shortage), how benefits should be distributed to retired persons, how immigration priorities should be assigned to people desiring to come to the United States, and how office space should be allocated among smokers and nonsmokers. Two of the situations, scholarship and rehabilitation, were constructed to evoke as clearly as possible the classic microjustice criteria of merit and need, respectively. The next two, gasoline and retirement, were seen as current issues in which both microjustice and macrojustice elements appeared to be salient. The last two, immigration and smoking space, were issues that did not seem to us to be handled at all well by familiar microjustice formulations. The spirit of our study, however, was to gain rather than to test or rule out ideas. Our first aim was simply to map out the primary incidence of micro- and macrojustice principles, and to demonstrate that macrojustice principles were indeed necessary to capture an important part of what people felt was embodied by justice.

In each case, we asked the respondents to consider a total of six possible principles of justice and to endorse as many of them as they felt would be relevant to a fair and just solution to the issue in question. Three of the principles were principles of microjustice: need, effort, and ability. Three were principles of macrojustice. The first was a minimum principle, that the smallest allocation be not too much less favorable than the largest allocation. (This is actually a relative minimum principle and was presented to the subjects in words rather than with any arbitrary numbers. Note also that in practice its effects are equivalent to those of a relative maximum or ceiling principle.) The second was a subgroup principle, specifying that there should be an appropriate balance kept between the average allocations to two different subgroups of the total population. The third was a principle specifying that the average or the total allocation within the particular domain in question take into account constraints placed by requirements of other domains or other systems, which we have termed an average or total principle.

Because the actual content of most of these principles is necessarily different in each situation, it seems advisable to provide readers with a complete listing of the items we used. This is done in Appendix A. It will be immediately clear that in many instances more than one item could have been written to embody a particular principle. Our choice in such cases was somewhat arbitrary. On the other hand, it will also be clear that in some situations, it was very hard to construct a compelling version of one or another of the principles. Thus, our assessment of the preference for micro- and macrojustice in any one situation is at best an approximation. We can attach more confidence to patterns appearing across situations.

The subjects in the study were 192 college students, 96 from Southern Methodist University and 96 from the University of Michigan. Half the subjects at each site were male and half were female. The male subjects at each site were individually contacted by male interviewers, and the female subjects were contacted by female interviewers. The subjects were approached in dormitories and asked to fill out a survey about public policy. As an additional incentive, the participants were promised a chance to win $20 in a lottery to be held at the conclusion of the study. Prizes of $20 were in fact won by five students at each site.

The order of situations was uniquely randomized for each subject. The order in which the microjustice and macrojustice alternatives appeared in each situation was counterbalanced, with half the males and females at each site receiving the microjustice alternatives first and half the macrojustice alternatives first. The subjects were asked not to sign their names to their questionnaires. (A list of names was kept separately for purposes of the lottery.)

In addition to evaluating the situations, the subjects were asked for demographic data on age, year in school, major, grade-point average, ethnic identification, sex, parental income, political views, party affiliation, and religious background.

The overall percentage of respondents endorsing each principle in each situation is shown in Table 9.3. As can be seen, the scholarship situation was the only one of the six in which all three microprinciples were more frequently endorsed than all three macroprinciples. Endorsement of the micro- and macroprinciples in the scholarship situation was compared with endorsement in each of the other five situations, using a 2×2 analysis of variance with situations and type of principle (micro versus macro) as the independent variables and mean frequency of endorsement of micro- and macroalternatives (0–3) as the dependent variable. The interaction contrast was significant at the .001 level in all five cases.

Table 9.3. Percentage of Respondents Endorsing Each Principle in Each Situation

Situation	Principle					
	Need	Effort	Ability	Minimum	Subgroup	Total
Scholarship	83	45	66	7	37	14
Rehabilitation	86	20	42	5	25	41
Gasoline	69	44	19	29	43	38
Retirement	76	31	18	20	57	35
Immigration	61	31	47	35	20	70
Smoking space	48	32	19	30	18	19

It is somewhat surprising that need was endorsed almost as frequently for the award of scholarships as it was for the award of money for the rehabilitation of accident victims. (Need was, in general, the most frequently endorsed principle, especially if our total principle is understood to refer to the need of another system.) Conversely, it is interesting to note that ability (in this case, the ability of patients to recover) was endorsed second only to need in the case of rehabilitation. The minimum principle was strikingly low in these two paradigmatic microjustice situations. It achieved a respectable frequency of endorsement in all the other situations.

The pattern for rehabilitation was, in general, much the same as the pattern for scholarship, with the exception that effort was abnormally low, lower indeed than it was in any of the other situations. In this case, more so than in most of the others (but consider also the retirement item), effort referred to a past behavior for which people might be blamed (carelessness). The respondents were clearly reluctant to employ this as a basis for denying funds for rehabilitation.

The gasoline and retirement items demonstrate a dramatic shift from the scholarship item. In both cases, all three of the macrojustice principles were more frequently endorsed than in the scholarship situation. With one exception, all of the microjustice principles were less frequently endorsed than in the scholarship situation.

There are at least three possible explanations for the greater endorsement of macrojustice principles in the gasoline and the retirement situations than in the scholarship and the rehabilitation situations. The first is that the gasoline and retirement situations were seen by the respondents as affecting essentially everyone, and it was this universality of impact that called forth a macrojustice orientation. Scholarships for medical students and rehabilitation funds for accident victims, on the other hand, are directly relevant to only a small minority of the population. A second, related explanation is that the gasoline and retirement situations were seen as ones in which individuals were, in general, less in control of, and hence less responsible for, their outcomes than the scholarship and rehabilitation situations. A third explanation, somewhat contradictory to the first, is that our student respondents were more familiar with and actually felt personally closer to the possibilities of getting scholarships to medical school or recovering from accidents than they did to the perhaps more remote issues of commuting to work or managing during retirement. Assessment of the relative merits of these explanations is a task for future research.

The immigration and smoking-space items likewise show less endorsement of microjustice principles than the scholarship item, and more endorsement of the minimum and total principles. The smoking-space question is especially illuminating. In general, principles of justice were seen as less relevant to this situation than to the other situations considered. Need was endorsed less frequently here than in any other case, and the endorsement of principles of effort, ability, subgroup balance, or limitation of total resources was also at or near the bottom of the ranges for these alternatives. By contrast, the minimum principle stood out. It was the only one in the smoking-space situation that was endorsed with a high degree of frequency relative to its endorsement in other situations.

The subgroup and total principles clearly depend on what subgroups are mentioned and what reasons are given for putting constraints on the total. For example, the endorsement of the subgroup principle was greatest in the retirement situation, where it referred to a balance between the sexes. It was reasonably high in the gasoline situation, where it referred to a balance of allocations among people of different income levels, and in the scholarship situation, where it referred to a balance of benefits given to minority and nonminority students. The subgroup principle was least frequently endorsed in the rehabilitation, immigration, and smoking-space situations, where it referred to differences between patients living in the North and South, differences between potential immigrants of different religious and racial groups, and differences between cigar, pipe, and cigarette smokers. Presumably the subgroup distinctions in these cases were less relevant to the subjects and, in their eyes, less relevant to the policy decisions at hand than in the other cases. Since the subgroups in this study were embedded within the situations—and were different for each situation— only future research can establish to what extent these differences were due to changes in the situation or the resource at stake versus changes in the identity or relevance of the subgroup.

The percentage of situations in which different types of respondents endorsed each principle is shown in Table 9.4. The raw data for Table 9.4 were the mean proportion of times each respondent endorsed a given principle over the six situations. Most of our demographic variables (all those not shown in Table 9.4) made no difference in the subjects' endorsement of different principles. The variables listed in Table 9.4, except for Michigan–Texas, all interacted with micro- versus macroprinciples at the .05 level when subject to 2 × 2 analyses of variance using the background characteristic and micro–macro as independent variables. They did not, on the other hand, interact with situations, that is, the effects in Table 9.4 were general across situations rather than due to idiosyncrasies of a particular item.

Females were more likely to endorse macrojustice principles than males, and social-science majors ($N = 23$) were more likely to endorse

Table 9.4. Percentage of Situations in Which Different Types of Respondents
Endorsed Each Principle

Type of respondent	Principle					
	Need	Effort	Ability	Minimum	Subgroup	Total
Male	69	36	33	17	27	35
Female	71	30	36	24	39	37
Physical sciences	71	36	37	21	25	36
Social sciences	65	33	31	25	39	37
Republican	68	39	40	17	38	34
Democrat	71	23	33	27	34	34
Texas	69	36	36	18	38	32
Michigan	71	30	33	23	27	39

macrojustice principles than physical-science majors ($N = 66$). The differences in both cases were more marked for the minimum and subgroup
principles than for the total principle. This sex difference fits with other
research indicating that females are more sensitive to the collective interests, and especially the socioemotional requirements, of a group (see
Brickman & Bulman, 1977, for some discussion of this literature). The
difference between social-science and physical-science majors seems to
be nicely explained by the fact that the former group is explicitly trained
to be aware of group-level interests and the requirements of human
collective action.

Students who described themselves as Democrats ($N = 49$) were
more likely to endorse a minimum principle than students who described
themselves as Republicans ($N = 72$), while Republicans were more likely
to endorse the principles of effort and ability. This seems to be less a
difference in micro- versus macro-orientation or system-level consciousness *per se* than in how members of these two groups perceive the
collective interests of their own subgroups. Democrats have historically
been more committed to government action on behalf of the disadvantaged, action that is ideally justified by a minimum principle. Republicans have historically justified opposition to such actions on the grounds
that effort and ability determine, correctly, what people get.

The Michigan–Texas pattern suggests a somewhat greater macrojustice orientation by Michigan subjects and a more individualistic orientation by Texas subjects except for the fact that the Texas subjects
were more likely to endorse the subgroup principle. Texas has a much

more multicultural society than Michigan. Texas politics is concerned with working out not only the relationships between blacks and whites but the relationships of each of these groups to native Americans, chicanos, and Mexicans. It should not be surprising that this experience may have sensitized Texans to the importance of subgroup principles. Our primary interest in including subjects from two different parts of the country, however, was not to assess regional differences, but simply to increase the generalizability of our overall description of preferences and their relationships.

We find it very encouraging that the pattern of our subjects' endorsement of micro- and macroprinciples of justice shifted, across situations and across respondent characteristics, in coherent and predictable ways. It should not be thought, however, that these subjects were simply choosing between microjustice and macrojustice principles *per se*, or that the microjustice and the macrojustice principles form unitary clusters. They do not. Indeed, if a person endorses one macrojustice principle, this may make it less, rather than more, necessary for the person to endorse another macrojustice principle in that situation. In general, the probability that any two principles would both be endorsed by a given respondent varied widely depending on the situation. In only two instances were the probabilities of co-occurrence above .60 and consistent across all six situations. The average probability of endorsing both effort and ability principles was .62. The average probability of endorsing both minimum and subgroup principles was also .62.

6.   DIRECTIONS FOR FUTURE RESEARCH

There is no more important question than the question of how the fairness with which individuals are treated sets limits to, and is in turn limited by, the fairness of the overall distribution of resources in a society. In the language of this paper, this is the question of the problematic relationship between microjustice and macrojustice. An overriding irony in the pursuit of justice is that the full attainment of one form may preclude the attainment of the other.

From now on, when we do research on distributive (or procedural) justice, we should keep in mind the question of macrojustice as well as the psychologically more salient question of microjustice. Attention to macrojustice may help us to solve certain problems, like the problem of equality, that are essentially intractable if considered at the microjustice level alone. It will help us to redress the imbalance of an individualistic and essentially incomplete equity theory. Finally, it will enable us to make better contact with the full set of rules that determine how questions of justice are fought out and resolved in the political arenas of the real world. The debate over affirmative action, for example, is largely a debate between micro- and macroprinciples of justice. In the *Bakke* case, newspaper articles in favor of admitting Bakke to medical school (and against the affirmative-action program that Bakke contended had caused him to be rejected) stressed the microjustice argument that Bakke was deserving: he had been a U.S. Marine and a hospital volunteer, he had studied premed during his off-hours, and his test scores were better than those of minority candidates who had been admitted (Jemmott & Tebbets, 1980). Articles against admitting Bakke and in favor of affirmative action stressed macrojustice arguments: society had discriminated against members of minority groups in the past, and society would benefit from having more minority doctors (Jemmott & Tebbets, 1980).

We are already in a position to formulate a number of specific hypotheses about the causes and consequences of a macrojustice orientation. The general cause is a high degree of awareness of the group as a collective whole, or a broad degree of identification with other group members in general (cf. Brickman, 1977). Collective awareness is more likely to be high, we may hypothesize, when the entire distribution of relevant resources in a group is public and available to all members rather than private or known only to a few or to none (cf. Brickman & Campbell, 1971). A heightened degree of collective awareness should also be more likely to occur under conditions of scarcity (cf. Heilbroner, 1974). Hence, we would predict greater salience of macrojustice principles for the distribution of necessities rather than luxuries, or under conditions of scarcity rather than abundance. Alternatively, it is possible that the salience of macrojustice concerns will be a curvilinear function of the amount of resources available to the society, with macroconcerns more salient when resources are either very scarce (in which case, people are forced to think about the group as a whole in pursuing their own individual needs) or very abundant (in which case, people are freed to think about the group as a whole by the satisfaction of their own individual needs).

The presence of class or categorical distinctions in a group should also heighten the salience of macrojustice concerns, especially subgroup principles (cf. Tajfel, 1978) and especially if the distribution of subgroup membership coincides with the distribution of important skills or rewards. Experimentally, this phenomenon could be studied by observing the response of mixed-sex or mixed-race groups to situations in which skills or rewards either differentiate or do not differentiate the sexes or the races at some initial point in time. Finally, as in the lottery study, we can predict that macrojustice principles in general will be more salient and more acceptable when their microjustice alternatives are perceived to be biased, invalid, or corrupt.

We cannot speak as clearly about the consequences of a macrojustice orientation. It is interesting to note, however, that Modigliani and Gamson (1979) have recently distinguished among three models of thinking about politics. The first treats politics as an extension of interpersonal experience. The second, which they call the "solidarity orientation," organizes political thinking around a salient set of group identifications. The third, called the "ideological orientation," views political discourse as primarily concerned with the production and distribution of collective goods. Modigliani and Gamson suggested that identical communications and events are reacted to very differently by people with each of these three orientations. The orientations appear to coincide quite well with a microjustice orientation, a subgroup macrojustice orientation, and a system-level macrojustice orientation, or at least this is a correspondence that could be empirically tested. If true, it suggests that a macrojustice orientation leads to an entirely different form of responding to political events than a microjustice orientation.

We also hypothesize that people with a macrojustice orientation are more willing to authorize third-party interventions in ongoing relationships or smaller units, especially when they see the pursuit of microjustice within these relationships or units as having become a social trap diminishing their overall viability. People with a macrojustice orientation may be less concerned with attributing responsibility or allocating praise for apparently meritorious behavior or blame for careless or reckless behavior. They may, for example, be more willing to endorse no-fault insurance. Finally, just as collective awareness is predicted to foster a macrojustice orientation, the elicitation of a macrojustice orientation should, in turn, foster a higher degree of collective awareness.

To dramatize the difference between micro- and macrojustice, we have highlighted instances in which the two are incompatible and people are forced to choose between them. In general, however, it seems likely that societies endorse some version of both micro- and macroprinciples (like the subjects in our second study) whether they realize it or not. A society with only microprinciples would risk an evolution of distributions that would be unprecedented, unacceptable to critical segments of the population, and hence unstable. A society with only macroprinciples would risk demoralization if the members felt their outcomes were not contingent on their behavior and hence that they had no control over their outcomes. Future research needs to map out precisely what combinations of micro- and macroprinciples are compatible, both logically and psychologically. We should find out whether compatibility is enhanced when the microprinciples are overt and the macroprinciples are covert (as may often be the case in American society), versus when the macroprinciples are overt and the microprinciples are covert, versus when both sets of principles are overt. And we need to understand not only the cost of endorsing only microprinciples or only macroprinciples, but also the conditions under which societies are allowed, or forced, to do this.

As we discover the conditions under which questions of macrojustice are salient, and the consequences of making them salient, we will expand our understanding not only of macrojustice but of microjustice and, indeed, of all of social behavior as well.

ACKNOWLEDGMENTS

We wish to thank Ronald Cohen, Lita Furby, Louise Kidder, Melvin Lerner, Melvin Mark, Dale Miller, Jeffrey Paige, and Margalit Tal for helpful feedback and suggestions. We would also like to thank Lisa Celucci, Thomas Chamberlin, Linda Fleming, Cindy Miller, John Mitchell, and Tom Robinson for their assistance in collecting and coding data.

APPENDIX   A

Microjustice and Macrojustice Alternatives in Social Policy Decisions

Which of the following should be taken into account in deciding how scholarship money should be distributed to applicants to medical school?

—Scholarship money should be given to applicants according to how much they need the award in order to attend school.
—Scholarship money should be given to applicants according to how hard they have tried to acquire experience and get acquainted with the field by working in health related jobs.
—Scholarship money should be given to applicants according to how able they are to do well in medical school, as predicted by such indicators as scores on the Medical School Admissions Test.
—The smallest scholarship awards should not be too much less than the largest awards.
—There should be an appropriate balance kept between the average scholarship awards given to minority and non-minority students.
—The total amount of money for scholarship awards should take into account the standard of living of people not in medical school.

Which of the following should be taken into account in deciding how government subsidies for rehabilitation should be distributed to accident victims?

—Money for rehabilitation should be given to accident victims on the basis of how much they need the subsidy to pay for treatment.

—Money for rehabilitation should be given to accident victims on the basis of how hard they tried to avoid the accident by behaving in a careful and responsible manner.

—Money for rehabilitation should be given to accident victims on the basis of how able they are to recover sufficiently to live a normal life.

—The smallest subsidy for rehabilitation should not be too much less than the largest subsidy given for rehabilitation.

—There should be an appropriate balance kept between the average subsidies available to patients living in the North and the South.

—The total amount of money available for rehabilitation should not interfere with the total amount allocated for medical care for non-accident victims.

If there were a severe oil shortage in this country, which of the following should be taken into account in deciding how gasoline should be allocated to commuters?

—Gasoline should be distributed to commuters on the basis of how much they need the gasoline in order to drive to work.

—Gasoline should be distributed to commuters on the basis of how hard they have tried to conserve energy by carpooling with other commuters.

—Gasoline should be distributed to commuters on the basis of how able they are to contribute to the economic welfare of the country by the important nature of their job.

—The smallest amount of gas given to a commuter should not be too much less than the largest amount given to a commuter.

—There should be an appropriate balance kept between the average gas rations available to commuters of different income levels.

—The total amount of gasoline for commuters should not interfere with the total amount of heating oil allocated to farmers.

Which of the following should be taken into account in deciding how benefits should be distributed to retired persons?

—Retirement benefits should be given to retirees according to how much they need the money in order to live.

—Retirement benefits should be given to retirees according to how hard they have tried to save money by investing in a pension plan.

—Retirement benefits should be given to retirees according to how able they will be to use the money in a responsible manner.

—The smallest retirement benefits given out should not be too much less than the largest benefits.

—There should be an appropriate balance kept between the average retirement benefits received by male and female workers.

—The total amount of money for retirement benefits should not interfere with the total amount allocated for unemployment benefits.

Which of the following should be taken into account in deciding how immigration privileges should be allocated to people desiring to immigrate to the United States?

—Permission to immigrate should be given to applicants on the basis of how much they need to come to the United States to escape persecution.

—Permission to immigrate should be given to applicants on the basis of how hard they have tried to help the United States by supporting free world interests abroad.

—Permission to immigrate should be given to applicants on the basis of how able they are to contribute to the United States by virtue of their professional skills.

—The longest delay between the submission of an application for admission and its consideration should not be too much longer than the shortest delay.

—There should be an appropriate balance kept between the number of persons of one religious or racial group given permission to immigrate and the number of persons of other religious or racial groups.

—The total number of immigrants should not be so large as to take jobs from American citizens.

Which of the following should be taken into account in deciding how office space should be allocated among smokers and non-smokers?

—Office space should be allocated according to how much the smoker needs to smoke and the non-smoker to avoid smoke.

—Office space should be allocated according to how hard the smokers tried to show consideration for non-smokers and the non-smokers for smokers.

—Office space should be allocated according to how able the smokers and non-smokers are to contribute to the group of which they are a part.

—Space available for smokers should not be too much less than the space available for non-smokers.

—There should be an appropriate balance kept between the average amount of space available for cigar, pipe and cigarette smokers.

—The amount of attention paid to smoking as an issue around the office should not interfere with the amount of attention paid to drinking as an issue.

REFERENCES

Adams, J. S. Inequity in social exchange. In L. Berkowitz (Ed.), *Advances in experimental social psychology*, Vol. 2. New York: Academic Press, 1965.

American smooths out the bumps. *American Way*, August 1978, p. 59.

Arrow, K. J. *Social choice and individual values*. New York: Wiley, 1951.

Brickman, P. Adaptation-level determinants of satisfaction with equal and unequal outcome distributions in skill and chance situations. *Journal of Personality and Social Psychology*, 1975, 32, 191–198.

Brickman, P. Preference for inequality. *Sociometry*, 1977, 40, 303–310. (a)

Brickman, P. Crime and punishment in sports and society. *Journal of Social Issues*, 1977, 33(1), 140–164. (b)

Brickman, P., & Bryan, J. H. Moral judgment of theft, charity, and third-party transfers that increase or decrease equality. *Journal of Personality and Social Psychology*, 1975, 31, 156–161.

Brickman, P., & Bryan, J. H. Equity versus equality as factors in children's moral judgments of thefts, charity, and third-party transfers. *Journal of Personality and Social Psychology*, 1976, 34, 757–761.

Brickman, P., & Bulman, R. J. Pleasure and pain in social comparison. In J. Suls & R. Miller (Eds.), *Social comparison processes*. Washington: Hemisphere, 1977.

Brickman, P., & Campbell, D. T. Hedonic relativism and planning the good society. In M. Appley (Ed.), *Adaptation-level theory: A symposium*. New York: Academic Press, 1971.

Brickman, P., & Stearns, A. Help that is not called help. *Personality and Social Psychology Bulletin*, 1978, 4, 314–317.

Cohen, R. L. On the distinction between individual deserving and distributive justice. *Journal of the Theory of Social Behaviour*, 1979, 9, 167–185.

Cole, N. S. Bias in selection. *Journal of Educational Measurement*, 1973, 10, 237–255.

Deutsch, M. Equity, equality and need: What determines which value will be used as the basis of distributive justice? *Journal of Social Issues*, 1975, 31, 137–150.

Eckhoff, T. *Justice: Its determinants in social interaction*. Rotterdam: Rotterdam University Press, 1974.

Folger, R. Distributive and procedural justice: Combined impact of "voice" and improvement on experienced inequity. *Journal of Personality and Social Psychology*, 1977, 35, 108–119.

Folger, R., Rosenfield, D., Grove, J., & Corkran, L. Effects of "voice" and peer opinions on responses to inequity. *Journal of Personality and Social Psychology*, 1979, 37, 2254–2261.

Fuchs, V. R. Redefining poverty and redistributing income. *The Public Interest*, 1967, 8, 89–94.

Goldman, A. H. *Justice and reverse discrimination*. Princeton, N.J.: Princeton University Press, 1979.

Heilbroner, R. L. *An inquiry into the human prospect*. New York: W. W. Norton, 1974.
Hofstadter, D. *Goedel, Escher, Bach: An eternal golden braid*. New York: Basic Books, 1979.
Jaynes, J. *The origin of consciousness in the breakdown of the bicameral mind*. Boston: Houghton Mifflin, 1976.
Jemmott, J. B., III, & Tebbets, R. Applying social cognition: A content analysis of the *Bakke* case. *Personality and Social Psychology Bulletin*, 1980, 6, 30–36.
Kanter, R. M. (Ed.). *Communes: Creating and managing the collective life*. New York: Harper & Row, 1973.
Langer, E. J. The illusion of control. *Journal of Personality and Social Psychology*, 1975, 32, 311–328.
Lerner, M. J. The justice motive in social behavior. *Journal of Social Issues*, 1975, 31, 1–20.
Lerner, M. J., Miller, D. T., & Holmes, J. G. Deserving and the emergence of forms of justice. In L. Berkowitz & E. Walster (Eds.), *Advances in experimental social psychology*, Vol. 9. New York: Academic Press, 1976.
Leventhal, G. S. The distribution of rewards and resources in groups and organizations. In L. Berkowitz & E. Walster (Eds.), *Advances in experimental social psychology*, Vol. 9. New York: Academic Press, 1976.
Mark, M. M. *Justice in the aggregate: The perceived fairness of the distribution of income*. Unpublished doctoral dissertation, Northwestern University, 1980.
Modigliani, A., & Gamson, W. A. Thinking about politics. *Journal of Political Behavior*, Spring, 1979 1, 5–30.
Nozick, R. *Anarchy, state, and utopia*. New York: Basic Books, 1975.
Okun, A. *Equality and efficiency: The big trade-off*. Washington: Brookings Institute, 1975.
Platt, J. Social traps. *American Psychologist*, 1973, 28, 641–651.
Rawls, J. *A theory of justice*. Cambridge, Mass.: Harvard University Press, 1971.
Rubin, Z., & Peplau, A. Belief in a just world and reactions to another's lot: A study of participants in the national draft lottery. *Journal of Social Issues*, 1973, 29, 73–93.
Runciman, W. G. *Relative deprivation and social justice*. London: Routledge & Kegan Paul, 1966.
Sampson, E. E. On justice as equality. *Journal of Social Issues*, 1975, 31, 45–64.
Seligman, M. E. P. *Helplessness*. San Francisco: Freeman, 1975.
Tajfel, H. *Differentiation between social groups*. London: Academic Press, 1978.
Thibaut, J. W., & Kelley, H. H. *The social psychology of groups*. New York: Wiley, 1959.
Thibaut, J. W., & Walker, L. *Procedural justice*. Hillsdale, N.J.: Erlbaum, 1975.
Thorndike, R. L. Concepts of culture fairness. *Journal of Educational Measurement*, 1971, 8, 63–70.
Thurow, L. C. The income distribution as a pure public good. *Quarterly Journal of Economics*, 1971, 85, 327–336.
Tversky, A., & Kahneman, D. Judgment under uncertainty: Heuristics and biases. *Science*, 1974, 185, 1124–1131.
Walster, E., Walster, G. W., & Berscheid, E. *Equity: Theory and research*. Boston: Allyn & Bacon, 1978.
Wortman, C. B. Some determinants of perceived control. *Journal of Personality and Social Psychology*, 1975, 31, 282–294.
Wortman, C. B., Hendricks, M., & Hillis, J. W. Factors affecting participant reactions to random assignment in ameliorative social programs. *Journal of Personality and Social Psychology*, 1976, 33, 256–266.

INSTITUTIONAL SETTINGS

Close Relations
Justice, Self-Interest, and Social Bonds

One might argue that at its simplest, most basic level, the key to finding constructive responses to scarcity and threat in our society is to discover as much as possible about the conditions that promote people's "caring for" one another. If so, what better microcosm is there in which to study this process than the dynamics that go on between a husband and wife, man and woman, who decide to share their lives on an enduring basis? In our society, as the contributors point out, that relationship begins, typically, with two adults' sharing the "bright illusion, we are one." In the natural history of these relationships, this stage ends and often fades into the bleak conviction "we are hopelessly different." At that point, the relationship either dissolves or continues in another form, usually involving degrees of conflict, compromises, mutual adaptations, and agreement to disagree (Berscheid & Campbell). At the very least, we can say with confidence that in almost every case, the initial caring for each other ends or takes a different form. Obviously, the more we know about these processes and what they have to do with the sense of justice or injustice, the closer we will be to generating principles or techniques applicable to the problems we face as a society. Of course, caring for one another is no insurance that a constructive solution will be found. Conceivably, where the caring is so intense, the most constructive response may be missed. For example, all choose to perish rather than face the prospect of surviving without the others. However, without the "caring," at least in its minimal form of respect for the decency and trustworthiness of the others, there is probably little hope of generating an adaptive or constructive solution.

Two questions might appear early in the reader's thoughts. What does the "caring for" that is typical of the initial stage of a close relationship with its "bright illusion, we are one" have to do with issues of justice and entitlement? Second, why do we consider the initial stage an illusion that must come to an end? In one sense, it is not an illusion but a genuine and valid experience for those involved. Do we consider it an illusion, then, because the factors that lead to its demise are an inevitable result of the subsequent events that every couple in our society must experience? Perhaps if we understand more about these subsequent events, we will be closer to understanding why, for example, under certain pre- or postdisaster conditions people who were relatively removed from one another often quite naturally care for and willingly aid one another as they would themselves or their own. But then, as has been observed, that illusion of "we are one" fades as the crisis recedes into the past. Why? What are the relevant factors? Is this phenomenon truly as mysterious as it appears to be inevitable?

In terms of the justice-motive theory model, in the initial stage the participants perceive each other in an "identity" relation. Tacitly, they have generated a stable view of the relationship so that it appears as a psychological given that they (both of the pair) are entitled to what would meet their needs—give them the greatest happiness. To the extent that any planned activities become relevant, the ultimate concern is for the other's as well as one's own welfare: the other person's welfare is not experienced or thought of as distinct from one's own.

Regardless of how enduring and intense the initial bond created by the sexual involvement with one another may be, most of the traditional cultures represented in our society have provided explicit prescriptions for the maintenance of this identity relation between husband and wife. These traditional roles have provided the procedures for each partner's caring for the other. It was clear what a good husband would do—usually it had something to do with being "the provider" of those external resources desired by the family members. The wife would care for the husband by performing her housewifely duties. As Berscheid and Campbell note, times have changed, and the old rules for caring for one another are no longer accepted fully by most couples. They are left sooner or later to work out how to (or even whether to) remain together. The social changes that have taken place have provided the members with alternative ways of viewing a marriage.

If Berscheid and Campbell are correct, then it is becoming more common and appropriate for the partners' identity relation to become a relatively trivial part of their encounters (possibly only in the bedroom, if there). Increasingly, they come to view each other as in a "unit" relation with clearly diverging goals (and thus norms of justified self-interest appear), or more commonly in a "nonunit" relation with at least temporarily common goals. The latter leads to each partner's bargaining and jockeying for the best deal for him/herself as the couple blend their efforts. As Holmes describes so well, however, this latter process is incompatible with a sense of identity or intimacy with the other, and eventually, given the weak social bonds and weakened constraints from the environment, many partners separate and take advantage of the new social forms of meeting their sexual and social needs.

How inevitable is this sequence, and what are the processes that determine its outcome? Kidder, Fagan, and Cohn see an inherent flaw in the structure of most husband–wife relations that is quite likely to engender the perception of a

*nonunit relationship. As long as there is a clear differential in the power that
each member has in the relation, then, regardless of accommodations that are
made on a temporary basis, there is a latent, but very real, difference in relative
status: superiority–inferiority. If this remains tacit or is acceptable to both parties,
then each member will feel that the arrangement is appropriate, that is, a just
one. However, according to these authors, the husband's access to greater economic resources gives him greater power than the wife in the exchange, because
his resources have general currency in the society, whereas what the wife contributes to the relationship, in terms of nurturant and affectional activities, is
specific to the particular relation. With that differential in power, the husband
has an inherently superior status, while the wife's in an inferior or dependent
one. If the woman also perceives herself as an equal—in a unit relation with her
husband—then the differential status and all that accompanies it, in terms of
decision-making power, etc., will obviously be seen and felt as unjust, and the
demand for change and/or conflict will arise.*

*According to Kidder, Fagan, and Cohn, until there is truly equal power
in the relationship, which can occur only when men and women share equally
in economic as well as other forms of power, it is quite natural for the relationship
to deteriorate into a nonunit one. They believe the identity relation must be
transformed into or at least blended with a true unit—or sense of equality—for
stability in close relations to occur.*

*Holmes's chapter adds an enormous amount of insight into these processes.
For example, he generalizes from H. H. Kelley's recent work to describe an
attributional process that could quite naturally lead to a reduction in the "caring-
for" orientation of the initial identity relation. Once the partner acts in a way
that appears to be self-interested rather than caring, that act can easily lead to
the imputing of a stable orientation to the other as uncaring or untrustworthy.
Thus, the initial identity relation is replaced by a unit or nonunit relation. The
"offended" partner, recognizing great areas of common interest with the other,
may respond to this new perception by attempting to institute procedures that
ensure that whatever occurs between them will be at least fair and just. Each
of them gets a fair share while the couple live together. In his chapter, Holmes
describes very persuasively how each of the strategies that people adopt from
their activities in other spheres of their lives—use of contractual agreements,
close monitoring of relative inputs and outcomes—eventually generates the explicit awareness of the psychological bases of these accommodative strategies. The
couple are self-concerned participants trying to work out the best deal for themselves or to protect their own interest against threats from the other's obviously
self-interested efforts. They are essentially in a nonunit relation with one another.*

*Holmes, however, points to an alternative to this degenerative sequence.
And it makes sense. If the members in the initial identity relation can begin to
view much of what they do to or for one another from a longer time perspective,
the identity relation can be maintained as they view their own and the other's*

acts in "positional" terms. What needs to remain as the essential sharing orientation in the relation is that whatever is happening "now," the dominant goal
is the enhancing of one another's welfare.

The various chapters have provided a great deal of information and understanding concerning what social-cultural and interpersonal events are likely
either to preclude or to destroy that initial "illusion." Kidder, Fagan, and Cohn
may be right. As long as there is radically differential "power" in a relationship,
then processes are set in motion that preclude the emergence of that common
orientation. Holmes, as well as Berscheid and Campbell, may also be right. As
the obvious signs of "caring for" are mixed with or replaced by other common
activities, then the identity relation will be replaced by the perception of the other
as a contestant or someone with whom one tries to strike a bargain for common
interests (nonunit relation). It is also possible that at this time in our society,
as Berscheid and Campbell point out, there are relatively few remaining social
constraints or incentives that would lead people to view one another in terms
of their overall welfare rather than in terms of immediate gains and losses.

In any case, these chapters will provide the reader with a considerable insight
into those factors that promote or inhibit people's "caring for" one another.

10

The Changing Longevity of Heterosexual Close Relationships

A Commentary and Forecast

ELLEN BERSCHEID and BRUCE CAMPBELL

1. INTRODUCTION

These are, indeed, "hard times for lovers" as the contemporary song proclaims. The chance of an adult heterosexual close relationship (CR) surviving "until death do us part" is slimmer than ever before. The rate of marital dissolution in the United States has reached an all time high (Levinger & Moles, 1976; Norton & Glick, 1976), and the current rate of dissolution among relatively high-commitment nonmarital CRs (e.g., cohabitation) is even higher (see Glick & Norton, 1977; Macklin, 1972). Approximately 40% of all current and potential marriages among women now in their late 20s may eventually end in divorce, according to Glick and Norton's (1978) estimates derived from the Census Bureau's Current Population Survey for June 1975.

The instability of adult heterosexual CRs is a recent phenomenon. Around 1960, the divorce rate turned sharply upward, more than doubling between 1963 and 1975 (Moles & Levinger, 1976; Glick & Norton, 1978), so that by 1976, there was approximately one divorce for every

ELLEN BERSCHEID and BRUCE CAMPBELL • Department of Psychology, University of Minnesota, Minneapolis, Minnesota 55455.

two new marriages contracted (Hunt & Hunt, 1979). And, ironically, at the same time that CRs have become substantially more vulnerable to disruption and dissolution than they were just a generation or two ago, CRs are now seen by most people as being the prime source of personal happiness (e.g., Freedman, 1978; Klinger, 1977).

Why is the discrepancy between the desire for a close and enduring relationship with a person of the opposite sex and its probability of fulfillment widening? Numerous candidates have been proposed for the role of villain in the widespread drama of CR dissolution. Most of these represent changes in societal conditions of one variety or another. Some social scientists, for example, have proposed that changes in economic conditions (e.g., increased affluence) are the major cause. Others have argued that changes in certain societal values (e.g., concerning sexual freedom and personal growth) are critically responsible for the decrease in the average life expectancy of a CR. Still others have pointed the finger at other societal changes (e.g., the erosion of religious beliefs and practice; increased social and geographic mobility). Each of these factors no doubt possesses some validity as a candidate. The many societal changes that have taken place in the past few decades have almost certainly played some role in bringing about the recent upsurge in the rate of dissolution of CRs.

But what role? Changes in societal conditions do not terminate relationships; people do. Somewhere in between the quantitative indicators of changes in societal conditions, on the one hand, and the separation and divorce statistics, on the other, lies a relationship between a man and a woman, one of whom (or both) decides to terminate rather than to maintain the CR.

Explanations of CR instability whose principal referents are global societal conditions are not wholly satisfactory from the social-psychological point of view, for they shed little light on the psychological processes that intervene between a specific societal condition and an individual's decision to terminate his or her relationship, thus adding another tally to the divorce side of the ledger. From the psychological perspective, any assertion to the effect that a given social change has affected the longevity of CRs implies, either explicitly or implicitly, that the social change in question has directly or indirectly altered the psychological factors that constitute the immediate causal antecedents of CR termination. Further, and from that perspective, specification of the ways in which a given social change has affected the longevity of CRs amounts to charting the direct and indirect causal effects of that social change on the immediate psychological antecedents of individuals' decisions to maintain or to terminate their CRs.

To understand current CR instability, then, and to prognosticate about the likely future of CRs, it is not enough to point to recent past social changes along a number of dimensions and to arm oneself with projections for the future along each. It is necessary, in addition, to understand the psychological factors that promote the stability of a CR and those that lead to its disintegration and to try to draw lines of correspondence between these and more distal sociological factors.

In this chapter, we focus on just one facet of CR dynamics that has been identified as critical to CR stability and dissolution: "external barriers" to relationship termination. We shall try to trace the causal pathways (a) between certain social changes and barrier reduction; (b) between barrier reduction and decisions to terminate a relationship through its direct and immediate effects on the psychological antecedents of such decisions; (c) between barrier reduction and decisions to terminate a relationship through its effects on other psychological factors critical to CR dynamics that in themselves alter the psychological antecedents of termination decisions and, thus, the indirect and delayed effects of barrier reduction on decisions to terminate a relationship; (d) between such decisions to terminate a relationship and certain social conditions; and, finally, (e) between such social change and further barrier reduction, which additionally weakens intact CRs, both newly formed and of longer duration, in the manner outlined above. The general, and circular, causal model described above is illustrated in Figure 10.1.

2. PSYCHOLOGICAL ANTECEDENTS OF CR TERMINATION

Drawing on the work of Festinger, Schachter, and Back (1950) and of Thibaut and Kelley (1959), Levinger (1976) has proposed that the stability of a CR is a function of its cohesiveness. The "cohesiveness" of a relationship is the resultant of the total field of forces acting on the two partners to remain in the relationship. Within this total field, three categories of forces, the net sum of which defines the pair's cohesiveness, are distinguished:

1. The *attractiveness of the relationship* itself to each partner. As suggested by Thibaut and Kelley (1959), each partner's attraction to the relationship is assumed to vary directly with the rewards that the partner perceives that he or she gains from the relationship and inversely with perceived costs. Thus, individuals are assumed to evaluate the attractiveness, or quality, of their interpersonal relationships by means of a subjective calculus in which the perceived rewards and costs experienced in the relationship are tallied, weighted, and algebraically combined to determine the net profit (rewards minus costs) yielded by participation

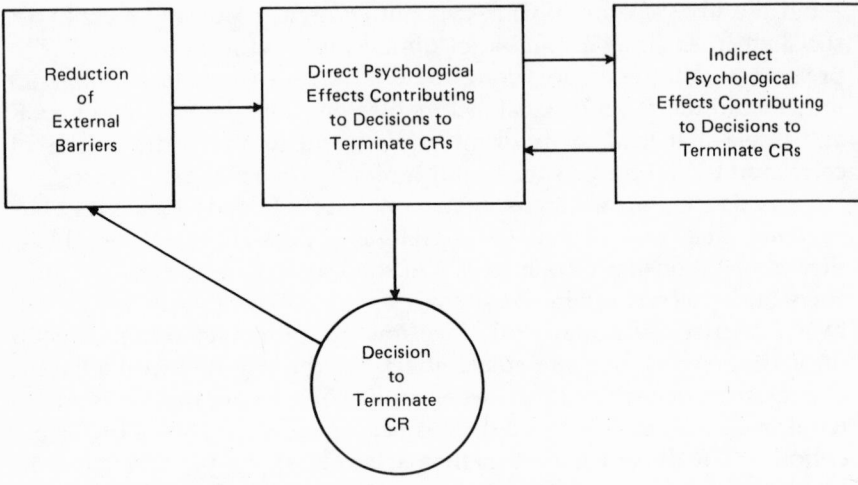

Figure 10.1. A general causal model of the effects of barrier reduction on decisions to terminate CRs.

in the relationship. Further, the individual's attraction to the relationship, or his or her satisfaction or dissatisfaction with it, is conceived to depend on the assessed profit of the relationship in comparison to the minimum level of profit that the individual expects to receive in the relationship (the individual's "comparison level," or CL). People are attracted to and satisfied with their CR to the extent that their assessed profits from it exceed their CL, and they are dissatisfied to the extent that their assessed profits fall below their CL. According to Levinger, and other things being equal, the higher the net attraction of the partners to the relationship, the higher its cohesiveness and the less subject to dissolution it ought to be.

2. The *attractiveness of* (actual or potential) *alternatives,* including especially alternative relationships but also the attractiveness of the alternative of the absence of any CR. According to Thibaut and Kelley (1959), although a person's satisfaction with a relationship depends on the amount of profit received from it relative to the person's CL, satisfaction does not directly mediate the decision to terminate a CR. Rather, it is the degree to which profits in the current relationship exceed (or fall below) profits that might be obtained in the best available alternative relationship (the "comparison level for alternatives, or CL_{alt}) that is critical for decisions to terminate. Thus, even if an individual is dissatisfied with a relationship (its profits fall short of CL), if the profits perceived

to be available in his or her best potential alternative are no better (or are worse), the individual would not be expected to terminate the relationship. The immediately precipitating cause of a decision to terminate a CR is, then, the perception that the profits available in an alternative relationship (or no relationship at all) exceed the profits to be gained by remaining in the current CR. Levinger theorized, then, that other things being equal, the greater the alternative attractions confronting each partner in a CR, the lower its cohesiveness and thus the more vulnerable to dissolution it ought to be.

3. The *external barriers* that act to contain the partners within the CR and prevent them from leaving the relationship. Any aspect or property of the physical or social environment is considered an external barrier to the extent that it leads a participant in a CR to anticipate sustaining costs of any kind—financial, social, psychological, emotional, or otherwise—should he or she voluntarily terminate the relationship. Barriers are conceived of as "psychological restraining forces," which affect a person's behavior only if he or she wishes to leave the relationship. External barriers, then, tend to prevent the termination of dissatisfying CRs even when attractive alternatives exist by making it costly for the individual to dissolve the relationship. Thus, according to Levinger, any reduction in barriers tends to decrease the cohesiveness of the relationship and to increase the probability that one or both partners will voluntarily choose to dissolve their CR.

2.1. The Direct Effect of External Barriers on Termination Decisions

People ought to become dissatisfied with their relationship whenever they perceive that their profits from it no longer exceed their CL, and they should then begin to contemplate terminating it. Such contemplation should include an assessment of the profits that might be available to them in potential alternative CRs as well as the costs of termination of the current CR. It is at this point, and especially when the potential profits within an alternative relationship are higher than those obtained in the current CR, that the presence or absence of strong barriers may make the difference between terminating the CR or remaining in it despite dissatisfaction. For when strong barriers exist, when "exit costs" are perceived to be great, an alternative must be extremely attractive indeed for the dissolution of the current CR to result in an overall increase in profits for the individual. Not only must the new relationship be intrinsically more profitable than the current one, but it must be more profitable by a wide enough margin to overcome the exit costs incurred in leaving the current relationship.

In Levinger's framework, then, barriers contribute to the longevity of CRs by acting as restraining forces that, if sufficiently strong, may prevent the individual from dissolving a CR even though he or she might otherwise wish to do so. The direct impact of external barriers is to prevent the termination of a CR that is unattractive (or unsatisfactory) to one of the participants where that participant has a more attractive alternative (but not attractive enough to overcome exit costs). Thus, reductions in external barriers, in proportion to the magnitude of their reduction, directly render a decision to terminate a CR more likely.

If this were the only effect of barrier reduction, its impact on CR stability would be significant wherever changing social conditions altered external barriers. But the effects of barrier reduction on CR stability are far more pervasive and insidious in contributing to the demise of CRs than it appears. We shall argue that not only has barrier reduction the direct effect on CR stability attributed to it by Levinger, but it indirectly affects the other two factors deemed vital to CR cohesiveness, and in addition, it indirectly contributes to further social changes that tend to perpetuate and further increase barrier reduction and thus increased CR instability. These arguments may be best made in the context of illustrations of social changes that have reduced external barriers to relationship termination.

3. RECENT SOCIAL CHANGES RESULTING IN BARRIER REDUCTION

For large segments of the population, recent social changes have brought about substantial reductions in external barriers to the termination of their heterosexual CRs. To document this point, we are generally forced to rely on data pertaining to marriage and divorce specifically, since adequate data concerning nonmarital CRs are not readily available. Although some of the changes that we discuss here, such as the easing of legal obstacles to divorce and the relaxation of religious constraints against marital dissolution, are relevant only to marital CRs, other recent changes, such as the increased educational credentials and economic independence of women, ought to have similar implications for the longevity of other forms of adult heterosexual CRs as well.

3.1. Reductions in Economic Barriers

One of the most important social changes that has resulted in the weakening of a previously formidable barrier to the termination of a CR is the significantly greater economic independence that women in our society have achieved in recent years. When employment opportunities for women were extremely limited, and when public assistance programs of the type we take for granted today were as yet unknown, there were perhaps few women, especially few with children, who could obtain a divorce or a separation without suffering severe financial hardship and a drastic reduction in their standard of living. As Weiss (1975) put it, "At one time the great majority of married women were financially dependent on their husbands. Unless they could count on alimony as a part of the divorce settlement, they could hardly dare to divorce" (p. 7).

Evidence that financial dependence on one's spouse does indeed constitute a powerful barrier to the termination of a CR may be seen in the impressive empirical relationships that exist between economic status and the incidence of marital dissolution. It is a well-established fact that the divorce rate is distinctly higher at low-income than at middle- or upper-income levels (Hunt & Hunt, 1978). This relationship may be interpreted as indicating that the economic losses involved in dissolving a high-income marriage are relatively greater than those resulting from the termination of a lower-income marriage, and hence the threat of economic loss constitutes more of a barrier to marital dissolution in higher-income marriages. As Levinger (1976) observed, "In the higher economic strata, where earning potentials between husband and wife differ widely, her alternative status is financially unattractive to her. In lower economic strata, wives have materially less to lose and more to gain from a divorce" (p. 41). Indeed, since the financial resources currently available to mothers of dependent children through public assistance programs may exceed, at least in the lower-income marriages, the earning power of their husbands, the question of whether or not programs such as AFDC (Aid to Families with Dependent Children) actually encourage the dissolution of low-income marriages has been frequently raised (e.g., Moles, 1976).

While the overall correlation between economic status and the likelihood of marital dissolution is negative when the entire income range is considered, there are revealing exceptions. For example, although marital stability is highest for men who have graduated from college or have gone to graduate school, the marriages of women who have had at least some graduate training show lower than average stability (Glick & Norton, 1978). And while divorce-proneness decreases with increasing family income for nonworking wives, divorce-proneness increases with the increasing personal income of wives who work. A woman earning $15,000 or more is three times as likely to be currently divorced as a woman earning under $3,000 (Hunt & Hunt, 1978), and although this statistic may partially reflect greater motivation to earn money among those women already divorced, it seems unlikely that the entire relationship can be accounted for in this fashion. In fact, in an examination of the fate of marriages following an application for divorce, Levinger (1976) found that while the husband's income was inversely related to divorce-proneness, the amount of the wife's income was positively related to the likelihood of divorce. Thus, even the exceptions to the general rule that divorce-proneness and economic status are negatively related suggest that with women's increasing economic independence from their partners, a potentially potent barrier to the termination of CRs has correspondingly diminished.

This economic barrier, which undoubtedly contributed importantly to CR longevity, continues to be vigorously assaulted. Over the course of this century, not only has the rate of employment of women in general shown a major increase (Pleck, 1978), but just between 1960 and 1975, the labor force participation rate of wives (collapsing across age groups) rose from 30.5% to 44.4% (McEaddy, 1978). The increase in the rate of employment of wives in the youngest age groups has been even more dramatic. In the five-year period between 1970 and 1975, the labor-force participation rate for wives aged 20–24 jumped from 47.4% to 57.1 percent (McEaddy, 1978). But perhaps most notable of all has been the upswing in the rate of employment of married women with school-aged or preschool children. The rate of employment of mothers of school-aged children roughly doubled between 1948 and 1972, while the rate of employment of mothers of preschool children almost tripled during the same period, and the rate of employment of women in these categories is continuing to climb steadily (Hoffman, 1978). In brief, there were approximately 38.8 million women in the United States work force in 1976 (Hoffman, 1978), and the husband is the sole wage-earner in only about a third of husband–wife households (Hunt & Hunt, 1978).

There are other, not unrelated, social changes that suggest that the economic dependence of women will decreasingly constitute an effective barrier against the dissolution of CRs. First, American women are becoming increasingly well educated. The number of women pursuing post-high-school or graduate degrees has increased dramatically in recent years. Between 1970 and 1975, for example, the number of married women under 35 years of age attending college increased by over 62% (Hunt & Hunt, 1978). And looking only at the older women in this group, the increase was even more remarkable. Specifically, the number of women aged 25–34 attending college increased some 94% during the same five-year period (Hunt & Hunt, 1978). Because of these recent trends, at least some post-high-school education is now the majority experience for women under 35 (Market Opinion Research, 1978). In addition, increasing numbers of women are continuing their education beyond an undergraduate degree. Between 1960 and 1975, for example, the number of women aged 35–54 who had pursued at least some graduate education more than doubled (Glick & Norton, 1978).

Second, more and more American women are delaying marriage. While the median age at first marriage for men has shown virtually no change since 1967, this indicator has risen gradually but steadily for women (Norton & Glick, 1976; Glick & Norton, 1978). One reflection of this trend is that the proportion of women aged 20–24 who are still single has increased from 28% in 1960 to 43% in current estimates (Glick & Norton, 1978). In 1976, in fact, slightly over 10% of women were still not married by age 29 (Hunt & Hunt, 1978). Thus, more and more women are evidently delaying marriage until after they have completed some higher education or have gained several years' experience in the labor market. They thus are generally entering marriage with greater occupational qualifications and with more extensive experience in a single, self-sufficient lifestyle than was typical heretofore.

It thus seems reasonable to conclude that whereas economic barriers confronting women may have contributed previously to the longevity of CRs, fewer and fewer women are, or will be, deterred from terminating their CRs by the prospect of overwhelming economic losses should they decide to do so. It should be noted that as a result of these changes in the economic status of women, barriers that formerly inhibited the freedom of men to terminate their CRs readily have also probably diminished. A man whose partner is quite capable of adequately supporting herself may feel less guilty about dissolving his marriage than he would if she faced real economic hardship as a result of his departure.

3.2. Reductions in Legal Barriers

Another recent social change that has undoubtedly resulted in the weakening of a once formidable barrier against the dissolution of CRs is the reduction of legal obstacles to divorce. Most of these have resulted from the enactment of "no-fault" divorce laws (where neither spouse is accused of any wrongdoing, either spouse can obtain a divorce simply by claiming that the marriage has suffered "irretrievable breakdown," and neither spouse can prevent the other from obtaining a divorce). These constitute, of course, a radical change from previously existing divorce codes, which sometimes made a demonstration of adultery mandatory for a divorce decree to be granted (Glazer-Malbin, 1978).

In 1970, California introduced the first no-fault divorce law, and other states quickly followed. By January 1974, 23 states had enacted no-fault legislation. Now more than half have adopted some form of no-fault divorce (Hunt & Hunt, 1978), and most of the remainder are currently attempting to incorporate the no-fault concept into their legal codes (Norton & Glick, 1976). Further, as a result of the recent reforms in divorce laws, the length of state residence and the duration of the predivorce separation period necessary to obtain a divorce have generally been shortened (Norton & Glick, 1976).

Reduction of the legal obstacles to divorce has reduced the emotional and psychological costs encountered in the process of obtaining a divorce, as well as much of the bitterness and disgrace that former grounds for divorce entailed. In addition, another effect of these changes has been to reduce the financial costs of obtaining a divorce. Thus, the legal obstacles that may once have discouraged the dissolution of CRs have diminished significantly over the past few years.

3.3.   Reductions in Religious Barriers

Religious barriers to marital dissolution have also undergone erosion in the tide of social change. Although belief in the sanctity of marriage and adherence to religious prohibitions against divorce may once have functioned as virtually insurmountable barriers for large numbers of the religiously devout, opposition to divorce on the part of major organized religions has diminished substantially in recent years. As Weiss (1975) has pointed out, even the Roman Catholic Church, once immovable in its opposition to divorce (and excepting the pope's pronouncements on his visit to the United States in 1979), has slowly become more permissive. The softening of opposition to divorce on the part of major religious institutions has been accompanied by noticeable increases in the frequency of divorce within major denominations that have traditionally had lower than average rates. While recent government data on the divorce rates of different religious denominations is not available, data collected by sociologists and public-opinion researchers suggest that the Catholic and Protestant divorce rates are converging (Hunt & Hunt, 1979).

Thus, available evidence seems to warrant the conclusion that barriers arising from religious precept, though once perhaps sufficiently strong to prevent the dissolution of large numbers of CRs, have weakened considerably in recent years. Moreover, since changes in the divorce laws effectively deny one spouse the power to prevent the other from obtaining a divorce because of religious convictions, perhaps few marriages that one or both partners find unsatisfactory can now be expected to endure for primarily religious reasons.

3.4.   Reductions in Barriers due to Children

The common-sense notion that the presence of dependent children in a marriage acts as a significant barrier against marital dissolution is supported by considerable empirical evidence. Hunt and Hunt (1979), for example, examined the effect of children on the rate of marital dissolution among women aged 35–39 who had been, or still were, married. They found that 12% of the childless women in this age group were currently separated or divorced, as compared with only 7% of those who had children. Similarly, based on the results of a national survey conducted in 1975, Glick and Norton (1978) found that of women who had obtained a divorce before the age of 30, only 13% had borne more than two children. Levinger (1976) also examined data pertaining to the effect of children on the durability of marriages and concluded:

> husbands and wives with minor offspring themselves feel more restraint against breakup than those with no offspring. Findings from various sources do show that, with length of marriage controlled, childless couples have generally higher separation rates than child-rearing couples, although the differences have decreased during recent decades. (pp. 39–40)

It would seem, then, that the presence of dependent children can act as a restraint on marital dissolution.

There are a number of factors, however, that have undermined the potency of the barrier that children constitute. The most important of these is that American women have acquired increased control over their fertility in recent years. As Hoffman (1978) has pointed out, less than 20 years ago the most effective contraceptive measures available today (principally the pill, contraceptive sterilization, and the IUD) were used by only a very small proportion of couples or did not yet exist. By 1973, however, roughly half the couples of reproductive age were using one of these methods, and when the U.S. Supreme Court legalized abortion on request in that year, the option to terminate an unwanted pregnancy first became widely available—an option that millions of American women have exercised since.

These recent improvements in control over fertility have, in turn, produced several interrelated effects. For example, American women are marrying later, having children later, and having fewer children when they have them (Hoffman, 1978). At the height of the post-World War II baby boom, the fertility rate reached 3.80 children per woman, but by 1976, it had dropped to an all-time low of 1.75. It now stands at almost precisely 2.00 (Hunt & Hunt, 1978). Not only the actual fertility rate but the desired family size reported by women has noticeably decreased over the past two decades (Hoffman, 1978). And in addition to wanting and having fewer children and having them later, women now stop having children sooner than they used to. While the average wife at the turn of the century did not have her last child until she was 33, today's average wife has her last child at the age of 30 (Hunt & Hunt, 1978).

Because of the declining birthrate and the tendency to delay having children, young married women are "having few if any children to complicate a return to 'singlehood' " (Glick & Norton, 1978, p. 190), if they choose to divorce or separate. Women with small families are also more likely to be in the labor force and thus tend to be more financially independent of their husbands (Norton & Glick, 1976). Furthermore, with smaller family size, women spend fewer years with preschoolers at home, which tends "to free the time of the potential divorcee for work outside the home" (Norton & Glick, 1976, p. 16). As a result, the rates of dissolution of childless and of child-rearing marriages have been converging.

There is at least one additional factor that may be responsible for the convergence. As Levinger (1976) has pointed out, when parents believe that greater damage to their children would result from a divorce or a separation than from continuing the relationship, the existence of children can be a potent source of psychological restraint, but when this is not the case, such restraint may be negligible. As divorce involving children becomes more and more commonplace, and examples of children who have adjusted well become more readily visible, and if scientific studies continue to suggest that divorce is not necessarily harmful to the welfare of children (e.g., Feldman & Feldman, 1975; Luepnitz, 1979), the formerly accepted belief that serious harm to children inevitably accompanies divorce will continue to dissipate.

It seems reasonable to conclude that while a barrier to the termination of a CR represented by the presence of children may once have provided the "glue" that held many subjectively unsatisfying CRs together over extended periods of time, these barriers have diminished in recent years to the point where fewer and fewer couples can be expected to remain together primarily (if not exclusively) "for the sake of the children."

3.5. Reductions in Social Barriers

One of the most widely discussed of all recent social changes, and one that has clearly resulted in the elimination of important barriers against the dissolution of CRs, is the relatively recent fading of the social stigma attached to divorce. Weiss (1975) observed that "Only 40 years ago divorce was scandalous, as novels and articles of that time demonstrate. . . . Just 25 years ago Margaret Mead commented on the fading but still visible image of the divorced person as selfish, irresponsible, or shady" (p. 6). Today, by contrast, "Men and women can divorce without stigma and without jeopardizing their jobs or careers even in such sensitive fields as politics or the ministry" (Hunt & Hunt, 1978, p. FC2), and most social scientists would agree with Norton and Glick

(1976) that "the phenomenal upsurge of divorce in this country during the last ten years has been stimulated by a growing acceptance of the principle that divorce is a reasonable, and at times desirable, alternative to an unhappy marriage" (p. 12).

The growing social acceptance of divorce may be seen in large-scale national survey data where the number of respondents who believe that "divorce is the answer to an 'ailing' marriage" increased from 53% in 1970 to 60% only four years later (Levinger & Moles, 1976). This small but visible increase (in addition to the high absolute level of endorsement of the item) attests to the growing acceptance of divorce in our society. Another indication of the increasing social acceptability of divorce is evident in the fact that first divorces increased twice as much between 1960 and 1970 as did divorce among those who had already divorced and then remarried. Thus, the divorce rate of those who had not already shown that they found divorce an acceptable option (i.e., first divorces) almost converged with the rate of divorce among those who had already demonstrated behaviorally (i.e., redivorces) that they had no fundamental objection to divorce (Weiss, 1975).

In addition to the fact that the stigma attached to ending a marriage is truly "fading into irrelevance" (Goldstine, Larner, Zuckerman, & Goldstine, 1977), the sheer prevalence of divorce in recent years virtually guarantees that almost anyone contemplating divorce will know someone who has already dissolved his or her marriage. As Hunt and Hunt (1979) put it, "Everyone knows—or at least rubs elbows with—someone who has been divorced. They can hardly avoid it since for every nine married people there is one FM (formerly married) plus three FM's who have remarried" (p. 7). Acquaintance with people who have already been through a divorce or separation may make it considerably easier for an individual contemplating the termination of his or her own marriage actually to go through with it. Seeing firsthand examples of others who have successfully survived the dissolution of their marriages may substantially reduce the fears and uncertainties of people contemplating their own divorces. Further, while ignorance of such apparently trivial matters as the name of a good divorce lawyer or the necessary legal steps involved may once have been sufficient to discourage some people from divorce, friends who have gone through a divorce themselves may prove to be eager and readily accessible sources of information about the "how-to-do-it" mechanics of dissolutions.

3.6. Summary

The evidence suggests, first, that the economic dependency of the vast majority of American women has substantially decreased in recent years and can be expected to decrease further in the future. Second, legal obstacles to divorce have greatly diminished. Third, religious opposition to separation and divorce has decreased. Fourth, constraints against dissolving CRs because of the involvement of children have weakened. And, finally, the social stigma of separation and divorce, and the accompanying community and primary-group pressures against CR dissolution, have all but vanished for most segments of the population. The net result of all of these social changes is clear. Under present social conditions, and more than ever before, the majority of men and women in our society face fewer and weaker barriers to leaving their partners should they become sufficiently dissatisfied with their relationship to contemplate its termination and should they perceive a more attractive alternative. In the next section, we shall discuss how barrier reduction may help ensure not only that they will become dissatisfied with their CR, but that they will find a more attractive (at least, temporarily) alternative.

4. THE INDIRECT EFFECTS OF BARRIER REDUCTION ON TERMINATION DECISIONS

4.1. Effects of Barrier Reduction on the Attractiveness of the Relationship

The great family sociologist Willard Waller (1930/1967) once observed that the stability of a marriage is not necessarily related to the sweetness of its contents. At the time that Waller made that statement, it was undoubtedly true, for (as the previous section documents) there were formidable external barriers to relationship termination that kept partners contained in marriages that had soured considerably. As the number and potency of external barriers have declined, however, Waller's observation has become increasingly invalid. Indeed, perhaps the most important effect of barrier reduction has been that as external barriers to leaving a CR have decreased and have failed to serve as the individual's reason for remaining in the relationship, the burden of purpose and justification for maintaining the relationship has increasingly fallen on the "sweetness" of its contents. It can be argued, in fact, that as barriers to leaving CRs have been reduced, the presence of rewards within the relationship, including the presence of positive emotions and feelings toward the partner, have become an increasingly important criterion by which large numbers of individual decisions to maintain or terminate a CR are made.

This shift in the *raison d'être* of CRs might not be so serious in impact were it not for the likelihood that when the heavy burden of justification of the relationship is placed on its internal contents, the contents themselves will change in such a way that their probability of becoming sour is increased. Thus, barrier reduction can be hypothesized to have a number of effects on the relationship itself and hence, indirectly, on CR termination.

4.1.1. Increased Focusing on the Internal Dynamics of CRs: "Taking the Pulse" of the CR

Where there are few barriers to leaving, participants in CRs are truly in the position of continuing to choose freely to remain in their relationship. But like other freedoms, the continued freedom to choose to stay or to go has a price. To have a perpetual choice means that one must choose not once, but over and over again. And to do so, one must continually expend time and energy in evaluating and reevaluating the wisdom of the choice.

When virtually insurmountable barriers surround a CR, making its termination almost unthinkable, little time and attention need be devoted to evaluating the alternatives that may present themselves from time to time, and there is also little need to devote much time and energy to calculating the value in one's personal reward–cost equation of the spouse's most recent behavior. But when the maintenance of a CR depends almost entirely on one's satisfaction within it, and a continued preference for it over potential alternatives, assessing one's satisfaction with the relationship and the nature and desirability of possible alternatives may come to absorb considerable amounts of time. The need to monitor and to assess and reassess one's satisfactions and alternatives may be seen as increasing the cost of maintaining the CR, thereby lowering its profitability and decreasing its attractiveness.

In addition to evaluating their own satisfaction and the alternatives, there is yet another reason that people may frequently feel compelled to take the pulse of their CR even when they themselves are satisfied and have no other attractive alternatives. When the partner's continued presence in the CR is not assured by the existence of powerful restraints against his or her departure, there ought to be increased vigilance for any signs that he or she might choose to leave. Thus, when the partner's continued availability (and the uninterrupted provision of whatever profits he or she supplies) is predicated on the partner's continued satisfaction with and preference of the CR over alternatives, the need to continually assess the partner's satisfaction with the CR, and to divine just exactly what the partner's alternatives are, becomes pressing. The need to continually assess the partner's satisfactions and alternatives, as well as one's own, further adds to the cost of the CR, lowering its profitability and attractiveness.

Only indirect evidence may be offered for the hypothesis that as the barriers to termination of a given CR decrease, "taking the internal pulse" of the relationship increases. However, the vast increase in "evaluate-your-marriage" rating scales, the large number of "formulas" and discussions of what ought to constitute a satisfactory marriage, and a general obsession with the subject in the popular press and other media help attest to the concern that people now show for evaluating the goodness of their CRs.

4.1.2. Increased Insecurity

The need to evaluate continually the sweetness of the contents of the CR is costly even if the results of the evaluation prove to be positive. There is reason to suspect, however, that the results are proving discomfiting to an increasing number of people. In particular, consideration of the partner's barriers to leaving, as well as the partner's alternatives, may now, more than ever before, lead to personal feelings of insecurity in the relationship.

Although a person's perceptions of the barriers and alternatives confronting his or her partner may often differ in important respects from the partner's own assessment (hence the surprise and shock that many people experience when their partner walks out unexpectedly), one would nevertheless guess that most people are cognizant of the presence (or absence) of constraints on the partner to terminate the relationship, and these contribute to feelings of personal security (or insecurity) within the relationship. For example, a man whose partner faces extreme economic hardship should she lose his financial support might well realize this and therefore spend little time worrying about whether she might terminate the relationship (and largely devote himself to assessing his own satisfactions and alternatives). But as the number of women who are economically dependent on their partners diminishes, the insecurity factor for men can be hypothesized to have become more potent in that fewer and fewer men can count on economic factors alone to ensure the longevity of their CR. Similarly, a woman whose spouse would never consider leaving because of strong religious convictions may realize this and may thus have little doubt that he will remain with her for life. But fewer people than ever before can rest assured that the religious convictions of their partners will effectively shield them from divorce or separation. In addition, and if only because of recent changes in divorce laws, men and women alike have lost considerable legal leverage with which to prevent their spouses from obtaining a divorce, and many live in fear of losing their spouses before having a chance to work out marital problems (Hunt & Hunt, 1978).

Thus, as a result of the social changes that have reduced barriers to CR termination, the security that participants in CRs formerly had may be vanishing. As Glazer-Malbin (1978) observed:

> Freedom to remake the interpersonal relationships between people is . . . magnificent and frightening. At once, the array of possible new relationships tempts, while we realize, however dimly, that the cost is a loss of the comfort of knowing *what* is *expected* with *whom*, and for *how* long. (p. 9, emphasis in the original)

If it is true that as a consequence of reduced barriers, individuals are typically more vulnerable to anxiety and insecurity about the permanence of their relationships, then this is simply another cost that can be expected to reduce the profitability of a CR and lower its attractiveness.

Besides being uncomfortable and costly in themselves, anxiety and insecurity are bound to have such effects as distrust and possessiveness, which will additionally increase costs within the relationship. In addition, insecurity should lead to especially strenuous (and costly) efforts to revive the partner's satisfaction with the relationship when that factor has fallen dangerously low in the face of attractive alternatives. Many of the strategies and mechanisms for resolving conflict, for increasing communication, for dealing with jealousy, and so on that are currently being recommended by marriage counselors (for whom business is booming), promoted in "self-help" books and articles, and taught in various seminars and lectures may be seen as attempts to provide individuals with the knowledge and skills to revive or maintain the satisfaction level within a CR. That such programs are paid for and subscribed to by increasingly large numbers of American couples bears witness to the fact that more and more couples feel a need to develop internal satisfaction safeguards to protect the longevity of their CRs (and thus their own security) and are willing to spend time, money, and energy in this pursuit. What is not frequently recognized, however, is that these constitute additional costs of maintaining a CR in contemporary society.

4.1.3. The Increased Importance of Positive Emotion and Feeling

It seems reasonable to propose that, largely as a consequence of the social changes we have discussed and the resultant barrier reduction and redefinition of the *raison d'être* of CRs, the factor that is now most closely linked to personal satisfaction or dissatisfaction with a CR is the nature of the emotional experiences that the person has in the context of the relationship. It is this factor that may be increasingly monitored by CR participants, for as Harever (1978) observed, "In periods when the family's economic stability was at stake, emotional gratification had less importance" (p. 35). But as economic and social conditions have changed, emotional experiences have apparently come to play an ever more important role as determinants of people's satisfaction with their CRs:

> The expectations of women and men have expanded from marital satisfaction which depends on rather solid, extrinsic rewards such as a clean house, a well-prepared meal, overt respect, good earnings, to a desire for ephemeral intrinsic satisfactions such as happiness, personal growth, sexual satisfaction, closeness, and the like. (Glazer-Malbin, 1978, p. 17)

And the high premium people place today on their personal emotional experiences is reflected in the fact that, according to survey results, both men and women claim to "put love above job, sex, money or power as a goal in life" (Hunt & Hunt, 1978, p. FC5). The notion of "falling in love" is still, as it was formerly, seen to be a necessary condition for the establishment of a marital CR (Kephart, 1967; Campbell, 1977), but now remaining "in love" may have been added to the criteria for its maintenence.

If it is true that men and women today expect their CRs to provide them with positive emotional experiences and believe that CRs that no longer do so should be dissolved, then the burden on the contents of the relationship to justify its existence is so heavy that the demise of the CR may be inevitable. Neither theoretical analyses of the temporal course of emotion, including its experience in CRs (Berscheid, in press), nor empirical data (e.g., Blood, 1967) suggest that intense and frequent positive emotion is likely to be sustained within a relationship.

Clinical data lead to the same pessimistic conclusion. For example, based on their extensive experience with couple therapy at the Berkeley Therapy Institute, Goldstine, Larner, Zuckerman, and Goldstine (1977) have proposed a three-stage model of the "life-cycle" of CRs; the emotional tenor of the relationship changes with each stage. Goldstine *et al.* stated that

> At the point in their relationship when two people discover that they love each other and wish to become partners, they tend to experience a harmony and mutual delight that we call Stage I. . . . This phase of a relationship is characterized by openness, optimism, and mutual engrossment. (p. 8–9)

The first phase of a CR, then, is typically characterized by all of the phenomena that the term *falling in love* implies. It tends to be a period of heightened and intensified positive emotional experiences perhaps unmatched by any other period in most people's lives. The partners, according to Goldstine *et al.*, tend to "find themselves filled with sweet and bittersweet emotions—with longing, passion, fondness, uncertainty—and wonder" (p. 149). They compose a kind of mutual admiration society where "entranced by the desirable qualities they discover in each other, they shrug off whatever shortcomings they find" (p. 152) and are tempted to believe that the relationship will realize all their fantasies and transform their lives.

But the delight of the first stage is not, and cannot be, sustained forever. As Goldstine *et al.* described:

> To a couple's distress, real life gradually impinges on their relationship. Conflict surfaces, failure intrudes, and boredom casts its pall. The bright illusion *We are one* fades into the bleak conviction *We are hopelessly different.* (p. 9, italics in the original)

In the process of coordinating their habits and routines, expressing and negotiating tastes and preferences in the making of joint decisions, dividing up duties and responsibilities, and with all the other realities of increasingly interdependent coexistence that come with actually living together, frustration, disagreement, conflict, and disappointment inevitably arise. Negative emotional experiences dramatically increase in frequency, and the joys and delights of the first stage begin to fade.

It is at this stage that the presence of external barriers may keep a couple contained in the relationship long enough for the conflicts to be worked out and for the couple to graduate to a third stage, described by Reik (1944/1972) as "the afterglow." In the third stage, the couple have resolved or laid to rest the major conflicts that arose during Stage 2 through compromise, mutual adaptation, "agreeing to disagree," and, perhaps not infrequently, through eliminating from their relationship spheres of interaction that were sources of friction and dispute. But because "Stage II follows Stage I with the relentlessness of death and taxes, no matter who the participants in the relationship are" (Goldstine, et al., 1977, p. 190), because few external barriers may keep people motivated to resolve their differences, and because the existence of extremely intense negative emotions within the CR may signal that it no longer has a reason for being, the CR may be aborted at this point.

Even CRs that successfully weather the storms of Stage 2 and reach the emotional placidity characteristic of Stage 3 may be vulnerable to dissolution, for they, too, fail to meet the criterion of providing frequent and intense positive emotion within the relationship. In fact, some CR partners may take the emotional tranquillity characteristic of the third stage to indicate that their emotional relationship has stagnated and died. They may then terminate the relationship because, as the Goldstine *et al.* (1977) client complained, "she can't come across with what I want . . . excitement, [and] passion" (p. 181).

4.2. Effects of Barrier Reduction on the Number and Attractiveness of Alternatives

Not only may barrier reduction indirectly contribute to the dissolution of CRs by altering the attractiveness of the relationship itself in such a way that a decision to terminate the CR is made more likely, but barrier reduction has also indirectly increased the number and attractiveness of alternatives to many current CRs. Probably the most important way in which reductions in barriers have resulted in increased alternatives to people in currently intact CRs is through the simple fact that each time a CR dissolves (e.g., via other effects of barrier reduction), two additional people have been added to the pool of possible alternative relationships for persons in that social network. And as the "field of eligibles" for alternative relationships expands for the individual who is contemplating the termination of his or her own CR, the probability necessarily increases that the individual will perceive at least one candidate in the field as offering sufficiently more profit than the current partner. Thus, the termination of any given CR indirectly weakens currently intact CRs by increasing the number of potential alternative partners.

Not only are there more alternative relationships possible now, but the same social changes that reduced the barriers (that helped produce an expanded field of eligibles) also help make the alternatives more visible and more tempting to people who are currently maintaining CRs. For example, increased contact between men and women in the absence of their partners necessarily accompanies the greater participation of women in the working world and in educational institutions. And perhaps not infrequently, relationships established between men and women in their common workplace or in a common educational setting may come to be seen as providing a more desirable foundation for a satisfactory CR than the current one provides, particularly if the individual has already become disenchanted with it.

In any event, it is clear that the dissolution of a marital CR no longer means the end of one's intimate life, as it once did. The chances of remarriage after divorce have grown almost as fast as the divorce rate itself, with roughly 80% of all divorced people, and virtually all in their 20s, eventually remarrying (Hunt & Hunt, 1978). One indication of the rate at which alternative relationships have increased is the fact that whereas in 1960 remarriages accounted for less than one-seventh of men's marriages, by 1975 remarriages following divorce accounted for almost one-fourth of all men's marriages (Glick & Norton, 1978).

Concurrently, the alternative of living together without marriage has been gradually accepted and has thus become an increasingly attractive alternative to many people contemplating replacing their current CR with another. It is estimated that roughly a million couples now live together unmarried, and while most of these couples are childless, one out of every five includes children of one or both partners by previous marriages (Hunt & Hunt, 1978).

Further, the increasingly positive social value accorded to the single lifestyle has made that alternative more attractive. While single adults were traditionally seen as something of a curiosity in our culture, a permanently single lifestyle has become socially acceptable and attractive, especially to women. As Levinger and Moles (1976) observed, "Today numerous women say that for them this alternative is more dignified than becoming or staying married" (p. 194). And in addition, the single life has become more attractive even to persons with children. As Brown, Feldberg, Fox, and Kohen (1976) noted:

> Single-parent status has traditionally been seen as transitional, a temporary status belonging to the difficult period after separation and ending when the single parent remarries. Those who remained single parents were seen as not quite normal. . . . However, despite a high rate of remarriage following divorce, single parenthood can no longer be viewed as temporary or deviant; it is a viable, even preferred life pattern for an increasing number of divorced mothers. (p. 119)

The statistics reveal that single parenting—most frequently on the part of women, but also increasingly on the part of men—is indeed becoming more prevalent. Orthner, Brown, and Ferguson (1978) have reported that according to U.S. Census Bureau data, there are now over 4.5 million families in the United States headed by a single parent with children under 18. Of these, 90% are headed by women, while the remaining 10% are headed by men. The attractiveness of single parenting to people considering it as an option to remaining married, in addition to having been increased by its greater visibility, has undoubtedly been enhanced by single parents' organizations that offer single parents and their children a variety of supportive programs and social events.

All of these increases in the number and attractiveness of alternatives, as they work their causal effect on intact CRs and lead to the dissolution of many of them, tend to perpetuate and fuel the social changes that helped reduce the external barriers that increased the number and attractiveness of alternatives in the first place. More single par-

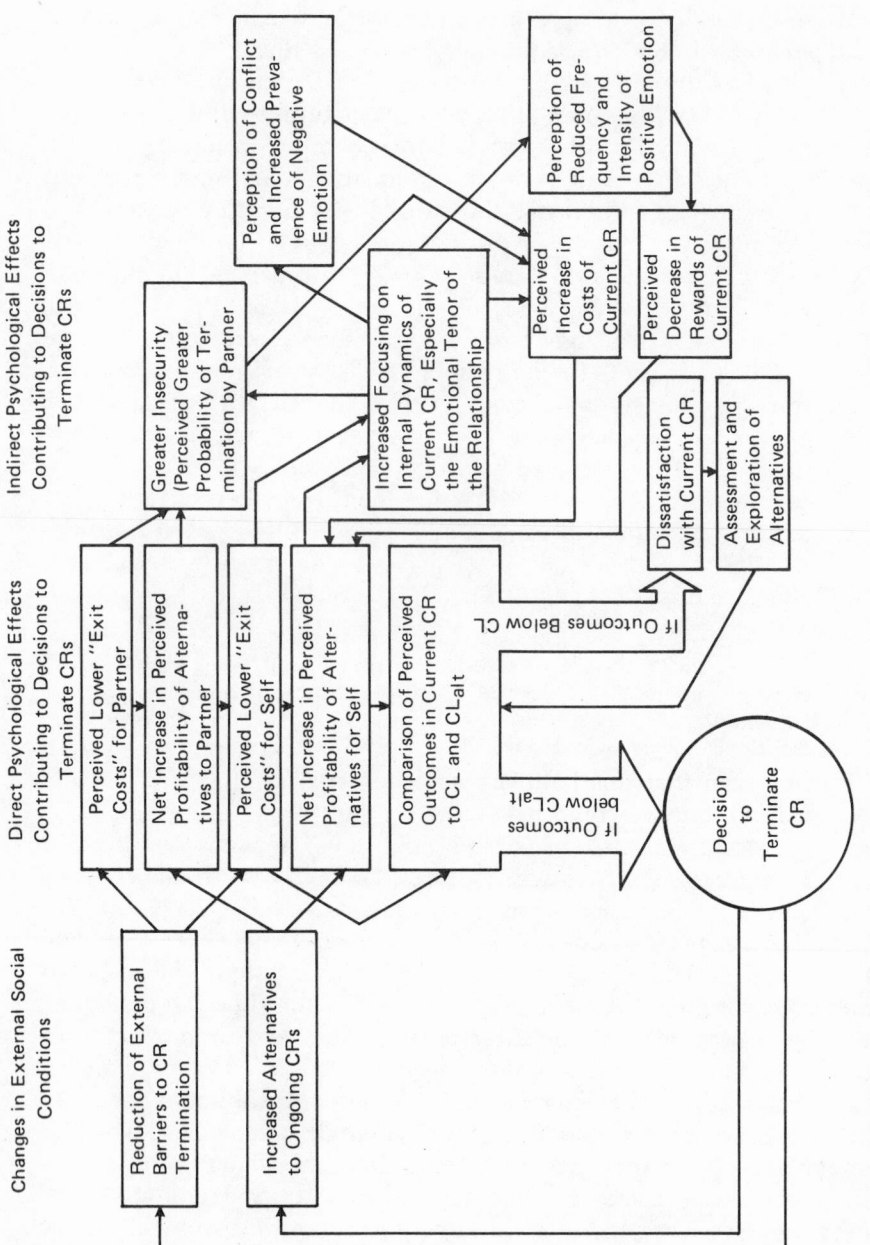

Figure 10.2. An elaborated causal model of the effects of barrier reduction on decisions to terminate CRs.

ents, for example, means more membership for organizations that support and encourage single parenting and hence better programs to facilitate this alternative to an unhappy marriage; it means greater social acceptance of this lifestyle, and it means greater demand for equal economic and educational opportunity for women; it means that for religious institutions to have relevance in modern-day society, particularly for children, traditional values must be altered; it means a greater demand for laws that will facilitate rather than hinder property settlement and child custody decisions. In short, it means more social change that will, as a by-product, further reduce external barriers to CR termination. The causal pathways we have traced are illustrated in Figure 10.2, an elaboration of the general model presented in Figure 10.1.

5. SUMMARY AND AFTERWORD

We have tried to show how individual decisions to terminate a CR lead to further changes in the external social conditions that are relevant to the longevity of CRs. Specifically, we attempted to indicate how reductions in external barriers, through their direct and indirect effects on decisions to terminate a given CR, contribute ultimately to the dissolution of many other CRs (see Figure 10.2). We discussed the ways in which reductions in barriers may bring about increases in alternatives and additional social changes that act to further decrease barriers (and it can be seen that this pathway has been specifically represented in Figure 10.2), but we could have entered the causal path at any point and in any of the recycles. For example, some of the barrier reductions we initially discussed, such as the weakening of the social stigma attached to divorce, might have been discussed as one of the consequences of certain prior barrier reductions (e.g., the growth in the economic independence of women), mediated by the effects that this prior event had on many CRs. Or it would in principle have been possible to start our discussion with the social changes that have had the effect of increasing alternatives, and to demonstrate how increased alternatives, through their direct and indirect effects on decisions to terminate a CR, may bring about both reductions in barriers and further increases in alternatives. But for our purposes, it is sufficient merely to acknowledge the theoretical existence and importance of feedback from individual decisions to terminate CRs to properties of the social context of CRs, which, as perceived by individuals in CRs, are crucial determinants of those very decisions—namely, the barriers to the termination of CRs that exist and the number and attractiveness of alternatives.

As we arrive at the conclusion of our analysis, three additional comments might be made. First, although recent social changes have reduced barriers for perhaps the majority of participants in CRs today, we would not expect all (or perhaps even most) of these social changes to have affected any given CR. For example, although we have argued that religious convictions play less of a stabilizing role in CRs today than they once did, we would still expect such beliefs to operate as potent deterrents to divorce in many individual cases. Similarly, although the social stigma of divorce has all but disappeared, we would still expect to find specific individuals for whom personal meanings attached to divorce would constitute a powerful barrier. Or, again, while we have suggested that many individuals may hold unrealistic expectations regarding the frequency and intensity of the positive emotions they should experience in their CR, we would not assume that all individuals do. Thus, our analysis does not allow us to make predictions about the probable longevity of a particular CR in the absence of specific information about the perceptions of those individuals, but our framework does suggest what kinds of information ought to be particularly important to gather in order to make such predictions.

Second, just as our analysis does not imply that barrier reduction has been equally dispersed across all CRs, neither do we suggest that there is a symmetry between partners in a given CR in this respect. There may be potent external barriers constraining one but not the other partner from dissolving the relationship. The implications of such asymmetry for the internal dynamics of the relationship are beyond the scope of this paper, but the plight of individuals caught in such situations (and the point for present purposes) is illustrated by the remark that one husband of our acquaintance made to his wife: "Yes, I know we haven't been getting along well lately, but what does that have to do with our marriage?" What it had to do with the marriage is that two months later they were separated and on the road to divorce, for it takes only one person to abort a CR, and it takes two to maintain it. Thus, it is the decision of the person with the fewest external barriers that is likely to prevail, further guaranteeing the potency of barrier reduction as a factor in CR longevity.

Finally, we want to stress that our analysis, despite its gloomy prognostications for CR longevity in the future, does not prejudge the social changes that have occurred as inherently "bad" or inherently "good." Only philosophers, and each of us as an individual, can decide whether the benefits of the social changes discussed outweigh their costs in terms of human happiness and suffering. The only conclusion that can be safely drawn is that whether an individual welcomes or deplores the revolutionary changes that have occurred in our society within a very short period of time, no one can escape their consequences, not even in the most intimate and personal aspects of life.

REFERENCES

Berscheid, E. S. *The problem of emotion in close relationships.* New York: Plenum (in press).

Blood, R. O., Jr. *Love match and arranged marriage.* New York: Free Press, 1967.

Brown, C. A., Feldberg, R., Fox, E. M., & Kohen, J. Divorce: Chance of a new lifetime. *Journal of Social Issues,* 1976, *32,* 119–133.

Campbell, B. *The perceived importance of romantic love as a determinant of marital choice: Kephart revisited ten years later.* Unpublished manuscript, University of Minnesota, 1977.

Feldman, H., & Feldman, M. The effect of father absence on adolescents. *Family Perspective,* 1975, *10,* 3–16.

Festinger, L., Schachter, S., & Back, K. *Social pressures in informal groups.* New York: Harper, 1950.

Freedman, J. L. *Happy people.* New York: Harcourt Brace Jovanovich, 1978.

Glazer-Malbin, N. Interpersonal relationships and changing perspectives of the family. In H. Z. Lopata (Ed.), *Family factbook.* Chicago: Marquis Academic Media, 1978.

Glick, P. C., & Norton, A. J. Marrying, divorcing, and living together in the U.S. today. *Population Bulletin,* Vol. 32, No. 5, Washington: Population Reference Bureau, 1977.

Glick, P. C., & Norton, A. J. Marrying, divorcing, and living together in the U.S. today. In H. Z. Lopata (Ed.), *Family factbook.* Chicago: Marquis Academic Media, 1978.

Goldstine, O., Larner, K., Zuckerman, S., & Goldstine, H. *The dance-away lover.* New York: Ballantine, 1977.

Harever, T. K. Family time and historical time. In H. Z. Lopata (Ed.), *Family factbook.* Chicago: Marquis Academic Media, 1978.

Hoffman, L. W. Changes in family roles, socialization, and sex differences. In H. Z. Lopata (Ed.), *Family factbook.* Chicago: Marquis Academic Media, 1978.

Hunt, B., & Hunt, M. What's happening to marriage and the family in America? *Family Circle,* September, 1978, FC1–FC8.

Hunt, M., & Hunt, B. *The divorce experience.* New York: New York American Library, 1979.

Kephart, W. Some correlates of romantic love. *Journal of Marriage and the Family,* 1967, *29,* 470–474.

Klinger, E. *Meaning and void: Inner experience and the incentives in people's lives.* Minneapolis: University of Minnesota Press, 1977.

Levinger, G. A social psychological perspective on marital dissolution. *Journal of Social Issues,* 1976, *32,* 21–47.

Levinger, G., & Moles, O. C. In conclusion: Threads in the fabric. *Journal of Social Issues,* 1976, *32,* 173–207.

Luepnitz, D. H. Which aspects of divorce affect children? *The Family Coordinator,* 1979, *28,* 79–85.

Macklin, E. D. Heterosexual cohabitation among unmarried college students. *Family Coordinator,* 1972, *21,* 463–472.

Market Opinion Research, National Commission on the Observance of International Women's Year. The life work of women. In H. Z. Lopata (Ed.), *Family factbook.* Chicago: Marquis Academic Media, 1978.

McEaddy, B. J. Women who head families: A socioeconomic analysis. In H. Z. Lopata (Ed.), *Family factbook*. Chicago: Marquis Academic Media, 1978.

Moles, O. C. Marital dissolution and public assistance payments: Variations among American states. *Journal of Social Issues*, 1976, *32*, 87–101.

Moles, O. C., & Levinger, G. Introduction. *Journal of Social Issues*, 1976, *32*, 1–4.

Norton, A. J., & Glick, P. C. Marital instability: Past, present, and future. *Journal of Social Issues*, 1976, *32*, 5–20.

Orthner, D. K., Brown, T., & Ferguson, P. Single-parent fatherhood: An emerging family life style. In H. Z. Lopata (Ed.), *Family factbook*. Chicago: Marquis Academic Media, 1978.

Pleck, J. H. The work-family role system. In H. Z. Lopata (Ed.), *Family factbook*. Chicago: Marquis Academic Media, 1978.

Reik, T. *A psychologist looks at love*. New York: Lancer, 1972. (Originally published 1944.)

Thibaut, J. W., & Kelley, H. H. *The social psychology of groups*. New York: Wiley, 1959.

Waller, W. *The old love and the new: Divorce and readjustment*. Carbondale: Southern Illinois University Press, 1967. (Originally published 1930–1958.)

Weiss, R. S. *Marital separation*. New York: Basic Books, 1975.

11

Giving and Receiving

Social Justice in Close Relationships

LOUISE H. KIDDER, MICHELE A. FAGAN, and
ELLEN S. COHN

Love and justice are an odd couple, the former presumably all emotional, the latter all rational. They are more closely tied, however, than romantic tradition lets on (e.g., Friedan, 1963). In this chapter, we examine the connecting links between justice in the calculable world of work and money and in the presumably incalculable world of love.

Of the many possible forms of intimate relations, we will examine only adult relations. This eliminates relationships among children and relationships among children and adults. We include same-sex and opposite-sex adult relationships. We include loving and liking, sexual and nonsexual relationships. We inevitably talk a great deal about marriage, but we also pay a lot of attention to friendships, particularly same-sexed or homosocial friendships.

1. VARIETIES OF JUSTICE IN CLOSE RELATIONSHIPS

Close relationships involve both distributive and procedural justice. *Distributive justice* refers to the fairness of the allocation of resources

LOUISE H. KIDDER ● Department of Psychology, Temple University, Philadelphia, Pennsylvania 19122. MICHELE A. FAGAN ● Department of Psychology, University of Chicago, Chicago, Illinois 60637. ELLEN S. COHN ● Department of Psychology, University of New Hampshire, Durham, New Hampshire 03824.

among people in a relationship. Studying love and friendship as relationships in which people exchange resources or exercise power may be anathema to readers who like to think in more romantic terms. Rubin (1973) acknowledged this in his book on *Liking and Loving:* "The notion that people are 'commodities' and social relationships are 'transactions' will surely make many readers squirm. Exchange theory postulates that human relationships are based first and foremost on self-interest" (p. 82).

To analyze distributive justice in close relationships, we must suspend commonsense romantic conceptions of love and friendship. This is no different from analyzing distributive justice in many other settings where the surface conception is different from the theoretical analysis. For instance, in a study of what good teachers do, Uriel Foa and his students analyzed students' perceptions of what they received (Foa, Megonigal, & Greipp, 1976). Good teachers gave not only information but also something that Foa called "love," that is, warmth and understanding. We should not be surprised to learn, therefore, that love and friendship involve the exchange of other resources that are different from love.

There may be times when people do not calculate the exchange of resources in their relationships. People in love may say that justice is the farthest thing from their minds (Levinger, 1979b). This means, however, not that it is irrelevant but that the rewards are so great and so obvious that they need no calculation. When love wanes, and a relationship begins to decline, the net balance of costs and rewards becomes obvious and salient (Levinger, 1979a).

Procedural justice refers to the fairness of the *process* whereby decisions are made and resources distributed (Folger, 1977; Thibaut & Walker, 1975). To analyze procedural justice in close relationships, we ask "who decides" about important issues, and this amounts to a study of social power. Procedural justice in close relationships is tied to the exchange of resources. Jesse Bernard (1973) wrote, "when wives enter the labor force their 'power' in the marital relationship tends to increase" (p. 143). The person who brings home the bacon may also be the person who has the final say (cf. Kipnis, 1976).

1.1. Distributive Justice: The Exchange of Resources in Close
       Relationships

Marriages in other times and other cultures are often based on clearly calculable exchanges. Bride price is paid by a prospective groom to the father of the bride before a marriage takes place in many African societies. In India, the bride brings money to the marriage in the form of gold jewelry, silk saris, silver bowls, and actual cash. These form her dowry and make her marriageable (Goody & Tambiah, 1973). In addition to monetary contributions, prospective mates offer one another a great variety of resources and attributes, some of which are worth advertising, as in the matrimonial ads from the Indian newspaper *The Hindu:*

> Alliance invited with horoscope from Non-Srivathsa, M.B.B.S., M.D. Iyengar grooms, 29–34 years, tall, smart, handsome, U.S. settled or bound, for smart, fair, beautiful doctor bride, 26 years, 160 cms. height, and Green Card holder in U.S. (*The Hindu,* Sunday, July 2, 1978)

Not so long ago, Americans also advertised for spouses with special skills or attributes. During the latter part of the westward movement, men chose mail-order brides from magazines that included a section like the matrimonials above. One reconstructed version of a mail-order request has become a contemporary joke:

> Wanted: Strong, healthy, young farm-woman who can milk cows, shear sheep, sew, and who possesses tractor. Send photo—of tractor.

Advertisements for dates, companions, and occasionally spouses in contemporary American magazines provide as particular a list of attributes as the matrimonial ads in *The Hindu:*

> Successful businessman, holds masters degree, enjoys sports, tennis, golf, movies, museums, gourmet dining, and art. Have traveled extensively in U.S. and Europe. Am 50 years old, 6 feet, slender, muscular. Desire Caucasion female companion who is attractive, intelligent, broad minded, passionate, affectionate. Photo appreciated. (*Philadelphia Magazine,* July 1979, p. 206)

> Tall, sophisticated, sensitive woman, 38, desires meaningful relationship with warm, sincere, communicative male. (*Philadelphia Magazine,* July 1979, p. 207)

Matrimonial advertisements, personal ads, mail-order brides, dowries, and bride price show that marriages and other alliances are made by assessing and offering rewards. They are based on self-interest rather than altruism. Such advertisements for relationships between spouses, friends, and lovers are very explicit about the exchange—what each person gives and receives. We agree with Rubin (1973) that "we must face up to the fact that our attitudes toward other people *are* determined to a large extent by our assessments of the rewards they hold for us" (p. 82).

Distributive justice in close relationships or any relationship can take several forms (cf. Leventhal & Michaels, 1969; Lerner, Miller, & Holmes, 1976; Lerner, 1977; Deutsch, 1975). Meritorious equity is justice based on each person's merits—one person may receive more rewards than the other provided the first also does more to deserve the rewards.

Compensatory equity is justice based on each person's needs: one person may receive more than the other because the first person has greater needs. Equality is the third form of distributive justice: it requires that both persons' rewards be equal, regardless of their contributions or their needs. We regard equality as the appropriate form of justice in close relationships. Equality calls forth neither competition nor charity, both of which we regard as inappropriate in intimate relationships.

Meritorious equity is appropriate where merit is an issue, such as in grading systems in schools. If one student's essay is better than another's, the former deserves a higher grade. Compensatory equity is appropriate where need is an issue such as in welfare programs. If one mother has two dependent children, she needs twice as many pairs of children's shoes as a mother with one dependent child. While there may be times in close relationships when one person needs more money or another deserves more recreational time, the form of justice we advocate is equality, and the relationships we are interested in are egalitarian.

In exchanges that have calculated costs and rewards—as with wage labor and scholastic grades—it is fairly easy to calculate the input–outcome ratios and to determine whether justice was done, be it meritorious, compensatory, or equal justice. When we try to measure exchanges in intimate relationships, it is more difficult because we have no calculus to equate the different contributions and rewards that people exchange. One instrument for measuring exchanges in close relationships lists 22 services, including affection, understanding, day-to-day maintenance, finances, physical attractiveness, and 17 more diverse qualities (Trauppman-Utne-Walster scales, 1978). How many units of "day-to-day maintenance" equal how many units of "understanding and concern" or "physical attractiveness"? If one person offers financial support in return for the other person's commitment or caring, it is difficult to know whether "justice" prevails; the two resources are noncomparable.

Our thesis is that egalitarian relationships arise when both persons make valued contributions, and to do this, they must each have access to a wide range of equally valuable resources. If one person has fewer resources or less valuable resources, that person can be exploited, even in an intimate relationship. For instance, if I have only silver pieces and you have all gold, you may extract many more pieces from me than I from you. If the exchange rate is two silver pieces for every gold one, and you continue to mine gold while I mine silver, I will have to work twice as long as you to engage in each exchange. The adage that a housewife's work is never done may arise from an imbalance in valuable resources. Estimates of the number of hours women work range to as high as 80 hours a week when wage labor and housework are added together (Myrdal & Klein, 1968; Vanek, 1973). If the division of labor seems fair, it must be that her work counts for less, and like the silver miner, she must work twice as long.

Egalitarian relationships are most likely to exist when both persons have access to a wide range of resources, so both can contribute and receive similar rewards. Neither person possesses all silver or all gold, and neither can be exploited on the grounds that "One of my contributions is worth two of yours." The distribution of resources within a relationship, therefore, determines the distribution of power. Our analysis of social justice in close relationships, therefore, requires that we also look at partners' economic relationships. The real and symbolic value of men's and women's contributions in the economic world determines what they can do and expect in their relationships at home. Distributive justice affects procedural justice.

1.2. Procedural Justice: The Social Power that Accrues to Those Who Have Resources

Men's and women's participation in the labor force in twentieth-century America is not representative of all times and places. In the preindustrial economy, men and women had different but more equally valued jobs. He tended the sheep and she spun the wool; he fed the cows and she churned the butter. Industrialization and modernization eliminated home-based economy and removed women from the economic labor force (Oakley, 1974). In societies where women's work has economic value, women have bargaining power in both the marketplace and the home.

Judith van Allen (1972) described Nigerian women's economic and political institutions before colonialism and how they influenced women's relationships with men. Before colonial administrators revised Igbo political life, Igbo women had access to some important resources. They were farmers and grew all crops except yams, and they were merchants, selling their crops at the market. As both farmers and merchants, they held meetings, called *mikiri* or *mitiri*, where they set prices and established rules about proper market behaviors and farming practices. Their rules applied to men and boys as well as to women and girls, and they enforced them as a group. For instance, men were responsible for tending the livestock and growing yams. Yams have a shorter growing season than the women's crops, however, and after the yams had been harvested, the men sometimes relaxed their watch and the livestock wandered into the women's crops. The women organized through their *mikiri* to enforce control over the animals, and in one village, they "swore an oath that if any woman killed a cow or other domestic animal on her farm, the others would stand by her" (van Allen, 1972, p. 171). As merchants, the women also established rules that applied to men:

> For instance, rowdy behavior on the part of young men was forbidden. Husbands and elders were asked to control the young men. If their requests were ignored, the women would handle the matter by launching a boycott or strike to force the men to police themselves, or they might decide to "sit on" the individual offender. (p. 170)

To "sit on" or "make war on" a man meant that the women

> gathered at his compound, sometimes late at night, dancing, singing scurrilous songs which detailed the women's grievances against him and often called his manhood into question, banging on his hut or plastering it with mud and roughing him up a bit. (p. 170)

Igbo women's demonstrations for justice extended beyond economic matters. They also used their *mikiri* to protest personal matters and were prepared to "sit on" a man for mistreating his wife. The women gained bargaining power and the right to demand social justice in their intimate relationships because they controlled economic resources. Political institutions, organized strikes, and demonstrations arose among women who were farmers and merchants. They used their political and economic resources to promote justice at home.

Another description of precolonial African women confirms van Allen's analysis. Women in Tanzania were important food producers and some held high political offices before colonialism. Then they were introduced to "gentility" by European missionaries, educators, and administrators. Gentility cost them their economic and political positions and affected their domestic lives (Brain, 1978): "The character of the marriage relationship in various tribes and the status of women in the community are closely related to the share which they take in the provision of food" (Hailey, 1956, p. 821).

The women in both of these societies gained power through their economic activities. That power permitted them to "sit on" or "make war on" a man to enforce what they considered proper behavior, whether that behavior be in the marketplace or in relationships at home. The European introduction of "gentility" did away with such power. It is not genteel to "sit on" or "make war on" a man. Though the phrases are figurative rather than literal, they suggest unladylike gestures. Promoting social justice in intimate relationships may mean relinquishing "ladylike" behavior.

Colonial administrators were not solely to blame for eroding African women's economic and political power. Even in societies not introduced to gentility, there is the notion that being female is incompatible with being powerful. Among the southern Bantu, a woman must become a "husband" if she wishes to exercise power (O'Brien, 1977). A female "husband" marries another woman with all the rites and legal procedures of any other marriage. She assumes the role of husband not to engage in sexual relations with her wife but to be a leader. Women who exercise power are "husbands" and not wives (O'Brien, 1977).

Women in preindustrial America also gained power and authority through their contributions of food and the other necessities of life. They worked at subsistence tasks for most of their adult lives, and their work often made the difference between affluence and poverty for a farm family (Ginzberg, 1966; Oakley, 1974).

In contemporary America, employed wives have more voice in family decisions than unemployed wives. Donald Wolfe (1959) analyzed decision making in families living in the Detroit area. He classified families on the basis of who decided where to live, where to go on vacation, whether to change jobs, and five other important family issues. He found that wives who had never worked lived in families he called "husband-dominant," with the husbands making most decisions. Wives who were currently working were in families that had shared decision-making. A different sample of families in another part of the United States showed the same pattern. Working wives had more influence in family decision-making whenever there were disagreements than did housewives (Heer, 1958). A plausible alternative explanation for these data is that the wives who are dominant to begin with are the ones who are also able to work outside the home. In the second study, the author tested this explanation and found it untrue. When he controlled for the factor of personality dominance, he found that it was still true that working wives had more influence—demonstrating that it was their work status rather than their personality that made the difference. These and other studies (e.g., Kligler, 1954; Kelley, 1979; Centers, Raven, & Rodrigues, 1971) demonstrate that persons who control resources other than love also possess power in love relationships.

2. MEN'S AND WOMEN'S CONTRIBUTIONS IN THE WORLD OF WORK

If doing productive work gives a person power in relationships, why are there productive workers who have little social power? We need another factor in our formula. Not all labor counts as work, or, more accurately, people may value some work more than other work.

2.1. Who Does What Work

Two researchers devised one objective way to count how much work men and women do throughout the world (Aronoff & Crano, 1975). They used data from 862 societies in the Ethnographic Atlas (Murdock, 1967). They tested the idea that there is universal task segregation and sex-role differentiation in the family, to see whether men did more of the productive labor necessary for subsistence than women. They scored each society on the degree to which it was dependent on each of five subsistence activities: gathering, hunting, fishing, animal husbandry, and agriculture. They then coded the degree to which males and females were independently involved in these activities. If a society depended on fishing for 30% of its subsistence, and if men contributed 90% to fishing, their contribution to subsistence through fishing was $.30 \times .90 = .27$. If another society depended on agriculture for 60% of its subsistence and if women constituted 80% of the agricultural work force, their contribution to subsistence through agriculture was $.60 \times .80 = .48$. Aronoff and Crano computed these subsistence-contribution coefficients for men and women in 862 societies and found that "the worldwide percentage of food contributed by women is 43.88%" (p. 17). The worldwide average (44%) was roughly half, which means that women *traditionally* have been productive contributors to their societies' subsistence even if we do not compute the economic value of "housework," as some authors have suggested (e.g., Peterson, 1978; Chong, 1969).

Aronoff and Crano concluded that there is

> no support for the theoretical position of a universal principle of task segregation and sex differentiation in the family. While these data do not bear on considerations of socio-emotional specialization, they do demonstrate clearly that women play a most significant instrumental role in the family. (p. 17)

They added:

> it is clear that women contribute a significant share to the total subsistence larder of their society—and this observation holds even though the present analysis does not include that share provided by mother's milk. (p. 18)

These cross-cultural data show that women can and do make significant contributions to the economic welfare of their societies. Why do we find, then, in modern industrial societies, that they often do not appear to do so? And why, even in subsistence economies where their contributions are economically essential, do they not always have the power to "make war" for what they believe is right? Although women's contributions may be objectively valuable, their subjective worth may fall short of their real value; perhaps women's work seems less important.

Cross-cultural evidence and North American statistics both tell us that large numbers of women do large amounts of work. In the United States, for instance, 48% of married women are employed, and 42% of the labor force is female (U.S. Bureau of the Census, 1979). We have also seen in Aronoff and Crano's (1975) analysis of 862 societies that women contribute about 44% to the subsistence of their societies. In spite of these objective data, women's *perceived* contributions are often much less and their salaries much lower than men's (U.S. Bureau of the Census, 1979).

2.2. Which Work Counts

Anthropological observations show how women's objective contributions may be undervalued. For instance, the Hazda of central Tanzania have a diet that is 75% vegetable. Hazda women grow the vegetables. Their food, therefore, is 75% woman-produced. The Hazda, however, *think* of themselves as meat eaters and value the men's contribution more than the women's even though the proportions suggest that they should do the opposite (Brain, 1978). The objective importance of women's work may be overshadowed by the symbolic importance of men's work. Brain (1978) elaborated on this idea:

> African men cut trees down, and women cut them up. Women could fell trees, but they do not. Perhaps the element of danger involved in defense, undertaken by males as part of our primate heritage, is responsible for the rationalization of the division of almost all occupational roles between the sexes. Whether this speculation is reasonable or not, the fact remains that frequently what is considered the dangerous role (e.g., felling trees) is a male prerogative . . . men are conditioned to undertake the more exciting tasks, which allow for more individual acquisition of public approbation, while females are prepared for the more mundane, repetitive, and dull tasks. (p. 697)

Even if the division of labor is objectively equal, or even if women assume great responsibility for subsistence, the symbolic division of labor may find men doing the important work and women the menial jobs.

Several writers have argued that as long as women are identified with nature, through childbirth and lactation, and men are identified with culture, through producing tangible and nonperishable goods, women's role will be unequal to men's (Chodorow, 1974; Ortner, 1974). These writers do not believe that biology is destiny, as Freud once argued, but they do say that as long as cultural symbols and conventions identify women with childbirth and child rearing, they will have a double jeopardy: they will be unavailable and seem unsuitable for the important work of a society (cf. Firestone, 1971; Brown, 1970; Nerlove, 1974).

Discrimination against women in the labor force is a topic for a chapter unto itself and is documented elsewhere (e.g., U.S. Bureau of the Census, 1979). We introduce it here to make two points: (1) Women can and do work for food or money in almost the same numbers as men (44% & 42% are the estimates from cross-cultural and American data), but their work is undervalued; they do not earn the same money or same public approbation. (2) Men's and women's decision-making power and equality within the relationship reflect their economic activities outside the relationship. When women bring home food or money, they have the right to "make war" for what they believe is just, and they have the right to make decisions that affect both persons. When they share the power to make economic decisions, they may forfeit their previous monopoly on the power to make decisions about child rearing. They may also have to relinquish their monopoly on other aspects of the private world of nurturance and love.

3. MEN'S AND WOMEN'S CONTRIBUTIONS IN THE WORLD OF LOVE

3.1. Who Gives the Love

We have examined one half of the stereotype about the sex-role division of labor: men bring home the bacon. We found that this is not universally true, that women contribute roughly half the economic labor in America and in 862 other societies. We also found that when women work and their work is recognized, they have more power and authority in their families. We now examine the other half of the stereotype—that women tend the heart and matters of love and affection. Novels, poetry, and some social scientists describe women as specialists in the world of love. Bales (1953) called them socioemotional leaders, responsible for making people in groups and families feel happy, secure, comfortable, and cared for. Men, Bales said, are instrumental leaders, seeing that people complete their tasks. Tönnies (1963) distinguished between the world of love, community, and family life, which he called the *Gemeinschaft*, and the world of work and commerce, which he called the *Gesellschaft*. He considered *Gemeinschaft* the feminine world (Sampson, 1979).

These global proclamations raise two questions for us: (1) Are women the socioemotional specialists, giving love and affection more than men? (2) Do families work this way?

To answer the first question, Melodie Wenz asked 32 people who volunteered to be interviewed what they gave and received in their close relationship with a person of the opposite sex (Wenz, 1978; Kidder, Fagan, & Wenz, 1978). She used Traupmann, Utne, and Walster's scale (Walster, Walster, & Berscheid, 1978) and asked each person how much he or she contributed and received of 22 specific resources, such as understanding and concern, sexual pleasure, intelligence, physical attractiveness, liking, and help with day-to-day tasks. She found significant differences in what men and women reported contributing to their relationships. Men reported contributing more than women did in three areas:

1. Finances: contributing income to a joint account
2. Intelligence: being an intelligent, informed person
3. Physical attractiveness

Women reported contributing more than men did in five areas:

1. Liking the other person and showing it
2. Committing oneself to the other person and to the future of the relationship
3. Remembering special occasions: being thoughtful about sentimental things, such as remembering birthdays, anniversaries, and other special occasions
4. Showing affection: being openly affectionate; touching, hugging, kissing
5. Day-to-day maintenance: contributing time and effort to household responsibilities such as grocery shopping, making dinner, cleaning, and car maintenance

The women's contributions are all nurturant, socioemotional activities—providing affection and caretaking services. Men describe themselves as instrumental—bringing resources from the wider world outside in the form of money and information. By these self-report measures, the division of labor in close relationships conforms to Bales's view: women do the socioemotional and men the instrumental work.

Research on interaction patterns in informal groups shows that women are more intimate than men in artificially created groups, too. Eliabeth Aries studied the interaction patterns of all-male, all-female, and mixed-sex groups (1974). The all-male groups developed hierarchies; leaders emerged; certain men quickly became the most active speakers and dominated in subsequent sessions. The all-female groups shied away from hierarchies. The women said that they "felt uncomfortable in the leadership positions . . . and in some sessions they drew out more silent members" (p. 4). In the all-male groups, if a member missed some sessions, there was no provision to make up for lost time when he returned. In the female groups, absentees were encouraged to make up for their absences by talking more in subsequent sessions. In general, the all-female groups had a more equal division of speaking time. Aries concluded the difference in dominance means that men have "more ways than women for expressing competition and leadership . . . while females develop more ways for expressing affection and interpersonal concern" (p. 4). This was true also of the content of the discussions. The women disclosed a lot about themselves, their feelings, families, and friends. The men talked very little about themselves or their relationships. In the mixed-sex groups, the dominant men still talked more than anyone else, but what they talked about changed. They were more willing to express their feelings and talk about personal issues. The presence of women permitted or encouraged men to disclose more of themselves, but in style, they still remained dominant. In the presence of men, the women spoke less, and they spoke less than the men did of achievement, power, and societal institutions.

In these groups, whether same-sex or mixed-sex, men and women played different roles and spoke of different things. Their behavior conformed to sex-role stereotypes and to the Parsonian distinction. In real families, however, Leik (1963) failed to find such a sharp division of labor. He studied both real and artificial families to see whether husbands are task leaders and wives socioemotional leaders. He formed artificial families of unrelated people told to play the roles of mother, father, and daughter. The play-acting fathers emerged as task leaders, and the mothers and daughters played socioemotional parts. In real, intact families consisting of a mother, a father, and a college-age daughter, there was little sex-typed behavior. The only difference was that the daughters exceeded their fathers in showing socioemotional support. Those who played at being a family acted as though they had stereotyped scripts; those who were in real families were less differentiated. Perhaps real families are backstage locations, where men and women do not display such differentiated behaviors. In the privacy of one's own home, some of the sex-role stereotypes may be discarded (cf. Kidder, Bellettirie, & Cohn, 1977). There is still a grain of truth in the stereotypes, however; men and women do behave differently in social groups, and they report giving different resources in their close relationships: women provide care and concern and men provide money and information. Women's behaviors are maximally tailored to the world of love, men's to the world of work.

3.2. The Origins of Division of Labor of Love

Girls and women are the socioemotional specialists because they have been carefully taught. Table 11.1 shows that girls are taught to be nurturant, responsible, and obedient, while boys are encouraged to be achieving and self-reliant.

In not one of the cultures surveyed were boys or men taught to be nurturant, and in none were girls or women taught to be self-reliant (see footnoted figures). Our own society is no exception. Men are punished when they fail to act like leaders and when they are submissive rather than assertive; women are punished for the opposite. Assertive women and submissive men were rated psychologically unhealthy and socially unpopular by their fellow group members (Costrich, Feinstein, Kidder, Marecek, & Pascale, 1975). Women are also taught obedience. Only 3% of the cultures described in Table 11.1 required obedience of men, while 35% required it of women.

Such sex-role division of labor and training in nurturance and obedience serves some purposes, but at a cost. An unknown Japanese woman recorded the impact of such training in her diary (Unknown Japanese woman, 1975). A few days after her marriage she wrote:

> Two or three days later, the father of my husband's former wife visited me and said: "Maniki-She is a good man—a moral, steady man, but as he is also very particular about small matters and inclined to find fault, you had better always be careful to try to please him." Now as I had been carefully watching my husband's ways from the beginning, I knew that he was really a very strict man, and I resolved so to conduct myself in all matters as never to cross his will. (p. 67)

Parsons believed that if women remained socioemotional specialists and did not compete in the economic marketplace, they would not com-

Table 11.1. Socialization Pressures on Males and Females across Cultures

Characteristics	Number of cultures	Percentage cultures showing greater socialization pressures		
		Toward males	Toward females	Toward neither
Nurturant	33	0[a]	82	18
Responsible	84	11	61	28
Obedient	69	3	35	62
Achieving	31	87	3	10
Self-reliant	82	85	0[a]	15

[a]From Barry, Bacon, and Child, 1957.

pete with their husbands. Jean Lipman-Blumen (1976) criticized the Parsonian view as a means to justify the status quo:

> Parsons's formulation of the normative articulation between the family and the occupational world involved a single linkage—between the husband's work role and his familial role. Parsons concluded that comparable occupational roles for husband and wife would introduce a negative, competitive component in the marital relationship. Parsons, in effect, was providing an *ex post facto* explanation of a homosocial pattern in which only the husband had direct access to resources of income and occupation, and the wife and children gained their status through their relationship with the male figure. (p. 19)

Parsons's version of the sex-role division of labor represents the Victorian ideal of motherhood (Bernard, 1974), with men performing the instrumental functions of bringing home the money and women the socioemotional functions, making husband and children feel good again after their brushes with schools and jobs.

4. MEN AND WOMEN IN RELATIONSHIPS: DIFFERENT CONCEPTIONS OF JUSTICE AND OF CLOSE RELATIONSHIPS

Perhaps men and women play different roles in close relationships because they have different conceptions both of what it means to be just and of what it means to be in a close relationship.

4.1. Sex Differences in Preferences for Equity or Equality

The accumulated literature on sex differences in reward allocation says that men and boys divide rewards equitably, according to merit, while women and girls divide them equally, regardless of differences in work, effort, or merit (e.g., Leventhal & Lane, 1970; Leventhal & Anderson, 1970; Benton, 1971). The exceptions to this rule appear when men and women make their decisions privately (Kidder, Bellettirie, & Cohn, 1977; Greenberg, 1978). In public, with face-to-face encounters, they maintain the difference.

Lerner's (1975) model of the forms of justice predicts that people will choose different forms of justice depending on whether they perceive other people as *persons* or as occupants of *positions*. Those who perceive others as persons opt for equality; those who perceive others as interchangeable occupants of positions choose meritorious equity. Do women more often regard others as persons, and men regard them as occupants of positions?

The resources that men and women reported giving in close relationships (Kidder, Fagan, & Wenz, 1978) show that men and women may accumulate different experiences in responding to others as either persons or occupants of positions. We can classify these data using the typology in Figure 11.1 developed by Uriel and Edna Foa (1974, 1975).

The universalistic–particularistic dimension describes the extent to which it matters whether the resource is exchanged with a particular person or whether it can be exchanged with anyone occupying the appropriate position. Money is the most universalistic resource because it usually matters very little whether we do or do not know a salesperson, or whether we like or dislike a bank teller. We exchange money with these people regardless of their personal qualities, so long as they occupy the right positions. Love is the most particularistic resource because it matters a great deal with whom we exchange it. We do not give or receive love indiscriminately just because a person occupies the "right" position. We respond instead to the person, and to his or her particular qualities.

The five contributions made by women in Wenz's interviews belong to the categories of love and services (Kidder, Fagan, & Wenz, 1978). They are particularistic contributions. The three contributions made by men belong in different categories. Finances (money in the Foas' [1974] scheme) are a universalistic resource, to be exchanged with virtually anyone. Being well informed and intelligent (information in the Foas' scheme) is also universalistic; it is something one does for almost anyone. Being physically attractive is more difficult to code; it is, however, a resource that could be offered to and appreciated by anyone. It need not be tied to a particular person. This typology, like Parsons's, reveals a sex difference: women give particularistic resources, directed toward particular *persons*, and men give universalistic resources, directed toward

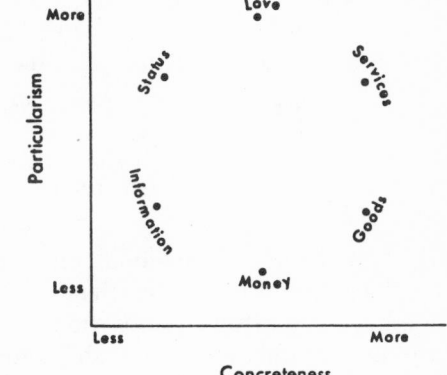

Figure 11.1. Resources classified along two dimensions: particularism and concreteness. From Foa and Foa, 1975, p. 5. Reprinted by permission of American Association for the Advancement of Sciences, © 1971.

an occupant of the appropriate *position*. This difference conforms with Lerner's prediction: women should prefer equality and men equity, for they give different resources, which permit them to perceive people differently.

These preferences for different forms of justice appeared in Aries's (1974) observations of interaction patterns in all-male and all-female groups. The women developed egalitarian structures, they eschewed leadership and distributed talking time equally. The men developed hierarchical structures; some became leaders and they did not seek equality. We may therefore find different conceptions of close relationships among men and women, differences that mirror their interactions in all-male and all-female groups.

4.2. Sex Differences in Close Relationships: An Examination of Homosocial Relationships

Homosocial or same-sex friendships provide a pure case for understanding men's and women's behaviors in close relationships. They show us how each sex behaves on its own, and not simply in reaction to the opposite sex.

4.2.1. Male Friendships

Men report more same-sex friendships than women do, but they are not intimate friendships (Lewis, Casto, Aquilino, & McGuffin, 1978). Men have more difficulty disclosing personal fears and feelings, as Aries found in the all-male groups, and consequently have more difficulties forming close friendships with other men (Lewis, Casto, Aquilino, & McGuffin, 1978). Men who have formed consciousness-raising groups to examine the sex role stereotypes which encourage competition and discourage self-disclosure among men lament the barriers to emotional intimacy between themselves (e.g. Pleck & Sawyer, 1974). Even men who are not self-consciously involved in that social movement agree:

> "Guys don't talk about things like that with each other. Me, I'm used to talking to women. I talked to my mother when I was a kid, not my father. When I got older, I talked to girls, not guys. And now I have a wife to talk to." (from a 31-year-old heavy equipment mechanic, interviewed by L. Rubin, 1976, p. 21)

Lionel Tiger's (1969) book on men in groups takes another point of view. He writes that male friendships and "male bonding" are strong, and he points to men's clubs, poolrooms, and bars as evidence that men can develop intimate relationships with one another. He does not demonstrate, however, that members of such groups engage in self-disclosure or provide emotional support for one another. This has led other writers to say that the "bonding" which Tiger identifies serves instrumental rather than emotional needs. The old joke that business deals made in the men's room exclude women from important decisions may not be a joke.

Lipman-Blumen (1976) wrote that male bonding, as it appears in men's clubs, bowling leagues, union halls, or corner groups, satisfies not intimacy needs but "intellectual, physical, political, economic, occupational, social, power, and status needs" (p. 16). Men's groups in the past have circulated their economic, political, occupational, and legal resources among the members and may have accomplished more for their careers than for their intimate relationships. The uneven distribution of such resources has also made men more interesting to both men and women (Lipman-Blumen, 1976). With a redistribution of resources, women are developing a new homosocial world that serves intimacy needs and occupational goals as well (Starker, 1979). Men are also beginning to experiment with new forms of close relationships among themselves. Research on men in groups shows that they do not become intimate without women present (Aries, 1974; Lewis et al., 1978). They require self-conscious commitment to socioemotional goals and are teaching one another to nurture and act in ways often reserved for women. The Berkeley Men's Center Manifesto describes one group's goals:

> We want to relate to both women and men in more human ways—with warmth, sensitivity, emotion, and honesty. . . . We are oppressed by this dependence on women for support, nurturing, love and warm feelings. We want to love, nurture and support ourselves and other men, as well as women. . . . We want men to share their lives and experiences with each other in order to understand who we are, how we got this way and what we must do to be free. (Pleck & Sawyer, 1974, p. 173)

4.2.2. Female Friendships

Women's problems are the mirror image of men's. They have long known how to be intimate; what they have not been able to do is serve instrumental needs for one another outside the home. They have also been caught between wanting to be instrumental and to achieve something for themselves and not wanting to compete with another person who could be a potential friend (Piliavin, 1976).

American women in the eighteenth and nineteenth centuries were close companions for one another. They provided emotional support and satisfied intimacy needs at a time when marriages were institutional rather than companionate (Bernard, 1973, 1976). They lived in a world "bounded by home, church, and the institution of visiting—that endless trooping of women to each other's homes for social purposes" (Smith-Rosenberg, 1975, p. 10). Women frequently spent more time with each other than with their husbands, both for company and for help with domestic chores, with sickness, and with child rearing (Bernard, 1976). When husbands traveled to distant places, women moved in together; and when a sister or friend moved to a distant town, letters served the same purpose that their visits otherwise would have. This excerpt from a letter shows what women looked for in their female friends:

> Suppose I come down . . . [and] spend Sunday with you quietly . . . that means talking all the time until you are relieved of all your latest troubles and I of mine. (Smith-Rosenberg, 1975, p. 14)

Professional American women today have not lost the need for such friendships. Married and single professional women said in interviews that even when they had an intimate relationship with a husband or a lover, they felt a need for a second and distinctly different type of close relationship with at least one woman in the immediate geographical area (Starker, 1979).

The female world of homosocial relationships became devalued in the early twentieth century and was replaced by an emphasis on heterosocial relationships. Both men and women deprecated female friendships (Bernard, 1976). When the home-based economy gave way to the industrial workplace, women did not possess the resources that were important in the public world of work: money, legal power, occupational skills, and status. "This uneven array of resources systematically made men more interesting to women, women less interesting and useful to other women" (Lipman-Blumen, 1976).

Experimental data show that women consider a task more interesting just because it is labeled "masculine" rather than "feminine." In a laboratory study, college men and women worked on two simple tasks: a marble maze and another game that required eye–hand coordination. The tasks were alternately labeled "masculine" and "feminine," with the rationale that one required fine motor coordination typical of women's skills and the other the large-muscle coordination typical of men's skills. In truth, the tasks were not sex-differentiated, and the sex labels were randomly assigned to the tasks. Nonetheless, the women said they enjoyed the "masculine" task more, regardless of which game had that label. What mattered was the label, not the task (Kipnis & Kidder, 1977). As women gain access to important economic and legal resources, they and their work may be revalued, and tasks labeled "feminine" may seem more interesting.

4.2.3. Homosexual Relationships

Homosexual relationships are the most intimate same-sex relationships. They also reveal sex roles and sex differences in forms of justice in close relationships. Women in lesbian couples show a preference for equality. In a survey of lesbian couples, Peplau and her colleagues (Peplau, Cochran, Rook, & Padesky, 1978) asked about the ideal and real balance of power within the relationship. Of this sample, 97% said that both partners *should* have "exactly equal say"; 64% reported that the balance in their current relationship *was* exactly equal. There was almost unanimous agreement that equality should prevail, but there was less than complete success in bringing it about.

On other questions about the relationship, there was not such unanimity. The women differed in their emphasis on dyadic attachment versus personal autonomy within their relationships. Personal autonomy was more important to the feminists and the radical lesbians, and dyadic attachment was more important to the women who held traditional sex-role attitudes. Both the radical and the traditional women said that they valued dyadic attachments, but the traditional women valued them more.

Another survey of gay and straight couples examined the partners' sex-role orientation and sex-role behaviors (Cardell, Finn, & Marecek, 1979). Members of lesbian couples were least sex-typed, and members of gay male couples were most sex-typed on the Bem Sex Role Inventory; heterosexual couples were between the two. Members of lesbian couples also had more similar sex-role orientations than members of heterosexual couples, and members of gay male couples had less similar sex-role orientations than members of heterosexual couples. If sex-role similarity and non-sex-typing are measures of equality, the lesbian couples in this study demonstrated a preference for equality. Members of less-role-differentiated couples were also most satisfied with their relationship, providing further grounds for preferring equal relationships. When the authors compared the satisfaction measures of members of each couple, however, they found that wherever there was sex-role-differentiated behavior, the person playing the more masculine role was also the more satisfied member. Equality produced happier couples, but masculine role behavior produced the happier individual.

Perhaps there is an inevitable conflict between maximizing individual happiness and maximizing couple satisfaction. Individual happiness may come from personal autonomy; couple satisfaction may require dyadic attachments. Individual happiness may thrive on meritorious equity; couple satisfaction may require equality. Individual happiness may come from performing the masculine role and being an instrumental leader; couple satisfaction may result from feminine activity and socio-emotional work. When men take on aspects of the feminine role, they may relinquish some individual happiness as they promote couple or group satisfaction; and women who take on aspects of masculine behavior may gain individual happiness at the expense of group or couple satisfaction.

If men and women reverse or share roles, they engage in a series of trade-offs. As men relinquish their control over economic resources and women relinquish their monopoly in the world of love, they each acquire skills and rewards in the opposite sphere. In addition, couples may choose to maximize either individual or couple satisfaction by emphasizing autonomy or dyadic attachment.

5.   A NEW ORDER OF INTIMATE RELATIONSHIPS: A PRESCRIPTION

Our prescription for social justice in close relationships calls for equal participation by men and women in both the world of work and the world of love. Legislation like the Equal Rights Amendment can go a long way toward providing for equal participation in the labor market. We cannot legislate equal participation in the private world of nurturance, although paternity leaves and flexible working hours permit men to share work in the home without penalty on the job. Neither legislation nor liberalized work time can guarantee an egalitarian relationship for any particular couple, however. What options remain for those persons who cannot negotiate equal participation in the world of work and the world of love?

The remaining option, which we regard as a requirement for social justice in any relationship, is the option to leave. The ability to leave an unjust relationship is essential to promote justice in close relationships. Examples of what happens when people are entrapped in relationships demonstrate our point. The majority of women in traditional China were imprisoned in nonegalitarian relationships. Joyce Jennings Walstedt (1977) described their problem and their solution:

> Except for those few who worked in the field, women were kept confined to their immediate living quarters, denied an education, and not permitted to own property or to work outside the home . . . men demanded premarital chastity for all women and punished female adultery by death or divorce. On the other hand, the wife had to be sexually available to her husband at any time but could not actively participate or show any sign of enjoyment. (p. 178)

Women were not permitted to initiate a divorce. The only way out was suicide, "an option used by thousands, especially during the newlywed years" (Walstedt, 1977, p. 178). Suicide is the desperate way out of an unsatisfactory relationship. Divorce and separation are, by contrast, humane. We are not advocating divorce or separation as desirable solutions; they are the solution of last resort if all other attempts fail.

We began this chapter by saying that egalitarian relationships between men and women develop when both persons possess a range of equally valued resources. Neither person can then exploit the other, wittingly or unwittingly. In addition to participating equally in the world of work, however, they must also learn to participate equally in the world of love. Men must learn to nurture as women give up their monopoly. Several writers with different philosophies have said that men must share the work of child rearing if we are to change the relationship between men and women. Dorothy Dinnerstein (1977) wrote from a psychoanalytic perspective:

> it is unreasonable to expect men to act as our brothers in the task of changing the sexual arrangements . . . until we stop being their mothers; until, that is, we stop carrying the main responsibility—and taking the main blame—for the early introduction to the human condition. (p. 90)

While we do not share Dinnerstein's psychoanalytic perspective, we agree with her conclusion: the arrangement between the sexes will not change until men become fathers in the best sense and share the work of raising children.

From a very different perspective, Elise Boulding (1976) reached the same conclusion:

> Marxist analysis failed to put its finger on one aspect of the oppression of women: the confining of breeder–feeder roles to women. It thought that by turning the state into the breeder–feeder all would be well. But human liberation depends on sharing breeder–feeder roles between women and men . . . There is no way out . . . but to confront parenthood, and no way for women but to confront sharing their centuries-old monopoly on the breeder–feeder role. (p. 15)

Boulding added, "Marx left love as woman's work by default. It simply could not be taken over by the state" (p. 115).

If love could be taken over by the state, we would all lose. If, instead, men and women both become proficient at working and loving, we will all stand to win.

ACKNOWLEDGMENTS

The authors thank their friends and colleagues for reading and commenting on several drafts of this chapter: Marilynn Brewer, Phil Brickman, Faye Crosby, Julia Erickson, Arnie Kahn, Bob Kidder, Dave Kipnis, George Levinger, and Jeanne Marecek.

REFERENCES

Aries, E. *Interaction patterns and themes of male, female and mixed groups.* Paper presented at the American Psychological Association Convention, New Orleans, August 1974.

Aronoff, J., & Crano, W. D. A re-examination of the cross-cultural principles of task segregation and sex role differentiation in the family. *American Sociological Review,* 1975, *40*, 12–20.

Bales, R. F. The equilibrium problem in small groups. In T. Parsons, R. F. Bales, & E. A. Shils (Eds.), *Working papers in the theory of action.* Glencoe, Ill.: Free Press, 1953, pp. 111–161.

Barry, H., III, Bacon, M. K., & Child, I. I. A cross-cultural survey of some sex differences in socialization. *Journal of Abnormal and Social Psychology,* 1957, *55,*327–332.

Benton, A. A. Productivity, distributive justice and bargaining among children. *Journal of Personality and Social Psychology,* 1971, *18,* 68–78.

Bernard, J. *The future of marriage.* New York: Bantam, 1973.

Bernard, J. *The future of motherhood.* New York: Dial Press, 1974.

Bernard, J. Homosociality and female depression. *Journal of Social Issues,* 1976, *32*(4), 213–238.

Boulding, E. Familial constraints on women's work roles. In M. Blaxell & B. Reagan (Eds.), *Women and the workplace.* Chicago: University of Chicago Press, 1976.

Brain, J. L. Down to gentility: Women in Tanzania. *Sex Roles,* 1978, *4*(5), 695–716.

Brown, J. K. A note on the division of labor. *American Anthropologist,* 1970, *72,* 1073–1078.

Cardell, M., Finn, S., & Marecek, J. Sex-role identity, sex-role behavior, and satisfaction in heterosexual, lesbian and gay male couples. Unpublished manuscript, Swarthmore College, Swarthmore, Pa., 1979.

Centers, R., Raven, B. H., & Rodrigues, A. Conjugal power structure: A re-examination. *American Sociological Review,* 1971, *36*(April), 264–278.

Chodorow, N. Family structure and feminine personality. In M. Z. Rosaldo & L. Lamphere (Eds.), *Woman, culture and society.* Stanford, Calif.: Stanford University Press, 1974.

Chong, S. P. The monetary value of a housewife: An economic analysis for use in litigation. *American Journal of Economics and Sociology,* 1969, *28,* 271–284.

Costrich, N., Feinstein, J., Kidder, L., Marecek, J., & Pascale, L. When stereotypes hurt: Three studies of penalties for sex-role reversals. *Journal of Experimental Social Psychology,* 1975, *11,* 520–530.

Deutsch, M. Equity, equality and need: What determines which value will be used as the basis of distributive justice? *Journal of Social Issues,* 1975, *31*(3), 137–149.

Dinnerstein, D. *The mermaid and the minotaur.* New York: Harper Colophon, 1977.

Firestone, S. *The dialectic of sex: The case for feminist revolution.* New York: Bantam, 1971.

Foa, U. G., & Foa, E. B. *Societal structures of the mind.* Springfield, Ill.: Charles C Thomas, 1974.

Foa, U. G., & Foa, E. B. *Resource theory of social exchange.* Morristown, N.J.: General Learning Press, 1975.

Foa, U. G., Megonigal, S., & Greipp, J. R. Some evidence against the possibility of utopian societies. *Journal of Personality of Social Psychology,* 1976, *34,* 1043–1048.

Folger, R. Distributive and procedural justice: Combined impact of "voice" and improvement on experienced inequity. *Journal of Personality and Social Psychology*, 1977, *35*, 108–119.

Friedan, B. *The feminine mystique*. New York: Norton, 1963.

Ginzberg, E. *Life styles of educated women*. New York: Columbia University Press, 1966.

Goody, J. R., & Tambiah, S. J. *Bridewealth and dowry*. Cambridge, England: Cambridge University Press, 1973.

Greenberg, J. Sex differences in the administration of punishment to deter, rehabilitate and restore justice. Unpublished manuscript, Tulane University, 1978.

Hailey, L. *An African survey*. London: Oxford University Press, 1956.

Heer, D. M. Dominance and the working wife. *Social Forces*, 1958, *35*(May), 341–347.

The Hindu, Madras, India. Sunday, July 2, 1978.

Kelley, H. H. *Personal relationships: Their structures and processes*. Hillsdale, N.J.: Erlbaum, 1979.

Kidder, L. H., Bellettirie, G., & Cohn, E. S. Secret ambitions and public performances. *Journal of Experimental Social Psychology*. 1977, *13*, 70–80.

Kidder, L. H., Fagan, M. A., & Wenz, M. *Sex differences in equity: A theoretical framework— Do Men Respond to Positions and Women to Persons?* Paper presented at the American Psychological Association Meetings, Toronto, August 1978.

Kipnis, D. *The powerholders*. Chicago: University of Chicago Press, 1976.

Kipnis, D., Cohn, E. S., & Catalano, R. *Power and affection*, unpublished manuscript, Temple University, 1980.

Kipnis, D. M., & Kidder, L. *Practice, performance and sex: Sex role appropriateness, success, and failure as determinants of men's and women's task learning capability* (Tech. Rep. No. 1). Philadelphia: University Science Center, May 1977.

Kligler, D. H. *The effects of the employment of married women on husband and wife roles*. Unpublished Ph.D. dissertation, Yale University, 1954.

Leik, R. K. Instrumentality and emotionality in family interaction. *Sociometry*, 1963, *26*, 422–435.

Lerner, M. J. The justice motive in social behavior: Introduction. *Journal of Social Issues*, 1975, *31*, 1–20.

Lerner, M. J. The justice motive: Some hypotheses as to its origins and forms. *Journal of Personality*, March 1977, *45*(1), 1–52.

Lerner, M. J., Miller, D. T., & Holmes, J. G. Deserving and the emergence of forms of justice. In *Advances in Experimental Social Psychology*, 1976, *9*, 134–161.

Leventhal, G. S., & Anderson, D. Self-interest and the maintenance of equity. *Journal of Personality and Social Psychology*, 1970, *15*, 57–62.

Leventhal, G. S., & Lane, D. Sex, age and equity behavior. *Journal of Personality and Social Psychology*, 1970, *15*, 312–316.

Leventhal, G. S., & Michaels, J. W. Extending the equity model: Perception of inputs and allocation of reward as a function of duration and quantity of performance. *Journal of Personality and Social Psychology*, 1969, *12*, 303–309.

Levinger, G. Marital cohesiveness at the brink: The fate of applications for divorce. In G. Levinger & O. Moles (Eds.), *Divorce and separation*. New York: Basic Books, 1979. (a)

Levinger, G. A social exchange view on the dissolution of pair relationships. In R. L. Burgess & T. L. Huston (Eds.), *Social exchange in developing relationships*. New York: Academic Press, 1979. (b)

Levinger, G., & Moles, O. *Divorce and separation: Context, causes and consequences*. New York: Basic Books, 1979.

Lewis, R., Casto, R., Aquilino, W., & McGuffin, N. Developmental transitions in male sexuality. *The Counseling Psychologist*, 1978, *7*(4), 15–19.

Lipman-Blumen, J. Toward a homosocial theory of sex roles: An explanation of the sex-segregation of social institutions. In M. Blaxall & B. Reagan (Eds.), *Women and the workplace: The implications of occupational segregation.* Chicago: University of Chicago Press, 1976.

Murdock, G. P. Ethnographic atlas: A summary. *Ethnology,* 1967, *6,* 109–236.

Myrdal, A., & Klein, V. *Women's two roles: Home and work.* London: Routledge, 1968.

Nerlove, S. B. Women's workload and infant feeding practices: A relationship with demographic implications. *School of Urban and Public Affairs Reprint Number 108.* Carnegie-Mellon University, Pittsburgh, Pa., 1974, pp. 207–214.

Oakley, A. *Woman's work: The housewife, past and present.* New York: Vintage, 1974.

O'Brien, D. Female husbands in southern Bantu societies. In A. Schlegal (Ed.), *Sexual stratification: A cross-cultural view.* New York: Columbia University Press, 1977.

Ortner, S. Is female to male as nature is to culture? In M. Z. Rosaldo & L. Lamphere (Eds.), *Woman, culture and society.* Stanford, Calif.: Stanford University Press, 1974.

Peplau, L. A., Cochran, S., Rook, K., & Padesky, C. Loving women: Attachment and autonomy in lesbian relationships. *Journal of Social Issues,* 1978, *34*(3), 7–27.

Peterson, R. D. Problems in estimating the value of household services. *American Journal of Economics and Sociology,* 1978, *37*(2), 145–148.

Philadelphia Magazine, Philadelphia: Municipal Publications, Inc., July, 1979.

Piliavin, J. A. On feminine self-presentation in groups. In J. I. Roberts (Ed.), *Beyond intellectual sexism: A new woman, a new reality.* New York: David McKay, 1976, pp. 138–159.

Pleck, J. H., & Sawyer, J. *Men and masculinity.* Englewood, N.J.: Prentice-Hall, 1974.

Rubin, L. *Worlds of pain: Life in the working-class family.* New York: Basic Books, 1976.

Rubin, Z. *Liking and loving: An invitation to social psychology.* New York: Holt, Rinehart and Winston, 1973.

Sampson, E. What would a *"feminine"* science be like? Paper presented at the meeting of the American Psychological Association, New York City, September 1979.

Smith-Rosenberg, C. The female world of love and ritual: Relations between women in 19th-century America. *Signs,* 1975, *1,* 1–30.

Starker, L. R. *Meeting intimacy needs: A study of professional women.* Paper presented at the Eastern Psychological Association, Philadelphia, April 1979.

Thibaut, J., & Walker, L. *Procedural justice: A psychological analysis.* Hillsdale, N.J.: Erlbaum, 1975.

Tiger, L. *Men in groups.* New York: Vintage, 1969.

Tönnies, F. *Community and society* (C. P. Loomis, trans.). New York: Harper & Row, 1963.

Traupmann-Utne-Walster (1977) scales. Published in E. Walster, G. W. Walster, & E. Bersheid, *Equity: Theory and research.* Boston: Allyn & Bacon, 1978, pp. 236–242.

U.S. Bureau of the Census, *Statistical Abstract of the United States,* 99th Edition, Washington, 1978.

Unknown Japanese Woman, Diary. In M. J. Moffat & C. Painter (Eds.), *Revelations: Diaries of women.* New York: Vintage, 1975.

Van Allen, J. "Sitting on a man": Colonialism and the lost political institutions of Igbo women. *Canadian Journal of African Studies,* 1972, *6,* 165–181.

Vanek, J. Keeping busy: Time spent in housework, United States, 1920–1970. Unpublished Ph.D. dissertation, University of Michigan, 1973.

Walstedt, J. J. The women of the People's Republic of China. In V. L. Lussier & J. J. Walstedt, *Women's lives: Perspectives on progress and change.* Newark: University of Delaware Press, 1977, pp. 177–189.

Walster, E., Walster, G. W., & Berscheid, E. *Equity: Theory and research*. Boston: Allyn & Bacon, 1978.

Wenz, M. Unpublished honors thesis, Temple University, 1978.

Wolfe, D. M. Power and authority in the family. In D. Cartwright (Ed.), *Studies in social power*. Ann Arbor, Mich.: Institute for Social Research, 1959, pp. 99–117.

12

The Exchange Process in Close Relationships

Microbehavior and Macromotives

JOHN G. HOLMES

1. INTRODUCTION

Following years of apparent neglect, the study of close relationships is emerging as an important focus for social psychologists. Until a decade ago, research on the topic was typically confined to identifying predictors of marital harmony or discord (cf. Barry, 1970). The demographic characteristics of couples, such as education, status, and family background, were examined in order to portray relationships that were "at risk." The large majority of studies used either self-ratings of satisfaction or divorce rates as reflections of troubled relationships. With the noteworthy exception of Levinger's contribution (see Levinger & Snoek, 1972), the research was marked by a lack of concern with considerations of both process and theory. Recent contributions have demonstrated an interest in the theoretical development of models of intimate exchange and "pair cohesiveness" (e.g., Kelley, 1979; Moles & Levinger, 1976), though evidence of an emphasis on the process of conflict and estrangement is meagre (e.g., Hill, Rubin, & Peplau, 1976; Kelley, 1979; Passer, Kelley, & Michela, 1978).

JOHN G. HOLMES • Department of Psychology, University of Waterloo, Waterloo, Ontario, Canada N2L 3G1.

Research on the formation and growth of intimate relationships also represents a recent trend. The focus of much of this work has been the selection of friends or intimates from the field of eligibles (cf. Rubin, 1973; Walster, Walster, & Berscheid, 1978). This emphasis is understandable, since it blends with the historical interest by social psychologists in first impressions, attraction, and exchange theory. But for the most part, it has also meant that research energies have been directed toward factors important in the formative stages, rather than toward processes that determine growth and development. This trend has also promoted the tendency to analyze events within a single time frame and to emphasize attributes (e.g., attractiveness, values) rather than dynamics (e.g., communication, decision making).

This chapter focuses on the dynamics of the exchange process in established close relationships, with particular emphasis on the relation between everyday interpersonal behaviors and the development of higher-order attitudes (macromotives) over an extended time frame. The term *macromotives* is used to describe the set of attributions that characterizes a person's feelings and beliefs about his partner, such as love, trust, and cooperation. It is intended to distinguish these longer-term interpersonal attitudes from short-term reactions to particular patterns of behavior.

2. THE DEVELOPMENT OF MACROMOTIVES

Overall, the evidence from various sources leads one to conclude that (conjugal) love and feelings of trust are critical determinants of the stability of close relationships (cf. Driscoll, Davis, & Lipetz, 1972; Hill, Rubin, & Peplau, 1976). On the other hand, the *process* that controls their development has received only scant attention. However, Kelley (1979) has recently presented a theoretical perspective that contributes in important ways to our understanding of the inference process involved in such interpersonal attitudes. He suggested that couples often evaluate an exchange in terms of its combined consequences; each person in a close relationship tends to be concerned with ensuring that *both* his own and the partner's outcomes will be as high as possible. Direct satisfaction may be gained by the knowledge that the other person's preferences are being served: "I want to do what will make you happy." When a person is responsive to the outcomes of the partner, his analysis of the rewards and costs attached to a behavior are modified to reflect this fact. If the choice of a particular behavior departs from what would best suit one's personal, self-serving interests, then he will be providing evidence of his feelings and attitudes toward his partner.

Kelley wished to stress the critical importance of behaviors that demonstrate sensitivity and responsiveness to a partner's needs and outcomes and that are at variance with egocentric models of the exchange process (see also, Lerner, 1977; Pruitt, 1972). The particular pattern of responsiveness that a partner displays provides the basis for inferring the person's "interpersonal dispositions." For example, assume that a person consistently places himself in a vulnerable position by disclosing personal information that could be exploited by a partner, and that this information helps the couple in their joint decision-making. This pattern of behavior could be taken as evidence of the person's attitudes of co-operation and trust in this particular relationship. Kelley believes that specific, everyday behavior is scanned and encoded in terms of its significance to such latent attributes. Thus, individuals in close relationships are held to have a deep concern with the "invariances" on which the relationship is based, the macromotives of their partners.

This conclusion was supported by the evidence that emerged from a series of survey studies on close relationships conducted by Kelley and his colleagues. In several studies, people were asked to describe specific behaviors that they wanted their partners to change or continue. Contrary to instructions, there was a strong tendency for answers to be expressed in terms of stable dispositions or attitudes, such as affection, understanding, and support. A similar trend emerged from investigations of attributional conflict, where disagreements on the interpretation of instances of negative behavior were studied (Orvis & Kelley, 1976; Passer, Kelley, & Michela, 1978). A multidimensional scaling of the causal explanations provided by each spouse revealed two important dimensions. First, the behavior was interpreted by both persons as reflecting a positive or a negative attitude toward the partner. Second, the causes communicated by the *offending party* formed an intentional–unintentional dimension. The offender appeared to use intention to justify the behavior by suggesting that the action was done for "good" reasons, with the partner's interests in mind. An explanation that stressed unintentional causes served to excuse the behavior by interpreting it in terms of external pressures or temporary internal states. On the other hand, the causes given by the *offended spouse* were best described on a dimension that reflected attributional tendencies to interpret the behavior in terms of the offender's enduring traits. Thus, one could infer that the evaluation of the negative behavior would be essentially "negotiated" if these divergent attributions were communicated: the issue in conflict is the relevance of the act to the enduring *interpersonal dispositions* of the actor.

Various aspects of Kelley's theory of interdependence are applied to the analysis of the exchange process presented in later sections of the chapter. The important point to be made here is that many of the events that comprise the day-to-day exchange process can be understood at the level of their direct, tangible rewards and costs, or they can be analyzed in terms of their impact on a person's attributions of the partner's interpersonal dispositions. An explicit microlevel accounting of a person's more concrete contributions and resources may fail to capture the essential nature of the quality of a close relationship.

3. THE LEVELS-OF-ANALYSIS PROBLEM

When one is guided by a desire to understand the dynamics of everyday exchange, there are obvious temptations to resort to a "microlevel" analysis of events. If the goal is to predict particular patterns of behavior in short-term time frames, then historical, molar variables, such as caring and commitment, seem only to cloud the issues. The measurement process is certainly aided by working with operational definitions of variables that are more objective, tangible, and molecular in nature. But the disadvantages of this approach are quickly drummed home if the "microbehavior" that is the focus of attention is controlled in important ways by "macromotives" that are ignored in the theoretical or experimental analysis of the events under study.

An experiment conducted by Gottman and his associates (Gottman, Notarius, Markman, Bank, Yoppi, & Rubin, 1976) illustrates some of the problems with research in this area. This research reflects the traditional emphasis in clinical psychology on communication deficits as a cause of marital distress (cf. Lederer & Jackson, 1968). In recent years, investigators have studied the behavioral interchanges of couples classified as distressed or nondistressed. Coding systems have been developed for classifying behaviors in dyadic interactions in order to estimate the degree of fit between social learning theories and the patterns of exchange characteristic of married couples.

Gottman et al. (1976) reviewed this literature and concluded that there is consistent evidence that distressed couples exhibit more negative behaviors and fewer positive behaviors in their interaction patterns than normal couples. These findings have been used as support for the notion that reciprocity in exchange characterizes the healthy marriage—this, in fact, is a critical assumption in the clinical literature on marriage counseling. Thus, results indicating a greater display of positive behavior in stable marriages are interpreted in terms of "social reinforcements," in a manner consistent with social learning theory (e.g., Birchler, Weiss, & Vincent, 1975).

Gottman et al. made the important point that high base rates of positive behaviors do not imply reciprocity. To demonstrate the occurrence of reciprocity, it is necessary to demonstrate a sequence of conditional states (in a Markov chain), so that the respondent modifies his behavior as a function of the preceding act by a partner. This contingent pattern of exchange had not been tested in previous studies. Gottman *et al.* examined the behavioral (communication) styles of distressed and nondistressed couples as they dealt with either high- or low-conflict tasks. Spouses in nondistressed marriages interpreted their partners' behavior as more positive and less negative than did spouses in distressed relationships, particularly on the high-conflict tasks. But there was no evidence of differential reciprocity between the two groups. The authors speculated that "perhaps it is precisely this *lack* of reciprocity in a context of high positive exchange that characterizes stable positive interaction in nondistressed couples" (p. 21). This speculation is in marked contrast to the point of view held by most researchers identified with the behavior modification approach to therapeutic intervention for distressed couples: "Concerning the resolution of problems, the focus should be on the exchange of positives, and specific training in reciprocity, that is, *an explicit process of 'quid pro quo'*" (italics ours; Birchler *et al.*, 1975, p. 359).

These studies and the emerging trend they represent are exemplary in several ways. Given the previous criticism in this chapter that process considerations and social interactions in close relationships have been virtually ignored (see also Birchler *et al.*, 1975), then these programs of research have made important contributions to the field. Also, a number of innovative procedures have been developed to describe behavioral sequences in the communication patterns of couples. "Improvisation" techniques (Raush, Barry, Hertel, & Swain, 1974) and "talk tables" (Gottman *et al.*, 1976) have made the study of interaction in close relationships more amenable to laboratory investigations. Birchler *et al.*, (1975) have made use of daily diaries for collecting information about communication patterns in the home environment. In spite of the strength of many studies on these dimensions, one can argue that there are critical weaknesses in this microbehavioral approach to the issues. These weaknesses will be illustrated by examining the assertion that distressed couples should be trained to engage in an "explicit process of 'quid pro quo,' " a belief that is widely held in the clinical literature.

3.1. Time Frame Considerations

There are important theoretical reasons to argue that contented relationships will be marked by fairness in social exchange, and that imbalances in the rewards and costs experienced will eventually result in estrangement (Kelley & Thibaut, 1978; Walster, Walster and Berscheid, 1978). Recently, Walster, Walster, and Traupmann (1978) have provided evidence in support of this general principle. Dating couples were asked to evaluate their contributions and outcomes in the relationship; they were then classified as underbenefited, equitably treated, or overbenefited. Men and women in relatively equitable relationships described themselves as more content and more happy; those who were underbenefited felt most resentful and angry about their position, while those who were overbenefited felt most guilty about theirs. The equitable couples also reported more frequent sexual relations, a finding consistent with Howard and Dawes's (1976) correlation between self-ratings of marital happiness and sexual intercourse. Finally, fair exchange appears to be associated with perceived stability in the relationships. Walster, Walster, and Traupmann's measures of equitable or fair exchange involved global perceptions of contributions that retrospectively summarized prior experiences over an extended period of time, rather than specific instances of behavior.

If we grant that fair exchange is a pivotal concept in close relationships, we must still address several critical questions that must be considered before the principle can be used to characterize the successful relationship. First, what is the optimal period of time for balancing or "equalizing" the outcomes of each partner? Second, how explicit should the evaluation and enactment of fair exchange be made? Does the conduct of a successful relationship require that a person adopt a conscious strategy of signaling or making salient his attempts to bring about a fair exchange? Finally, what resource dimensions are included in these considerations—does a quid pro quo perspective with exchange "in kind" result in increased satisfaction, or is exchange across different resource classes a viable approach? At the cost of oversimplification, the prevailing view appears to argue for the efficacy of an interpersonal strategy or style that requires establishing equity in an explicit way over very short periods of time (days or even interactions), within a particular resource class (i.e., exchange "in kind").

4. THE ADDRESS FRAME FOR SOCIAL EVALUATION

This perspective on exchange in close relationships rests on a number of assumptions. First, it is often presumed that the principles of social exchange that pertain to less intimate or casual relationships also promote growth and satisfaction in close relationships. If reciprocity, exchange in kind, and explicit rules and agendas result in evidence of effective, fair exchange in our experimental studies of transactions, do not similar principles apply to the realm of our more intimate social lives? Various authors have questioned this assumption, as we shall see later. Second, the time frame of exchange does not have a central place in this perspective. This omission has several implications that have not been systematically discussed in the literature. The application of certain general principles of exchange such as equality may in fact be deemed appropriate in certain contexts by participants in a close relationship, and it may lead to perceptions of fairness and satisfaction in the short run. This does not mean that the longer-term consequences of such an exchange will be similarly beneficial and will contribute to the health of the relationship. It will be argued that the time perspective of the involved parties and the particular nuances and messages associated with an exchange determine its ultimate effectiveness. Also, the psychological processes or motives that have been postulated to govern exchange in these microlevel episodes are typically correspondent in form. That is, there has been a strong tendency to assume that the parties to an exchange have short-term, transitory goals, or micromotives. This belief contrasts with the possibility that microlevel behavioral sequences are organized around coherent, long-term goals and motives. If such macromotives have a critical role in the development of healthy close relationships, short-term principles of exchange may be sacrificed in the service of more extended goals.

The major theme of the present chapter is that this perspective for evaluating a close relationship—this "address frame" (see Sampson, this volume)—is a prescription for conflict and estrangement. It will be argued that evidence of such interpersonal strategies in a relationship could be used effectively to construct a set of markers or symptoms to *identify* distressed couples or to predict the onset of impending strain and conflict. A tentative list of such indicators is described below. Each symptom can be regarded as reflecting the strategies or rules adopted by a person in planning and executing exchanges and/or the cognitive framework used to evaluate the fairness of an exchange:

1. Behavioral exchanges are monitored over short-term time spans and are examined for equivalence. Thus, relatively microlevel interchanges often emerge as the unit of evaluation: specific behaviors are more likely to be viewed as having "meaning" and symbolic value.

2. Interactions typically involve a large proportion of cases in which the idea of an exchange is implicitly or explicitly at issue. Explicitly, the conditional nature of a positive contribution is made salient to the other person. Implicitly, members of a relationship assume that benefits are given with the expectation that similar benefits will be returned.

3. Equivalence or fair exchange is best established by returning a benefit from the same or a similar resource class: exchange in kind. Benefits that are tangible or objective or that have established normative currency have greater value. Thus, features of social exchange that promote clear social comparison of outcomes in the dyad have more influence on perceptions of fairness.

4. The relationship tends to be rule-governed, with a relatively explicit social contract. The rules tend to stress procedural aspects of justice.

The term *contraction* could be applied to this set of conditions because (a) it captures the reductionistic, circumscribed view of the relationship, and (b) it emphasizes the sharply defined constraints placed on the variety of interaction rules that are viewed as satisfying the role demands associated with fair exchange. Each of these conditions is discussed here in detail to explicate the hypothesis that they tend to produce an increase in discord and discontent in a close relationship.

The present analysis depends rather heavily on theories and findings from the literature on social conflict, on content areas, such as negotiations, that are at a superficial level quite different in manifest features from close relationships. This approach is necessitated by the paucity of research in the latter area. At the same time, a strong argument can be made that the major themes in social conflict theory have a direct bearing on the topic, one that is, *a priori*, as defensible as the application of principles of reinforcement. For whatever reasons, theories of conflict have seldom been applied to an analysis of the exchange process in close relationships.

The dynamics of social conflict are important in this context because it is hypothesized that a circular, causal relationship exists between a microlevel perspective and experiences of conflict and dissatisfaction in close relationships. If individuals tend to adopt this perspective, an increase in conflict and discontent results. On the other hand, experiences of conflict or dissatisfaction reinforce and exacerbate the tendency to adopt the microlevel address frame. Thus, an amplifying causal loop between this exchange perspective and social conflict is viewed as descriptive of the process of social exchange in close relationships.

There are conceptual parallels between the form of this hypothesis and the notion of social traps (Platt, 1973). A social trap occurs when an individual makes a decision that furthers his short-term, personal gains but produces increasing costs in the longer run. This process has most often been related to the case where individual "rationality" ultimately conflicts with the welfare of social aggregates (See Brickman, Folger, Goode, & Schul, this volume; Schelling, 1978). In contrast, the present hypothesis suggests that dyadic, mutual concerns about short-term fairness may result in processes that undermine the relationship in the longer run.

4.1. Time Perspectives

If one were to plot the sequence of outcomes that each member of a relationship experiences, the graph would depict a pattern marked by its variability. A person's outcomes would appear volatile, with peaks and valleys and substantial fluctuation, given the intrusion of the host of potential causes of satisfaction or dissatisfaction. If the graph for the partner were overlaid, the profile would show large discrepancies in the relative outcomes of the two persons at any particular point in time. This variability and the size of the discrepancy between the outcomes would typically be decreased substantially if the profiles were aggregated or averaged over longer periods of time: the longer the time frame, the greater the *apparent* stability of outcomes and fairness of exchange. In contrast, when a short-term time perspective is used as the unit of analysis, these conditions appear to be exacerbated. They should appear this way not only to the scientist or the behavior therapist but to the actors who are involved in monitoring and evaluating their own relationship.

The apparent variability that results from a short-term focus should act to increase feelings of insecurity and anxiety about the relationship. Another cost of this perspective involves the decreased representativeness of the sample of behaviors that is used as the basis for inferences about the attributes and affections of one's partner. Small samples of behavior are likely to result in distorted or unreliable estimates: attributions tend to be "overdrawn" and more symbolic meaning tends to be attached to behaviors than is warranted.

Perceptions of inequitable or unfair exchange (i.e., outcome discrepancies) are thus more likely to result from a contracted perspective. This consequence is seen as relatively orthogonal to the particular rule of exchange that is adopted, whether it be equity or equality. These perceptions often cause overt conflict, which results in communicated attributions that serve to redress or rationalize the particular grievance (Kelley, 1979). These processes, in turn, make the experience more vivid and salient and increase the impact of what may once have been a potentially innocuous exchange.

The person's time perspective is likely to alter not only the fabric of the exchange, but also its texture. A constricted time frame causes a person to focus his attention on single issues or behaviors. The term *issues* is used here to refer to the content of the frequent decision-making activities required in any close relationship, decisions about services, finances, preferences and activities, tastes and plans, and so on. These implicit negotiations are likely to constitute the basis for a substantial portion of communication in the dyad. There are important costs associated with decision-making styles that compartmentalize complex exchanges into a set of single issues that are treated independently.

Erickson, Holmes, Frey, Walker, and Thibaut (1974) have shown that an orientation toward problem solving that involves "partitioning" multiple issues tends to result in increased conflict and less effective solutions than an "integrative" strategy that involves trade-offs and balances within the whole set of issues. Often, even issues that are initially seen in black-and-white terms can be broken down into components or extended in time, so that fair exchanges can be located. Negotiations that focus on a single issue highlight the conflict of interest between the two parties. The outcomes for *each* solution tend to be negatively correlated, so that the discussions appear to resolve into win–lose contests. This process often results in competitive, power-oriented strategies that produce hostility and distrust. If either party becomes invested in the conflict, simple chores like washing the dishes can take on substantial symbolic value.

The correspondence or correlation between the two partners' outcomes can usually be increased dramatically if they attempt to optimize their preferences across a set of different issues. The importance that two people attach to an issue usually differs, permitting package solutions that integrate individual preferences across issues. This strategy of integrative problem-solving typically demands a long-term time perspective (Walton & McKersie, 1965). It has the potential for increasing the outcomes of both parties as well as for reducing the intensity of conflict (Kelley, 1966). However, the success of this venture hinges on the concept of *trust*. Each person must be willing to forgo a fair exchange on certain issues and give the other the lion's share of benefits, with the hope or expectation that such a concession will be reciprocated on other issues or at a later time.

The manner in which couples deal with complex sets of issues is almost a litmus paper for the strength of their relationship. Walton and McKersie (1965) and Kelley (1966) have argued that for optimal solutions to emerge, each partner must adopt a trusting attitude, show sensitivity to mutual interests, and engage in open, honest communication. Short-term sacrifices and temporary imbalances are often necessary to secure growth in the benefits derived from a relationship (Holmes & Miller, 1976). In the extreme, one partner may agree to a clearly inequitable contribution in the belief that both will ultimately benefit, as in the case where one spouse accepts a relatively unrewarding job to support the other's graduate or professional training. This type of arrangement depends on a high level of trust by the underbenefited person, the belief that the other party acknowledges and shares similar principles of fairness or social justice.

In contrast to this picture, a desire to maintain short-term equity or equality should restrict attention to fewer issues and inhibit the development of integrative agreements. The noncorrespondence of outcomes that is typically associated with decision making on particular issues makes conflict of interest the salient perception. A sensitivity to opposing goals and dissimilarities should be reinforced over time. A couple may be hard put to resist adopting competitive strategies in these circumstances. And once aversive tactics enter into the equation—power, threats, distrust, and closed communication—they typically amplify the conflict process (Deutsch, 1973). The failure to extend one's time perspective also causes one to forfeit the increased benefits that accrue to integrative solutions and inhibits the development of trust. Trust requires the acceptance of risk and vulnerability, while short-term equivalence is designed to minimize risk and unilateral sacrifice (Holmes & Miller, 1976).

In summary, the desire for fair exchange in the short run may produce a paradoxical effect: it may result in perceptions of injustice and conflict of interest and may diminish the potential rewards available through effective cooperation.

There is little direct evidence to support these ideas in the domain of close relationships. Traupmann (reported in Walster, Walster, and Berscheid, 1978) examined the reactions of friends and strangers to inequitable treatment by a partner on an evaluation task. Friends appeared to be less distressed by the incident and attempted to negotiate a fair resolution, while the strangers appeared unwilling to tolerate momentary imbalances. This finding is consistent with Blau's (1964) argument that "too quick or too strict a concern with reaching equal balance in an interpersonal exchange appears to characterize nonfriends more than friends" (p. 148). Also, Driscoll et al. (1972) found that the degree of trust in a close relationship was directly associated with the passage of time and that love and trust became more strongly correlated over time. Thus, there is some evidence to support a claim that increasing attachment is related to less concern with microlevel fairness and that considerations of justice come to involve longer-term perspectives.

The reader should be reminded of the nature of the amplifying causal loop that is being proposed for these processes. It is suggested that distress, conflict, or dissatisfaction in a close relationship leads the members to monitor or evaluate their experiences more closely and to be less willing or capable of engaging in long-term or complex exchanges that put a stress on risk and trust. This contracting address frame then further exacerbates conflict and dissatisfaction. But note that a couple that adopts this "justice perspective" is hypothesized to be "at risk" irrespective of its present state of health, since the style itself is predicted to result eventually in an impaired relationship.

4.2. Explicit, Conditional Exchanges

Earlier in this chapter, the theory of interpersonal dispositions was briefly described (Kelley, 1979). Interpersonal dispositions are those relatively stable qualities of a person that are inferred (attributed) by observing the degree to which his decisions reflect sensitivity to his partner's outcomes, and his considerateness in taking account of them. A disposition is viewed as a stable tendency to "go beyond" or transform the personal rewards and costs of a decision in ways that are directly responsive to increasing the outcomes of the other person. The happiness of one's partner becomes an important motive, so that decisions are shaped by a concern for the couple as a unit, for "we" rather than for "you" and "I."

Several other theorists have emphasized the idea that persons in a close relationship show evidence of being responsive to the needs and outcomes of their partners, with potential costs in terms of immediate, personal gains. Blau (1964) noted that the pleasure derived from gratifying the needs of an intimate partner may be intrinsically rewarding if such a motive has become internalized or if strong identification produces vicarious affect. Lerner (1977) argued that a person who has a strong "identity relation" with another will respond directly to his needs, without consideration of the other's inputs. Pruitt (1972) suggested a norm of "mutual responsiveness" that is based more on extrinsic considerations. A person may forfeit personal gain to benefit a partner, with the understanding that fairness will eventually be restored. Similarly, Huesmann and Levinger (1976), in their incremental exchange model, argued that a partner's outcomes are given more weight if a person forecasts increased future benefits from such a strategy.

An important aspect of Kelley's (1979) theory is that it describes the process by which attributions about the other's dispositions and affections are drawn, based on the particular pattern of deviations from self-centered motives. Thus, this theory links the structure of an exchange to its attitudinal consequences for each partner.

The social contexts that permit clear inferences on the "responsiveness" of one's partner are limited, because his behavior must in some way stand in contrast to that which would be expected if he were acting solely in accord with his egocentric concerns. For instance, a decision that involves a low correspondence of outcomes (high conflict of interest) provides the opportunity to display one's responsiveness, while one that involves a high correspondence of outcomes does not. Any evidence that a partner's behavior can be adequately explained by the personal rewards that he expects to receive undercuts the inference that he is responding to the other partner's needs and outcomes.

A participant in a relationship faces a difficult strategic decision when he considers whether to make explicit the conditional nature of an act. If you have received a benefit from your partner, you may feel obligated to return a comparable benefit: a debt has been incurred. But if the contingent nature of your subsequent behavior is then made salient to your partner, its informational value is discounted in terms of any interpersonal disposition (affection, caring) that you might wish to express. And the value of your partner's initial act is somehow "cheapened" because you are acting as if he were anticipating such a benefit, that is, as if the action were predicated on such an expectation. Similarly, if one announces the social contracting nature of a benevolent behavior (if I do this, I expect you to do that), then an important aspect of its value to the other person is lost.

The results of a recent experiment by Clark and Mills (1979) illustrate that an act of explicit reciprocation carries a message about the intended scope of a relationship even between strangers. When it was clear that the relationship of a male (subject) and a female (confederate) would be restricted to a task-oriented interaction, the female was liked more when she explicitly returned a favour on receiving a benefit from the male. If there was some possibility that the female co-worker might be interested in establishing a friendship, then she was liked more when she did not reciprocate the benefit. Thus, in an "arm's-length" or positional relationship, it is appropriate to expect reciprocity in order to establish fair exchange. When the potential for a more personal relationship exists, one that goes beyond the participants' roles on the experimental task, the violation of the reciprocity "norm" seems to have been interpreted not as a failure to restore equity, but as an indication that the female is willing to consider the relationship a friendly, more social one. The important point is that there seems to be an understanding, even among strangers, that reciprocity demands are appropriate for impersonal transactions but not necessarily for more personal ones. (The authors suggested that the results demonstrate that reciprocity does not apply to "communal" relationships, in which each person is concerned with the needs and welfare of the other. To the present author, this conclusion does not seem warranted, given the context of an interaction between strangers.)

The Clark and Mills (1979) study is consistent with the idea that when the conditional nature of the exchange process is made explicit by either party, then a message is communicated about the relationship itself, or at least those domains relevant to the interchange. It makes salient the instrumental, extrinsic basis of a transaction and may thereby play a role in *defining* the perspective that is relevant for that type of interaction. As Lerner and Whitehead (1980) suggested, the emphasis is put on the exchange *process* and the rules for acceptable allocations, rather than on the goal of promoting the *relationship* itself.

One could speculate that in the early stage of romantic involvement, each person is centrally concerned with the state of the relationship and the partners' respective feelings for each other. An explicit focus on the

process of fair exchange and justice is not likely at this stage; if it did
occur too early, it would be interpreted as an attempt to define the
relationship in nonromantic, less personal terms. But as the level of
interdependence increases, it becomes necessary to negotiate the terms
of the relationship and to clarify role requirements. Driscoll *et al.* (1972)
reported that negative interpersonal behaviors and distrust increase at
this stage, irrespective of the level of intimacy that is expressed. This
result may be based on the necessity of focusing on the exchange process
itself and on an explicit accounting of each person's contributions.

It has been hypothesized that when the conditional nature of exchange is made salient, it serves to undercut the expression of certain
feelings (i.e., interpersonal dispositions) by providing sufficient, extrinsic explanations for one's contributions and positive behaviors. It also
inhibits the development of trust and perceptions of commitment, since
these processes require evidence of self-sacrifice and responsiveness to
another person's needs. This evidence depends on the display of behavior that cannot be interpreted primarily in terms of one's egocentric,
immediate rewards. Finally, when the exchange process is stressed,
there is pressure on each person to monitor the other's behavior and to
make a more explicit accounting of the value of his actions, in order to
defend one's own interests in the "bargains" that are struck. Not only
could this phenomenon result in invidious comparisons, but it forces
each party to adopt an evaluative perspective that focuses attention on
matters of power and relative contributions.

On the other hand, it is important to identify the benefits associated
with a microlevel exchange perspective, as well as its costs, if we are to
understand why it so frequently becomes the preferred interpersonal
strategy. First, when a partner is made aware that a behavior is explicitly
a part of an exchange, then reciprocity or "fairness" is more clearly
established. It ensures that one receives "credit" from one's partner
when credit is due. This credit is most likely to be seen as necessary
when the fairness of exchange is at issue; in turn, this circumstance is most
probably associated with domains of the relationship that have involved
prior conflict or unresolved insecurities. (Paradoxically, these are the
very circumstances in which the less said, the better, since the opportunity exists for clearly displaying one's trust and responsiveness.)

A couple's desire to verbalize and label an exchange and to clarify
its elements and procedures is thus postulated to vary directly with the
degree of risk and vulnerability that is perceived to be involved in the
issues. Typically, long-term arrangements and complex issues increase
risk and the associated desire to make explicit the conditions of an exchange. The degree of perceived risk also hinges on the trust and commitment that exists in the relationship. These psychological elements

are related to the strength and stability of the relationship, its "pair cohesiveness" (Levinger, 1976), including the barriers that inhibit dissolution (Berscheid & Campbell, this volume). Thus, when cohesiveness and "security insurance" are low, explicit exchange is evident over a wider range of issues and is extended over shorter periods of time. However, even healthy relationships may require "metarules" to govern long-term understandings, particularly on issues that are threatening and that tend to make personal insecurities surface.

It is interesting to note here that Barry (1970) concluded from a review of research on marital satisfaction that the best predictor of strain in close relationships is a weak identity or self-concept in the male partner. A husband with a frail self-concept would be less trusting and more vulnerable and sensitive to temporary insecurity and inequity. Thus, he would be expected to manifest a desire for both explicit reciprocity in the short run and clear role requirements. These tendencies are likely to exacerbate any existing strains in the relationship.

In summary, an excessive concern with interpreting one's interdependence with another in terms of explicit rules of exchange has important potential costs. The approach forecloses on opportunities for the growth of intimate feelings in the relationship by obscuring their expression. The attribution of interpersonal dispositions requires that a person show evidence of self-sacrifice and responsiveness to the outcomes and needs of his partner. To the extent that one communicates that one's benevolent acts can be explained by the laws of economic transactions, this process will be undercut and inhibited. And if such tendencies are widespread and become generally apparent to one's partner, they carry a message about the state of trust that exists in the relationship. Of course, it is also possible that each party will come to view their *own* behavior in this light. If a person becomes accustomed to gauging his contributions and behaviors in a close relationship according to the direct benefits expected in return, and this process is salient, then his perceptions of his own motives may be affected. The emphasis on extrinsic rewards could diminish his estimation of intrinsic emotions such as love and affection (see Seligman, Fazio, & Zanna, 1980).

4.3. Resource Equivalence and Substitution

In the discussion of symptoms of contraction, it was argued that chronically or acutely distressed couples would tend to restrict their exchanges to those that involve resources from the same class: exchange "in kind." There is little direct evidence to support this principle. Foa and Foa (1974) have concluded from their own research program that people prefer exchanges within a particular resource class and that the value of a resource for restoring equity is directly related to its similarity to the specific benefit received. But this evidence is a summary of experiments conducted with strangers in narrowly defined contexts. Walster *et al.* (1978) have argued that casual relationships are usually conducted in a single context—among classmates or tennis partners, for instance. They believe that close friends, on the other hand, interact in a wide variety of contexts and freely exchange one type of resource for another: "Intimates spend much of their time negotiating the values and exchangeability of various behaviors: negotiating the 'terms' of their relationship" (p. 152).

Admittedly, the narrow contexts of casual relationships tend to confine exchanges to particular resource classes. But there are other important reasons for this phenomenon that have not been given sufficient stress. First, if a person reciprocates a benefit with a resource from a different class, the effects may parallel those from the Clark and Mills (1979) study. That is, the behavior is likely to be interpreted as communicating a message about the relationship itself—a message or invitation to extend it to a different, closer footing. This notion is based partly on the findings of Foa and Foa (1974) that normal, more impersonal transactions are based on reciprocation in kind; a departure from this rule creates a disjunction that imparts information about a person's motives.

Second, exchange in less intimate relationships tends to involve resources that are relatively concrete and universalistic in value (money, goods, services, information) (Walster *et al.*, 1978). The currency of such resources has usually been calibrated in the public domain; they are less subjective and represent tangible, "hard evidence" of a contribution that is less dependent on the qualities of the particular doner or the perceptions of the recipient. This hard evidence permits each party in an exchange to gauge its fairness with a minimum of communication and trust, facilitating clear social comparison. The value of more subjective resources like affection and respect not only is more difficult to code and calibrate but depends on the history and intimacy of the relationship itself. Thus, exchange across widely divergent resource classes can be taken as presumptive evidence of the closeness and integration of a relationship; it requires the trust necessary to tolerate the ambiguities of "soft evidence" and the communication necessary to establish "equivalence classes."

There is a certain amount of theoretical ambiguity related to this question. Equity theorists suggest that parties to a close relationship are most content if fair (equitable) exchange has been achieved. But at what level? Within resource classes or in terms of total contributions? For instance, Kidder, Fagan, and Cohn (this volume) note that husbands believe that they contribute more instrumental resources (money, information) than their wives. To what extent does the equity formulation apply to this disparity? Note that women believe that their socioemotional contributions are greater than their husbands', so that an argument can be made that equity (or equality) is restored across resource classes. (Research studies have typically focused on "total" contributions, as in the Walster, Walster, and Traupmann [1978] study described earlier, though there is some evidence of "compensation" across specific resource classes when social commodities such as attractiveness are unequal [Berscheid, Walster, & Bohrnstedt, 1973]). The point to be made is that certain partners may strive for equity within resource classes, even in close relationships. This differentiation of the relationship into spheres of power would seem to reflect a lower level of integration than compensation across resource classes. For instance, do men who contribute greater instrumental resources expect to achieve more benefits that are tied to this resource class? Do they expect to control the finances and have greater input into family decision-making, or do they look for compensation elsewhere in order to achieve fair exchange?

Equity theory states that a person would be content if his contributions (overall or within classes) were greater, as long as the benefits he received were correspondingly higher. However, other theoretical perspectives view equality of contributions and benefits as the critical allocation rule in a close relationship, when a person views it in "personal" rather than "positional" terms (Lerner, Miller, & Holmes, 1976; Lerner, 1977). Note that the distinctive contributions of men tend, in fact, to be rather "positional" in nature—the resources are "universalistic" and relate to their external role—compared with the more "personal" contributions of women (e.g., affection, emotional support).

An equity perspective would offer certain advantages to men, since the value of their resources would be easier to establish. Also, the value of women's resources, such as affection, depends more directly on the husbands' acceptance and evaluation of it. Thus, a husband would have a large element of fate control in the negotiations over the value of each person's contributions. Since a strict application of equity across resource classes could provide the basis for unequal power in the relationship, these negotiations would have substantial impact. And it is hard to believe that a competitive orientation would not be introduced by such a process (Deutsch, 1975).

Several aspects of this analysis should be emphasized. First, the process of social comparison of inputs and outcomes both within the dyad and with salient outside alternatives is more evident in distressed couples and less intimate relationships. In such cases, principles of fairness are a central concern, and interpersonal comparisons are particularly salient. Second, principles of social comparison appear to be more critical to spheres of a relationship that are governed by equity considerations, rather than by equality, and in particular, need considerations. When the spectre of unequal power comes to the surface in an equity perspective, the comparison of relative contributions is most explicit. Social comparison is facilitated by resources that have more objective status, and one might speculate that the partner who contributes more of these resources will find it more advantageous to stress equity principles.

Other motivations could act to minimize the degree to which each person vies for a position of relative power based on his advantageous inputs. The critical ingredient in close relationships may be the strength of the interpersonal dispositions that are inferred, particularly qualities such as love and loyalty (Berscheid & Walster, 1978, pp. 39–45; Kelley, 1979). These dispositions are inferred from evidence of responsiveness to the unique needs of one's partner, and their development requires that each person respond to the other in "personal" terms rather than in the "positional" terms relevant to an equity perspective. One would expect that other resources could be substituted for love and affection only to the extent that they impart indirect evidence of such dispositions: "I did it because I love you." Thus, interpersonal dispositions such as love may constitute contributions that are different in kind from other resources, as many writers have long insisted! The strong desire for equal contributions on this dimension would inhibit tendencies to achieve equity by compensating with other resources—at least, in healthy relationships.

The complex role structure of a close relationship almost ensures that unequal contributions will emerge in certain domains. A woman with young children may commit substantial energies to her role, without any expectation of reciprocation in kind from her husband. To the extent that the value of her services must be explicitly negotiated, her husband may contend that they represent the "normal" role requirements of her position. An emphasis on role requirements makes salient the "positional" nature of a relationship, and it promotes the application of equity principles (Lerner, Miller, & Holmes, 1976). Also, defending the value of one's services as a resource emphasizes the extrinsic purpose of the activity. For both reasons, the interpretation of her efforts in terms of her "personal" dispositions, such as love, caring, and commitment to the family unit, is undercut. In a healthy relationship, the calibrating of such contributions is more subtle, and efforts are made to avoid an explicit analysis and defense of their virtues. But equity (power) considerations or episodes of conflict may institute these processes.

In summary, exchange in kind typifies casual relationships, particularly when tangible, objective resources are at issue. Healthy close relationships are marked by a subtle balancing of contributions across resource classes, with the exception of more personal inputs, such as love and affection. Equality is stressed for contributions that are closely tied to the domain of interpersonal dispositions. Equity concerns are less pronounced in these areas because they tend to emphasize competitive negotiations on the value of resources and to make salient the relative positions or power of each person. Distressed couples regress in the sense that they tend to behave in a manner more similar to that of casual friends. Social comparison within such dyads increases, and concerns with fairness often center on more objective resources, which can be easily calibrated without complex understandings or agreements as to their value. There is increased evidence of equity considerations both within and across resource classes, and differences in perceived spheres of power are more salient.

4.4. Rule-Governed Behavior

An argument can be made that fewer norms and role constraints are imported into present-day marital relationships than at any other point in our history. This change has equivocal value. Norms and roles reduce attributional clarity since they often provide sufficient explanation for a person's behavior. Rule-governed behavior thus tends to obscure a person's motives. This can be a benefit if the behavior in question is a negative one, but a loss if it is positive. That is, role requirements and rules tend to inhibit conflict, by removing the necessity of using "raw" power (Thibaut & Kelley, 1959) and by blocking the attribution of hostile motives (Holmes & Miller, 1976). But at the same time, they inhibit the growth of intimacy, since inferences about positive behavior (i.e., interpersonal dispositions) are also impaired (Holmes & Miller, 1976).

Norms and social "contracts" are most likely to develop in the areas of a relationship that involve unresolved conflicts or recurrent problems (Thibaut & Faucheux, 1965). Typically, social contracts are designed to cope with points of stress in the interactive system by diminishing the need for either party to resort to the use of personal power and threats. Rather, the contracts specify role requirements and/or a relatively fixed set of procedures for arriving at a solution to a dispute. Behavior becomes scripted, and the interaction is depersonalized by emphasizing the person's role or "position" (Holmes & Miller, 1976). These role requirements prevent clear personal attributions and tend to inhibit emotional responses. Thus, social contracts and norms serve an adaptive purpose by stabilizing relationships in domains that could produce severe strain or conflict. Rules also serve a coordinating function in close relationships. As the level of interdependence grows, numerous routines are developed to reduce the costs of spontaneous adjustments to everyday coordination problems. In the earlier stages of a relationship, the social contract is not sufficiently articulated to provide rituals for dealing with minor frictions or to provide more complex rules for settling serious disputes. As we noted earlier, Driscoll et al. (1972) found that negative interpersonal behaviors, disagreements, and distrust exemplified the interaction of couples in the period in which their interdependence was being negotiated, irrespective of the strength of love that was expressed.

If fewer norms and rules are imported into modern relationships or are imposed from without, what consequences will result? The social contract is more likely to represent an idiosyncratic creation that is sensitive to the requirements of a particular relationship. One would expect that it could then be tailored to be more responsive and functional for the specific needs of the couple. However, there are some hidden costs associated with these changes. First, if rules and contracts, in fact, develop as a means of crisis management, in areas of dispute and conflict, then the "testing ground" for the development of interpersonal dispositions is attenuated. One can most clearly express one's attitudes of love and responsiveness in contexts where conflict of interest exists; only then can one demonstrate that one's self-interest is being sacrificed to respond to the needs of the other person. If the social contract becomes overly elaborate in areas of potential conflict, a person has difficulty finding the opportunity to express his feelings clearly or to learn about his partner's.

In a similar fashion, rules designed to smooth the flow of routine interactions and increase coordination may result in certain positive behaviors' being "taken for granted." Also, rules that prescribe specific behaviors that fall under the umbrella of one's role contributions may cause dissatisfaction by facilitating the detection of instances of noncompliance. Particular cases of deviance from a rule are more likely to be noticed and remembered. More ambiguous standards would lead people to rely on their impressions of their partner's fealty to the intent of the rule structure. In long-term relationships, there is the risk that rules will proliferate over time, and that they will not be discarded when they have outlived their purpose. Thus, the problems described above may be exacerbated at a stage when strong emotions already have a tendency to fade (Berscheid & Campbell, this volume).

When norms are imposed from without, by cultural dictates, they are not likely to carry a message about the state of a *particular* relationship. (However, rules such as those in the legal structure may affect general beliefs about one's fellow citizens, as Peachey and Lerner [this volume] have suggested.) On the other hand, if a large number of rules or procedures are developed extemporaneously within a relationship to deal with areas of dispute, then their existence may affect a person's beliefs about his partner. When it becomes salient that a complex set of rules is necessary to protect one's interests, the responsiveness and motives of the other person are called into question. The basis of trust is that one's partner can be relied on to look out for and respond to one's needs, without the necessity of rules to ensure one's security.

In summary, the development of explicit rules and procedures to deal with areas of conflict can prove beneficial in reducing stress and resolving disputes. Social contracting is probably popular among therapists for this reason; they are usually faced with providing counsel to couples who are suffering from the strain of considerable interpersonal conflict and who see it as the "presenting problem" that requires a cure. The cure may be short-lived, however, if an extensive reliance on rules and norms blocks the expression of interpersonal attitudes; the rewards that constitute the basis for attachment will be dissipated, and intimacy and emotional intensity will be eroded. This tactic for distancing or avoiding conflict is thus a mixed blessing. Evidence that a couple has learned to rely on it across a wide range of issues would augur future difficulties.

5. CONCLUSION

The fabric of close relationships has changed dramatically in recent years, and it is likely to evolve even further in the future. The direction of change appears to be toward a concern with short-term satisfaction and fairness. Structural factors that helped to ensure longer-term stability have diminished in strength. In marital relations, for instance, externally imposed role definitions and norms have become less evident, and the conditions that created "security insurance" for a couple have been eroding. The concept of commitment appears to have less psychological force. As the barriers to dissolution are weakened, an increased emphasis is placed on the internal cohesiveness of the unit, on the direct benefits and attractions that a relationship provides. Imported norms are being replaced by rule structures designed specifically to cope with the role definitions and conflicts of the particular couple. In summary, the emerging social climate is likely to put substantial stress on close relationships by focusing attention on principles of exchange and role structures that are more relevant to coping with short-term issues than to developing long-term intimacy.

The potentially destructive impact of applying a microlevel perspective to principles of exchange was illustrated by a critical analysis of a set of assumptions typically adopted by behavior therapists in the area of marital counseling. This criticism is unfair on a number of counts and was exaggerrated for reasons of exposition. The value of examining microlevel behavioral sequences is not being questioned; indeed, this perspective has helped to identify important "disturbances" in marital interaction, such as aversive control cycles (Patterson & Hops, 1972). The critical point is that an understanding of macrolevel motives and agendas increases our ability to interpret and predict specific behavioral sequences within restricted time frames. And conversely, it is necessary to have a reliable picture of the empirical laws in the latter domain if we are to comprehend the development of macrolevel motives such as love and trust.

It was proposed that couples experiencing chronic or acute distress are more likely to adopt an analytical and behavioral style based on a short-term, microlevel perspective on fairness. A short-term address frame promotes microlevel evaluation, explicit and conditional exchange, reciprocity in kind, and rule-governed behavior. These conditions undermine the growth of intimacy in close relationships. The effects are predicted to occur irrespective of the couples' choice of equity or equality as the preferred rule of exchange. By failing to recognize the limitations and consequences of their implicit "theory," they fall prey to a cycle in which their behavior only exacerbates the initial, precipitating causes of strain and dissatisfaction. The contraction of their perspective causes them to take less account of the macrolevel motives that play a central role in maintaining an intimate relationship.

ACKNOWLEDGMENTS

The author is indebted to Harold H. Kelley and Mel Lerner for their insightful comments on many of the ideas presented in this chapter.

REFERENCES

Barry, W. A. Marriage research and conflict: An integrative review. *Psychological Bulletin*, 1970, *73*, 41–54.
Berscheid, E. and Walster, E. *Interpersonal attraction* (2nd ed.). Reading, Mass.: Addison-Wesley, 1978.
Berscheid, E., Walster, E., and Bohrnstedt, G. The body image report. *Psychology Today*, 1973, *7*, 119–131.
Birchler, G. R., Weiss, R. L., & Vincent, J. P. Multimethod analysis of social reinforcement exchange between maritally distressed and nondistressed spouse and stranger dyads. *Journal of Personality and Social Psychology*, 1975, *31*, 349–360.
Blau, P. M. *Exchange and power in social life*. New York: Wiley, 1964.

Clark, M. S., & Mills, J. Interpersonal attraction in exchange and communal relationships. *Journal of Personality and Social Psychology*, 1979, *37*, 12–24.

Deutsch, M. *The resolution of conflict: Constructive and destructive processes*. New Haven, Conn.: Yale University Press, 1973.

Deutsch, M. Equity, equality, and need: What determines which value will be used as the basis of distributive justice? *Journal of Social Issues*, 1975, *31*, 137–156.

Dion, K. L., & Dion, K. K. Love, liking and trust in heterosexual relationships. *Personality and Social Psychology Bulletin*, 1976, *2*, 187–190.

Driscoll, R., Davis, K. E., & Lipetz, M. E. Parental interference and romantic love: The Romeo and Juliet effect. *Journal of Personality and Social Psychology*, 1972, *24*, 1–10.

Erickson, B., Holmes, J. G., Frey, R., Walker, L., & Thibaut, J. W. Functions of a third party in the resolution of conflict: The role of a judge in pretrail conferences. *Journal of Personality and Social Psychology*, 1974, *30*, 293–306.

Foa, U. G., & Foa, E. B. *Societal structures of the mind*. Springfield, Ill.: Charles C Thomas, 1974.

Gottman, J., Notarius, C., Markman, H., Bank, S., Yoppi, B., & Rubin, M. E. Behavior exchange theory and marital decision making. *Journal of Personality and Social Psychology*, 1976, *34*, 14–23.

Hastie, R., & Kumar, P. A. Person memory: Personality traits as organizing principles in memory for behaviors. *Journal of Personality and Social Psychology*, 1979, *37*, 25–38.

Hill, C. T., Rubin, Z., & Peplau, L. A. Breakups before marriage: The end of 103 affairs. *Journal of Social Issues*, 1976, *32*, 147–168.

Holmes, J. G., & Miller, D. T. Interpersonal conflict. In J. W. Thibaut, J. T. Spence, & R. C. Carson (Eds.), *Contemporary topics in social psychology*. Morristown, N.J.: General Learning Press, 1976, pp. 265–308.

Howard, J. W., & Dawes, R. M. Linear prediction of marital happiness. *Personality and Social Psychology Bulletin*, 1976, *2*, 478–480.

Huesmann, L. R. and Levinger, G. Incremental exchange theory: A formal model for progression in dyadic social interaction. In L. Berkowitz & E. Walster (Eds.), *Advances in experimental social psychology*, Vol. 9. New York: Academic Press, 1976.

Kelley, H. H. A classroom study of the dilemmas in interpersonal negotiations. In K. Archibald (Ed.), *Strategic interaction and conflict*. Berkeley: Berkeley University of California Press, 1966, pp. 49–73.

Kelley, H. H. *Personal relations: Their structure and process*. Hillsdale, N.J.: Erlbaum, 1979.

Kelley, H. H., & Thibaut, J. W. *Interpersonal relations: A theory of interdependence*. New York: Wiley, 1978.

Lederer, W. J., & Jackson, D. D. *The mirages of marriage*. New York: Norton, 1968.

Lerner, M. J. The justice motive: Some hypotheses as to its origin and forms. *Journal of Personality*, 1977, *45*, 1–52.

Lerner, M. J., Miller, D. T., & Holmes, J. G. Deserving and the emergence of forms of justice. In L. Berkowitz & E. Walster (Eds.), *Advances in experimental social psychology*, Vol. 9. New York: Academic Press, 1976, pp. 134–162.

Lerner, M. J., & Whitehead, L. Procedural justice viewed in the context of justice motive theory. In G. Mikula (Ed.), *Social Justice*. New York: Springer-Verlag, 1980.

Levinger, G. A social psychological perspective on marriage dissolution. *Journal of Social Issues*, 1976, *32*, 21–47.

Levinger, G., & Snoek, J. D. *Attraction in relationships: A new look at interpersonal attraction*. Morristown, N. J.: General Learning Press, 1972.

Moles, O. C., & Levinger, G. (Eds.) Divorce and separation. *Journal of Social Issues*, 1976, *32*, 1–5.

Orvis, B. R., Kelley, H. H., & Butler, D. Attributional conflict in young couples. In J. H.

Harvey, W. J. Ickes, & R. E. Kidd (Eds.), *New directions in attribution research*, Vol. 1. Hillsdale, N. J.: Erlbaum, 1976, pp. 353–386.

Passer, M. W., Kelley, H. H., & Michela, J. L. Multidimensional scaling of the causes for negative interpersonal behavior. *Journal of Personality and Social Psychology*, 1978, *36*, 951–962.

Patterson, G. R., & Hops, H. Coercion, a game for two: Intervention techniques for marital conflict. In R. E. Ulrich & P. Mountjoy (Eds.), *The experimental analysis of social behavior*. New York: Appleton-Century-Crofts, 1972.

Platt, J. Social traps. *American Psychologist*, 1973, *28*, 641–651.

Pruitt, D. G. Methods for resolving differences of interest: A theoretical analysis. *Journal of Social Issues*, 1972, *28*, 133–154.

Raush, H. L., Barry, W. A., Hertel, R. K., and Swain, M. A. *Communication, Conflict and Marriage*. Washington: Jossey-Bass, 1974.

Ross, M., & Sicoly, F. Egocentric biases in availability and attribution. *Journal of Personality and Social Psychology*, 1979, *37*, 322–336.

Rubin, Z. *Liking and loving: An invitation to social psychology*. New York: Holt, Rinehart, and Winston, 1973.

Schelling, T. C. *Micromotives and macrobehavior*. New York: Norton, 1978.

Seligman, C., Fazio, R. H., & Zanna, M. P. Effects of salience of extrinsic rewards on liking and loving. *Journal of Personality and Social Psychology*, 1980, *38*, 453–460.

Thibaut, J. W., & Faucheux, C. The development of contractual norms in a bargaining situation under two types of stress. *Journal of Experimental Social Psychology*, 1965, *1*, 89–102.

Thibaut, J. W., & Kelley, H. H. *The social psychology of groups*. New York: Wiley, 1959.

Turner, J. L., Foa, E. B., & Foa, U. G. Interpersonal reinforcers: Classification, interrelationship and some differential properties. *Journal of Personality and Social Psychology*, 1971, *19*, 168–180.

Walster, E., Walster, G. W., & Berscheid, E. *Equity: Theory and research*. Boston: Allyn & Bacon, 1978.

Walster, E., Walster, G. W., & Traupmann, J. Equity and premarital sex. *Journal of Personality and Social Psychology*, 1978, *36*, 82–92.

Walton, R. E., & McKersie, R. B. *A behavioral theory of labor negotiations: An analysis of a social interaction system*. New York: McGraw-Hill, 1965.

Justice in the Marketplace
Allocation of Scarce Resources

Justice concerns related to allocation situations must be examined from two perspectives: that of the allocator–decision-maker and that of the recipients. In some cases, of course, the recipients themselves as a group make the allocation decision, thus functioning in both roles. Most commonly in our larger society, allocation decisions are made by those in the public and private sectors who are regarded as having a legitimate right to do so, either because they have been delegated the power or because they legally control the resources to be allocated. From the perspective of the allocator (or the group-as-allocator), the primary justice concern is how to arrive at a principle of allocation that will be "seen to be just" by the recipients. Otherwise, there is a risk of social unrest, and the right of the allocator to perform that function may be called into question. With regard to the recipients, the central concern is to understand the variables that determine which types of allocation procedures and outcomes are preferred, which are acceptable, and what people's responses are to unacceptable allocations.

As Greenberg notes in the first chapter of this section, allocation of resources is a ubiquitous process that raises serious justice-related questions. Under conditions of scarcity, allocation procedures and outcomes assume special significance, since scarcity, by definition, means that not all can have their needs and wants satisfied—and people pay closer attention to who gets what and to how allocation decisions are arrived at. It is generally agreed that if some people come to feel that they are being unjustly deprived of a valued good, there is a strong possibility that they will move to rectify the situation, often in ways that can be destructive to the well-being of the group as a whole.

Drawing on both sociohistorical and experimental research, Greenberg provides an insightful introduction to the special conditions that scarcity imposes on those who make allocation decisions. When group survival is a necessary consideration, need *and* efficiency *are shown to be the allocation principles of* choice, replacing in favor the equity and parity strategies that have prevailed generally in our society in times of abundance. Greenberg also addresses the important questions raised for allocators by the fact that people's expectations and desires are constantly adjusted upward, creating feelings of "felt insufficiency" even as resource shares increase. He notes Rescher's suggestion that in times of abundance, this type of insatiability could have a positive effect on the general welfare by forcing allocators to distribute goods more widely in response to new definitions of acceptable minima. The question remains of how well people can or will adapt their expectations to a period of declining resource availability.

It is this question that concerns deCarufel in his chapter, which deals in some detail with relative deprivation. He outlines the most common types of situations in which individuals and groups feel unjustly deprived and provides a useful overview of the effects that conditions of scarcity are likely to have on the perceptions and behavior of people in these situations. As do many of the authors in this volume, deCarufel assesses the role that self-interest plays in how people decide what is unjust and how they respond to feeling unjustly treated or to the threat of such an outcome. After describing a number of recent experimental studies that have examined the interrelationships among relative deprivation, justice concerns, and allocation decisions, deCarufel turns to a consideration of how the individual's experience of relative deprivation evolves into various forms of collective behavior designed to protest or redress the felt injustice. Finally, he turns his attention to the allocator, with an analysis of the problems created by allocator self-interest or power disparity between allocator and recipients, and of the possibilities for allocators to develop innovative allocation strategies in times of scarcity.

Deutsch's chapter has as its focus the necessity of developing innovative societal responses to economic hard times. He delineates the part that justice concerns play in how a group copes with economic hardship, and he sees such concerns as only one aspect of a complex set of factors that determine the quality of the group's response. In his lucid discussion of these factors, Deutsch emphasizes the importance of two in particular: how conflicting views of the correct way to distribute loss are reconciled and whether this conflict takes place destructively or constructively. In elaborating on the latter, he relies on his own detailed analysis of the differences between cooperative and competitive group processes, relating this point directly to the imperative for creative problem-solving presented by "crunch" situations. Thus, Deutsch places the problem of arriving at acceptable allocation principles squarely within the larger context of the need to develop more participatory, human-scale social units.

Cook and Pearlman offer a different view of the problems inherent in the cyclical nature of our present economic system. They note that a common response to lower economic growth rates has been an acceleration of efforts to promote *growth, and that there is good reason to believe that such a strategy—derived from the "trickle-down" theory—increases income differentials in a society. They examine this strategy, together with its opposite number, based on the "bubble-up" theory, in an effort to determine whether either approach to allocation provides both growth and increased equality. This chapter is, in essence, a highly original policy analysis that attempts to assess the effects of different allocation decisions on the relative economic conditions of various groups in the society. But the authors go beyond this point in their concluding discussion of the role of justice concerns in shaping public response to increased economic inequality and the resultant temptation for decision makers to engage in manipulative symbolic politics. Cook and Pearlman offer an appropriate final perspective on the nature of primary allocation processes as most of us experience them in real life, that is, as remotely sited, impersonally implemented policy decisions that we understand poorly and can at best react to, often in justice terms and sometimes with destructive consequences.*

13

The Justice of Distributing Scarce and Abundant Resources

JERALD GREENBERG

1. INTRODUCTION

1.1. Allocation as a Ubiquitous and Profound Phenomenon

Questions as to the fairest way of allocating resources arise in all social activities, and the way these ubiquitous questions are answered can have critical, widereaching impact on the individuals concerned and on society in general. The profundity of such questions comes into clear focus when the supply of resources in question is inadequate to meet all claims. Examples range from seemingly ordinary matters of allocating journal pages to authors (Latané, 1979) or deciding which little leaguers will get to play baseball, through more far-reaching concerns of how gasoline will be rationed (Pauly & Walcott, 1979) or whether veterans should be given preferences for civil service hiring (Labich, LaBrecque, & Camper, 1979). Consider also the tragic choices of who will be drafted into the armed services (U.S. National Advisory Commission on Selective Service, 1967), which women should bear children (Berelson, 1974), or who will be given access to hemodialysis machines and other scarce medical resources (Katz, 1973; "Scarce Medical Resources," 1969).

These examples highlight both the ubiquity and the profundity of

JERALD GREENBERG ● Faculty of Management, Ohio State University, Columbus, Ohio 43210.

resource distribution questions. Moreover, this diversity of allocation situations suggests some of the special kinds of problems that can arise when resources are in limited supply. This chapter identifies and discusses many of these special concerns and contrasts them with situations involving the distribution of abundant resources.

1.2. Scarcity: A Neglected Social-Psychological Issue

At the onset, it seems noteworthy to point out that issues of allocation are not uniquely within the province of the experimental social psychologist. Such issues have been the concern of scholars in such diverse fields as law (e.g., Calabresi & Bobbitt, 1978), moral and social philosophy (e.g., Rescher, 1966; Bowie, 1971), economics (e.g., Becker, 1965), history (e.g., Moore, 1978), anthropology (e.g., Gudeman, 1978), political science (e.g., Singer, 1974), and environmental science (e.g., J. N. Smith, 1974). The matter of resource scarcity has been a recurrent theme in much of this work.

With this background, it is curious that the social psychologists who have studied reward allocation—both psychologists (e.g., Leventhal, 1976a) and sociologists (e.g., Berger, Zelditch, Anderson, & Cohen, 1972)—have paid very little, if any, attention to the crucial matter of scarcity. This is especially surprising in view of the more comprehensive efforts based on the related topic of relative deprivation (e.g., Crosby, 1976; Runciman, 1966). Yet it is understandable for American social psychologists, weaned under what Tyler (1976) called "the assumption of abundance," to give greater attention to phenomena rooted in cherished American principles, such as how people respond to personal injustices, than to matters of scarcity, from which they have been made to feel insulated and immune. This assumption of abundance in America sharply contrasts with the more pessimistic belief in some other cultures that all resources are in permanent short supply (Foster, 1965). Of course, with "a chicken in every pot" as the norm, it is not surprising that comfortable Western social psychologists have turned their attention to the loftier, more patriotic matters of the "Great Society." With high governmental officials such as Vice President Mondale discounting the eventuality of gasoline rationing ("Fuel Cutbacks Discounted," 1979), it is often difficult for social scientists to give serious attention to such matters.

Nevertheless, our attention is focused by the events of recent history on the realization that the myth of inexhaustible abundance is just that: mythical. The changes in behavior and values needed to adjust to finite reality are apparent in the case of the world's nations' rushing to the aid of famine-struck East Bengal in 1971 (for a discussion of the moral implications of this action, see Singer, 1972). More personally involving for most Americans is the recent collective memory of waiting in long lines to purchase gasoline in the winter and spring of 1974, and again in the summer of 1979. As this is being written, in 1979, the threat of gasoline cutbacks and rationing once again remains a gloomy prospect as unrest in the Middle East has lowered the supply of oil, and has drastically raised the price of such derivative products as plastic goods and gasoline. Features currently appearing in the popular press (e.g., Labich, LaBrecque, & Camper, 1979) reiterate the dynamics of gasoline rationing discussed little over the five years previously (e.g., Wallich, 1973).

Unlike some of our environmental psychology counterparts (e.g., Bartell, 1976; Ferber, 1977), who used the "energy crisis" as a rallying point on which to focus their research, social and interpersonal justice researchers, regrettably perhaps, have failed to receive professional inspiration from the impending prospect of having to allocate scarce resources. Certainly, this failure has not been due to lack of time—the hours spent queuing up for gasoline may have been better spent conceptualizing than grumbling. Yet, responding to a call to action by studying the obvious problem of natural-resource allocation would have hindered our attempts to deny personally the threat of "the end of affluence" (cf. Ehrlich & Ehrlich, 1974).[1]

Although "the problem" went away for a while, and the lines returned to normal at the gasoline pumps, it is once again very clear that abundance is falling to scarcity, and that this fact has implications for our conceptions of justice. In a very real sense, this volume and the conference from which it was derived serves as a testimony to the recognition among justice theorists that the scarcity problem is indeed real, and the book should provide an outlet for our research, both in generating new theoretical problems, and in assessing the external validity of our current conceptualizations. Hopefully, this chapter will prove useful toward this end.

2. SCARCITY

Conditions of scarcity exist whenever the available supply of a resource is inadequate to satisfy all demands for it. Such would be the case, for example, among a group of three persons stranded on a desert

---

[1] This sentiment is at least as self-incriminating as it is accusative, for in December 1973, Gerald Leventhal and the author discussed but failed to implement plans to study the perceived justice of various gasoline-rationing policies.

island with just enough food to keep any two alive until they were rescued.

Purely from the perspective of equity theory (Adams, 1965) or distributive justice theory (Homans, 1961), each person's share should be proportional to his or her contributions. Thus, given that all three are equally deserving, it would be considered fairest for each of our persons to receive two thirds of a unit of the available food. However, if each person requires one whole unit of food to stay alive, the allocation of two thirds of a unit could be considered "tragically just." Accordingly, in the case of dire insufficiency, catastrophe prevention may require abandonment of justice principles based on merit in favor of other important considerations, such as need and efficiency.

This point has been dramatically underscored by Rescher (1966) in stating that

> An *economy of scarcity* is, by definition, one in which justice (in a restricted sense of the term) cannot be done, because there is not "enough to go around": if everyone is given a "share proportional to his claims and desert" then someone—or everyone—is pressed beneath the floor of the minimally acceptable level. (p. 96; italics in the original).

Similarly, in referring to the state of "splendidly equalized distribution," Hospers (1961) stated that "There is a bottom level of instrumental good . . . below which equality is useless because it is equality in nothingness" (p. 428).

These statements do not strictly imply, of course, that justice *cannot* be achieved under conditions of scarcity. Instead, they imply that under these conditions, justice requires that claims for resources be made on the basis of factors other than the usual one of contributions. These philosophical arguments suggest that other factors such as need and efficiency, are essential elements of fair allocations under conditions of scarcity. Our research evidence bears this out. Allocation decisions based on need and efficiency, when implemented, effectively impose the *utilitarian* standard of bringing the most good to the greatest number. The implementation of such utilitarian strategies inherently rejects egalitarian principles, since *un*equal distributions of the total good may maximize the overall benefit to all (see Bowie, 1971). The popular acceptance of such a utilitarian doctrine was facilitated by philosophers' recognition that insufficient resources threaten life itself (Halévy, 1928). As Mill (1957) suggested, equality of treatment is to be suspended when required by "some recognized social expediency" (p. 78).

Given that some persons' claims have to go unfulfilled for the good of the greatest number, the question is raised as to what criteria should be used to determine which individuals should receive the resources in question. Some philosophers would apply Bentham's formula ("Everybody to count for one, and nobody for more than one") by employing a lottery, thereby giving each individual an equal opportunity of being saved or sacrificed. Our interest in this issue is not to engage in a moral dialogue; rather, it is to examine how persons actually confront these questions.

To this end, our analysis focuses on two distinct types of evidence: sociohistorical accounts of allocations of scarce medical resources, and direct experimental evidence derived from laboratory and survey studies. This evidence effectively validates the importance of two factors— need and efficiency—in allocating scarce resources. Interestingly, these are factors that ethical philosophers have long recognized as essential components of serious theories of distributive justice (e.g., Ross, 1939).

2.1. Need

The importance of need as a determinant of allocation behavior has already been highlighted in previous theoretical accounts (e.g., Deutsch, 1975; Leventhal, 1976a,b). Only recently, however, have theorists shown an appreciation of the extent to which scarcity is instrumental in making needs salient as an allocation determinant. For example, passing mention is made of the scarcity–need link by Leventhal, Karuza, and Fry (1980), and by Cook and Parcel (1977), who speculated that "there may be greater consensus on the legitimacy of a particular distribution rule (such as distribution according to need) under conditions of scarcity than under conditions of abundance" (p. 86). The evidence to be discussed will lead us to accept this statement as a very reasonable conclusion.

2.1.1. Sociohistorical Observations

Calabresi and Bobbitt (1978) have provided an interesting cross-cultural account of the procedures devised in various nations for allocating hemodialysis machines to kidney patients, a paradigmatic tragic decision. The concern about patients' needs, although paramount in all the nations studied, is most readily apparent in the case of Italy, where "the first in need gets the good." In fact, some Italian kidney patients who were known to be terminally ill from additional diseases unrelated to their kidneys have been given access to artificial kidney machines over less ill patients.

The important point is that despite some ostensible disparities in allocation practices between nations, the slogan "Kidneys for all who need them" was the common cross-cultural theme (Calabresi & Bobbitt, 1978), making the concern about need preeminent. In fact, need was clearly given preemptive priority over more customary egalitarian practices, such as dispensing resources by lot or in accordance with order of survival. The random assignment of patients to scarce medical treatments has been found to be negatively viewed by the public, even for purposes of scientific investigation (Hillis & Wortman, 1976). Similarly, the tendency to give greater weight to needs as an allocative decision rule is fairly common in other medical situations, such as distributing emergency room services (Schwartz, 1975).

Interestingly, other methods of allocating scarce medical resources that do not inherently favor concerns of need have been suggested, but they have failed to gain wide approval. For example, an "all-or-nothing" plan that would deny artificial kidneys to all patients if all worthy recipients could not be accommodated has been suggested (Cahn, 1955), and it has been implemented in at least one hospital ("Scarce Medical Resources," 1969). However, such plans have not caught on, and allocation plans that favor need seem to predominate when distributing scarce medical resources (Calabresi & Bobbitt, 1978).

2.1.2. Experimental Research

2.1.2a. Previous Studies. Unfortunately, there have been very few studies that directly examine the effects of scarcity on allocation behavior. What little experimental research there has been has dealt with matters more circumscribed and less monumental than the tragic issue of artificial kidney allocation. This evidence also makes it clear that considerations of equity become less important as the resources to be allocated become more scarce (Lewis & Leventhal, 1976).

A series of studies by Lane, Messé, and their associates (e.g., Coon, Lane, & Lichtman, 1974; Lane & Messé, 1972; Katz & Messé, 1973) provided our first suggestion of the way people allocate insufficient resources. The subjects in these experiments were college or elementary-school students who divided monetary payment between themselves and a co-worker after performing a task. The total amount of payment available for allocation was assumed on the basis of pretesting to be either insufficient, sufficient, or oversufficient relative to the prevailing external standards for the type of work performed. It was found that adult males and children of both sexes kept greater proportions of the reward when the supply was insufficient (i.e., less than the prevailing pay standards) than when it was sufficient (i.e., consistent with the prevailing pay standards). (They also kept a greater proportion of the reward when the supply was oversufficient, but this result will be discussed in the section on abundance.) These findings have been offered as support for the contention that allocation decisions are made on the basis of appeal to internal standards of justice as well as of comparison with others. Another conclusion suggested by these studies is that the subjects gave priority to their own needs when the available supply was inadequate to meet both their own and their partner's needs.

A similar concern about need reflected in allocation behavior is found in a study by Karuza and Leventhal (1976). The subjects in this experiment allocated units of food among hypothetical children for whom food was either plentiful (low need) or scarce (high need). Not surprisingly, the subjects favored the needier recipients in making their allocations. An analogous finding has been reported by Leventhal and Weiss (1975), whose subjects divided monetary team earnings between themselves and a (fictitious) co-worker. The subjects were also informed of the degree of extraneous monetary resources that their alleged partners had via a bogus index that provided the impression that their partners' economic resources were either relatively scarce or abundant. It was found that the subjects recognized their partners' greater need and favored these individuals to a significant degree, a response that the subjects recognized as being based on their desire to help.

2.1.2b. New Evidence. The most direct evidence of the perceived fairness of basing the allocation of scarce commodities on need is provided by some of the author's own previously unpublished data (Greenberg, 1979c).

A very simple preliminary survey was taken from a sample of 155 undergraduate students (74 males and 81 females) at a midwestern university; the subjects were asked to rate the fairness of two schemes for allocating natural resources. Two different resources were considered—coal and oil—which, at the time the data were collected (the spring of 1977), had been publicized as being highly *un*equal in their availability. The subjects were reminded that coal was considered plentiful and that oil was scarce. (Ironically, a year later, a strike by the United Mine Workers made coal scarce, too.) Two allocation schemes were described: *equality*, in which the available supply would be divided equally among all customers; and *need*, in which the customers with greater dependence (such as hospitals) would be favored at the expense of less dependent customers.[2]

The means displayed in Table 13.1 clearly reveal the striking divergence in the perceived fairness of using need versus equality as a decision

---

[2] Admittedly, this operationalization of need was vague since the nature of the dependence was not clearly specified. However, our purpose was simply to compare the subjects' beliefs about the fairness of allocations made *without* consideration of external criteria (equality) with those made on the basis of *some* dependence criteria, however defined (need). This contrast was made salient by having the subjects evaluate both cases.

rule in allocating scarce natural resources. The overall interaction between allocation scheme and resource scarcity was highly significant, $F(1, 153) = 332.31$, $p < .001$. Post hoc tests using the Tukey HSD procedure revealed that this interaction was attributable to the perceived fairness of allocating a scarce resource according to need and the unfairness of allocating it equally. In the case of abundant resources, as we discuss later, the subjects did not perceive one scheme as more or less fair than the other. Distributions of abundant resources based on need and equality were *both* perceived as being neither particularly fair nor particularly unfair. This study provides the clearest evidence yet that considerations of need are perceived as fair determinants of allocating scarce resources. Moreover, under such conditions, considerations of equality were perceived as being patently unfair.

2.1.3. Conclusion

On the basis of the evidence from the diverse experimental procedures employed in these studies, it seems safe to conclude that under conditions of scarcity, considerations of need predominate in making allocation decisions. The evidence has shown both that scarce resources tend to be allocated in a manner that favors needier recipients and that such allocation practices are perceived as being fair. This conclusion appears warranted in light of both the sociohistorical observations and the experimental evidence collected. On this basis, we concur with Eckhoff's (1974) conclusion: "It goes without saying that the criteria of *need* must play a prominent part when the aim [of allocation] is to improve conditions for the recipients" (p. 248; emphasis in original).

Table 13.1. Perceived Fairness of
Distribution Schemes as a Function of
Resource Scarcity[a]

Distribution scheme	Resource scarcity	
	Scarce (oil)	Abundant (coal)
Need	7.41_a	4.96_b
Equality	2.36_c	5.06_b

[a] Responses could range from 1 (very unfair) to 9 (very fair). Means not sharing a common subscript are significantly different beyond the .05 level according to the Tukey HSD procedure. In each cell, $n = 155$. (Data from Greenberg, 1979c.)

2.2. Efficiency

A second important determinant of allocating scarce resources is efficiency. Obviously, an efficient distribution of resources, one that minimizes waste, becomes an important aim when the resource pool is very limited. With few resources available, each individual resource unit becomes important and should not be wasted. At the time this is being written, in fact, some major airlines have announced cancellation of their less popular flights—an efficiency-based move, doubtlessly motivated by the economic reality of not attracting enough customers to offset the rising cost of jet fuel.

2.2.1. Sociohistorical Observations

As would be expected, the available evidence suggests that wastefulness is far more prevalent among citizens of the affluent, developed countries, the "haves," than among citizens of the poor, underdeveloped, "have-not" nations. Consider, for example, Morrison's (1976) conclusion: "The billion or so people who live in the developed countries use more—and probably waste more—of the natural resources that are the focus of current concern over scarcity than are used by the other three billion combined" (p. 294).

A concern about efficiency also comes across in Calabresi and Bobbitt's (1978) account of artificial kidney allocations, particularly in Great Britain. In England, artificial kidneys are allocated with a simple, mechanistic, efficiency-determined egalitarianism: they are allocated so as to achieve the highest possible rate of success. By excluding any candidates who, for whatever reason, are questionable medical risks for recovery (e.g., the aged, laborers), the British attempt to yield the most life-years out of the limited number of available machines.

Schwartz's (1975) analysis of persons queuing for services provides still another example of the stress put on efficiency in providing access to scarce services. For example, he noted that holders of relatively scarce skills, such as doctors, can command more waiting time from their clients. In fact, "the greater the scarcity of a service, . . . the less is a server compelled to reduce the waiting time of clients" (p. 15). Waiting time is one of the costs borne by clients in seeking the services of providers of scarce skills. Lining up clients for servers facilitates attempts to distribute the server's benefits more efficiently, and it reaffirms the power that society bestows on holders of scarce skills (Blau, 1964).

The same thing would seem to apply to the distribution of scarce goods. Merchants can reduce their costs by having clients come to them and wait in line for goods, as witnessed by recent news stories of long lines for extensively discounted airline tickets. Popular accounts of long lines for scarce, but popular consumer goods in the Soviet Union provide additional evidence (H. Smith, 1976).

2.2.2. Experimental Research

There are two experimental studies that also clearly highlight the heightened concern about efficiency associated with allocating scarce resources. In the first, Leventhal, Weiss, and Buttrick (1973) had subjects allocate a fixed number of rolls of film between two fictitious interviewees under conditions that either encouraged minimizing waste (as is imperative under scarcity), or failed to mention waste, and in which the interviewees were believed to be high or low previous film users. Not surprisingly, the subjects dissuaded from wasting film gave more to the high users, those who were believed to be more likely to use the film before it spoiled.

Some additional experimental evidence that underscores the importance of efficiency under conditions of scarcity comes from some additional items in Greenberg's (1979c) survey study previously described. For a scarce resource (oil) and for an abundant one (coal), the respondents had to rate the perceived fairness of each of three pricing schemes: increasing rate with higher use, decreasing rate with higher use, and same rate regardless of use.

The means in Table 13.2 reflect a significant interaction between source of scarcity and pricing scheme, $F(2, 306) = 96.21$, $p < .001$. As shown, scarce resources were seen as being most fairly priced according

Table 13.2. Perceived Fairness of Pricing Schedules as a Function of Resource Scarcity[a]

Pricing schedule	Resource scarcity	
	Scarce (oil)	Abundant (coal)
Increasing rate with higher use	7.04$_a$	5.19$_b$
Decreasing rate with higher use	3.13$_c$	4.99$_b$
Equal rate regardless of use	4.99$_b$	6.66$_a$

[a]Responses could range from 1 (very unfair) to 9 (very fair). Means not sharing a common subscript are significantly different beyond the .05 level according to the Tukey HSD procedure. In each cell, $n = 155$. (Data from Greenberg, 1979c.)

to an increasing sliding scale, and least fairly priced according to a decreasing sliding scale, with intermediate fairness for equal pricing. This pattern stands in sharp contrast to the findings for abundant resource conditions, in which both unequal pricing schemes were seen as being intermediately fair, and equal pricing schemes were perceived as being more fair.

What makes these responses interesting is that they suggest that the subjects had some internalized sense of basic economic principles and that their perception of fairness was related to the assumed effects of pricing schemes on consumption. Most intriguing in this regard is the comparison between the fairness of the increasing and the decreasing schedules in the scarce condition. Increasing prices with greater use would penalize high users and encourage conservation; decreasing prices with higher use would discourage conservation and do little to penalize wastefulness. Since, of course, in scarce conditions, each unit is valuable, reckless waste is to be discouraged, and careful, conservative use is to be encouraged.

It is quite intriguing that the subjects reflected their awareness of this dynamic in their fairness judgments. It seems reasonable to suggest that what the subjects seemed to be saying in making their fairness judgments was "What is fair is what brings about good for all." In the case of scarce resources, what is fair is what would help alleviate the scarcity, or minimize its consequences for society. This is what Rescher (1966) referred to as "justice in the wider sense." It is interesting to note the highly utilitarian cast attached to this reasoning.

The reader should be cautioned, however, that the idealism expressed in these responses may, to some extent, be the result of the respondents' lack of involvement in the questionnaire. Most college students probably do not pay utility bills, and they may be free to recommend more idealistic policies than those, such as their parents, for whom the matter would have a definite monetary external referent. Although more work needs to be done in this area, it seems reasonable to conclude that concern about efficiency becomes salient as resources become scarce.

2.3. Need and Efficiency Interrelated

It should be pointed out that matters of need and efficiency are not necessarily completely independent of each other. In many cases of allocation of scarce materials, favoring recipients with the greatest need may also be the most efficient course of action (Eckhoff, 1974). As examples of such a possibility, consider the cases of allocating heating fuel to hospitals as opposed to theaters, or gasoline to ambulances as opposed to private autos. The utilitarian societal priorities of caring for the infirmed at the expense of diminishing pleasurable opportunities for others may well represent one of the basic societal values that are highlighted by the conflict imposed by scarcity.

The interrelationship between need and efficiency is particularly apparent in the World War I notion of *triage*, caring for those who can be saved by receiving treatment, and withholding treatment from those whose conditions make cure unlikely and those who will recover without medical attention. This principle permits a distribution of resources that is both efficient and sensitive to needs. Its popularity is attested to by its extension from trauma situations in military medicine to problems of world population (Ehrlich, 1968).

It is also feasible for need and efficiency to be in conflict. For example, although the British system of allocating artificial kidneys described by Calabresi and Bobbitt (1978) is ostensibly concerned with patients' needs, that country's single-minded efficiency orientation can actually subvert the priority of need. Such would be the case when healthier, less needy patients receive treatment at the expense of more seriously ill patients since the result is an increase in the total number of life-years gained per machine. This is what Eckhoff (1974) has referred to as the "fitness" criterion: allocating according to the ability to use resources best. In medical examples, the neediest are usually the least fit; thereby, the two criteria are pitted against each other. Even in highly needs-oriented Italy, the opportunity to subvert the medical system (Calabresi & Bobbitt, 1978) makes it possible for fitness to override need in actual practice. Since fitness is an efficiency-oriented criterion, the simultaneous application of need and efficiency criteria may sometimes be difficult.

3. ABUNDANCE

Abundant conditions are represented by states in which the available resources are in excess of the number needed to satisfy each recipient's demands for them. In an economic realm, this situation exists in the affluent societies, those in which the level of production exceeds the minimum requirement for standards of decency (Bowie & Simon, 1977). Among the conclusions drawn regarding the distribution of abundant resources in this section is that the concerns identified as paramount in distributing scarce resources—need and efficiency— are decidedly subordinate. By definition, basic needs become relatively unimportant under abundance because they are easily fulfilled. Moreover, wastefulness is readily tolerated because the surplus is large.

This phenomenon is, perhaps, clearest under conditions of what Rescher (1966) called "superabundance," where all wants and all needs are satisfied. Under such conditions, he wrote, the matter of distributive justice fails to arise. Adumbrating this idea was the eighteenth-century preutilitarian, Hume (1740/1964), who stated, "if nature supplied abundantly all our wants and desires . . . the jealousy of interest, which justice supposes, could no longer have [a] place" (p. 267).

Consider air, for example, which is so abundant that the matter of its distribution, and the validity of claims thereto, is trivial. Only as "breathable," high-quality air becomes scarce through environmental pollution does the issue of distribution become paramount. The same thing can be said for water, another plentiful natural resource. In regions of the world in which water abounds, its distribution is not at issue: all who claim it can receive it. However, in arid or polluted regions, the questions of who will or will not receive supplies of water is a crucial one (Thomas & Baker, 1976). Likewise, with the cost of kidney dialysis machines now greatly reduced, the intricately delicate issue of claims to them is relatively unimportant (Calabresi & Bobbitt, 1978). Similarly, the point has been made that the need for deciding treatment priorities among emergency room patients is markedly low when staff availability exceeds patients' demands for services (Schwartz, 1975).

3.1. Review of Experimental Evidence

The experimental studies reviewed earlier that suggest that need and efficiency are important considerations when distributing scarce resources also suggest that these considerations are less important under conditions of abundance. Good evidence for this is derived from Greenberg's (1979c) survey data summarized in Table 13.1. As previously noted, respondents failed to differentiate between the fairness of allocations based on need and of those based on equality. Both schemes were perceived as being neither particularly fair nor particularly unfair—an intermediate response suggesting the lack of importance of the matter of distribution when supplies are plentiful. Moreover, in pricing abundant resources, as shown in Table 13.2, an egalitarian scheme was preferred to plans that discouraged or encouraged conservation, suggesting the relative perceived unimportance of efficiency as a criterion. As Hume (1740/1964) said, "Encrease to a sufficient degree . . . the bounty of nature, and you render justice useless, by supplying its place with much nobler virtues" (p. 267).

In this connection, it is also useful to recall the findings of Messé and his associates (Lane & Messé, 1972; Katz & Messé, 1973) in the conditions in which their subjects distributed oversufficient rewards.

Compared with male allocators distributing an amount of reward consistent with internal standards of pay sufficiency, those allocating oversufficient amounts kept a greater portion for themselves. We have already explained a similar self-favoring response in the undersufficient condition by suggesting that subjects put their own needs ahead of their co-workers'. In this case, a similar response may be given a different interpretation by virtue of differences in the sufficiency of the available reward. Specifically, it would appear that subjects in the oversufficient condition in these studies left their co-workers a smaller portion of a larger initial reward, one that may have approximated an amount commensurate with external standards of fairness. The subjects' own more selfish responses seem to represent an attempt to keep the remainder left after both parties' internal standards of fairness had been satisfied.

The strategy represents a variation of Rescher's (1966) conception of allocating windfalls according to a "devious application of the proportionality principle," or what Komorita and Kravitz (1979) called the "equal excess" norm. That is, once the basic minima have been met, the excess would be distributed evenly among all recipients. However, the findings of Coon, Lane, and Lichtman (1974), Lane and Messé (1972), and Katz and Messé (1973) show that allocators kept the remainder for themselves rather than distributing it equally. Similarly, Komorita and Kravitz (1979) found that when prize values were high (the available points were plentiful), bargainers more frequently followed the more profitable equality norm than the equal excess norm. This evidence appears to represent a case of what Lerner (1974) has called "justified self-interest," the normative obligation to take advantage of the opportunity to benefit oneself.

However, since monetary rewards were at stake in these studies, it *cannot* be implied from this evidence that subjects will capitalize on the abundancy by favoring themselves. The problem lies in the fact that while money may be considered more or less abundant within the experiment, it is *not* an abundant commodity for most subjects outside the experimental context. Thus, conditions of abundance may not actually have been created. However, on the basis of the evidence presented thus far (and some additional evidence to be presented later), it appears that truly abundant rewards are allocated without regard for need or efficiency. Needs satisfied by abundant resources are, by definition, low; the capacity to provide the resource is so high that all demands can be met, and neediness cannot result.

3.2. Abundance-Breeds-Scarcity Paradox

A different reaction to allocating abundant resources is to raise the perception of what constitutes the basic minimum allocation, that is, to "escalate the minima" (Rescher, 1966). As societies become more prosperous, concerns shift from maintaining basic, minimum substinence, to maintaining more comfortable standards (Bowie & Simon, 1977; Rainwater, 1974). Thus, what is considered the basic "utility floor" is revised upward (Rescher, 1966).[3] The emphasis is upgraded from maintaining a minimum standard of living (i.e., just staying alive) to maintaining a minimum standard of decency (i.e., enjoying a purposeful life with a sense of self-respect) (Bowie & Simon, 1977). In fact, Ryan (1942) cited extensive theological precedents denoting that the improvement in living conditions is the basic duty underlying the distribution of "superfluous wealth."

As previously noted, this is what Rescher (1966) referred to as justice in the wider sense of promoting the general good. He argued convincingly that as resource abundance increases and a share in "the good life" becomes possible for all, it is not only plausible but *imperative* to elevate conceptions of what constitutes a minimum share of that good. This argument is consistent with Brickman and Campbell's (1971) notion of "hedonic relativism," according to which the subjective standards by which pleasurableness is gauged rise as the environment becomes more pleasurable. The inherently paradoxical nature of this line of reasoning is noteworthy. As the conception of the basic minimum is raised, a state of scarcity—or, more properly, "felt insufficiency"—is again created. Rising expectations of a "satisfactory life" keep people in a state of felt insufficiency.

3.2.1. Married Women in the Work Force

An interesting base of empirical support for this process is derived from reference to the literature on changing norms regarding women in the work force (for a review, see Scanzoni, 1972). In this connection, the work of Parke and Glick (1967) is noteworthy. These investigators suggested that divorce rates may decline because women's entry into the work force will raise access to economic resources on an *absolute* level, thereby minimizing one potential source of marital disharmony. However, it has also been noted that such an outcome may do little to minimize the equally important feelings of *relative* deprivation, since new comparisons vis-à-vis other working women may raise the level of expectations (Roszak & Roszak, 1969). This kind of comparison effectively maintains the irritating effects of felt insufficiency, albeit at a higher absolute level. This sociological example nicely demonstrates the para-

---

[3] Another possible consequence of affluence may be a reduction in the collective willingness to delay gratification (Kahn, Brown, & Martel, 1976), although this remains an untested possibility.

dox identified by philosophers: conditions of abundance lead to raised expectations, which result in further perceived scarcity.

3.2.2. The New International Economic Order

Further evidence of this process is suggested by a political–economic example: the creation of a new world economic power base in the Middle East. Pursuant to our earlier analysis, the issue of justice would sink to irrelevance if the world relied on a superabundant source of energy such as the sun. However, the issue of justice is highly relevant to the distribution of oil because (a) it was once believed to be abundant but has recently become recognized as dangerously scarce; (b) the traditional powers have grown increasingly dependent on it; and (c) its control is in the hands of nations that have traditionally wielded much less power.

Some environmentalists (e.g., Hardin, 1974) have advanced the "lifeboat" analogy to prophesy the dangers of uncontrolled resource depletion, leading to the ultimately tragic necessity of deciding who will live and who will die. However, others (e.g., Borgstrom, 1975) have likened the situation more to a "luxury-liner" analogy, noting that what is really threatened is the highly skewed level of affluence throughout the world. What comes to mind is the California drought of 1977. Although more lawns than people suffered (not considering fire dangers, of course), the limited availability of water proved inconvenient to the customarily affluent way of life (Fraker & Cook, 1977). Similarly, the Organization of Petroleum Exporting Countries (OPEC) cartel, it has been said, really threatens only certain affluent lifestyles; it hardly threatens basic survival (Morrison, 1976).

The OPEC actions are often justified in the name of equity rather than greed. Thus, the reason for their cornering control of this resource is to reestablish justice within the world, where their previous attempts at economic development were thwarted by inequitable returns for their petroleum. This sentiment has been expressed by Venezuela's President Pérez (1974) in a full-page letter appearing in the New York Times, and it is indicative of the burgeoning inequity theme in international relations (Brown, 1973). As Morrison (1976) noted, the OPEC move is indicative of the widespread movement in the developing nations in their "revolution of rising expectations" and their call for a "New International Economic Order" (cf. Barraclough, 1976), all in the name of justice.

3.2.3. Conclusion

The foregoing analyses provide further evidence in support of the philosophical paradox we have been speaking of: abundance leads to increased expectations, which lead to feelings of scarcity. The "haves" are threatened by the "have-nots' " attempts to control an essential resource because equitably sharing the world's riches threatens the "haves' " affluent lifestyle.

Thorstein Veblen, in his classic book, *The Theory of the Leisure Class* (1899) noted that accumulation serves consummatory value among the poor, and status value among the wealthy. Only the future can tell whether the justice of the "New World Economic Order" will facilitate redistribution of the world's riches so that one person's status is not maintained at the expense of another's life.

4. SPECIAL PROBLEMS AND ISSUES FOR FUTURE RESEARCH

4.1. Self-Enhancing Views of Justice

Several theorists have hypothesized the existence of a self-maximizing justice motive. This comes across in Walster, Berscheid, and Walster's (1973) first corollary of equity theory, that individuals will try to maximize their outcomes, even if this means behaving inequitably. A similar selfishness is inherent in Törnblom's (1977) second assumption of distributive justice, that people are more motivated to avoid injustices to themselves than to others. Likewise, Lenski (1966) adumbrated these sentiments by stating that when individuals are forced "to choose between their own, their group's interest, and the interests of others, they nearly always choose the former" (p. 30). Even earlier, Homans (1961) stated that in determining justice, people are likely to emphasize the worth of the characteristics and the behavior in which they are themselves most outstanding. Taken together, these propositions suggest not only that justice is in the eye of the beholder but that perceived justice is what benefits the perceiver.

4.1.1. Previous Research

Some evidence attesting to the operation of such a self-serving bias is provided by Karsh and Cole's (1968) interesting report of preferences for various pay systems in Japan. Specifically, a preference for traditional pay systems favoring seniority was expressed by older workers, while a new system favoring individual productivity was preferred by younger workers. In both cases, the workers favored the arrangements that would bring themselves the greatest benefit.

Additional evidence of a preference for self-serving allocation practices is provided by the finding that nontenured university professors more strongly favor equity-based (i.e., merit) pay increases than their tenured colleagues (Lewis & Leventhal, 1976, Study 1). This finding has been interpreted as reflecting the nontenured professors' desire to gain recognition. Since favorable recognition may enhance a professor's chances for being granted tenure, such preferences may clearly be viewed as self-serving.

Lewis and Leventhal (1976, Study 1) failed to find that the allocation preferences of tenured and nontenured employees was qualified by the amount of money available for raises. However, it remains an interesting possibility that self-serving allocation preferences would be stronger when the resources distributed are scarce. Probably very few salaried employees ever feel that their supply of money is completely sufficient; therefore money is almost always perceived as being a scarce resource. For this reason, Lewis and Leventhal's nonsignificant interaction between tenure and the amount of money available for raises was not particularly surprising. However, if money (or any resource, for that matter) were truly superabundant, it seems less likely that such self-serving responses would be made.

4.1.2. New Evidence

Such a possibility is nicely demonstrated in some previously unpublished telephone-survey data collected by the author. The survey reported the attitudes toward various gasoline-rationing plans expressed by 210 adult drivers who were selected at random from the telephone directory of a large southern city. After being asked what the usual rate of miles-per-gallon was on the car they usually drove, they were asked to select which of three gasoline allocation schemes would be the *fairest* to implement. These involved dividing the available supply (a) equally according to the number of registered vehicles; (b) in proportion to the number of gallons usually used; or (c) on a first-come–first-served basis. These alternatives were such that they provided relatively greater benefit to drivers of highly gasoline-efficient cars (a), inefficient, low-mileage cars (b), or neither (c). A random half of the respondents ($n = 110$) were asked to imagine that gasoline was very scarce, with only 25% of the amount available to satisfy all demands; the remaining half were asked to assume that the supply was plentiful, with 100% availability. (See Table 13.3.)

As expected, it was found that the fairest scheme for allocating scarce gasoline was related to the gasoline efficiency of the respondent's vehicle; the scheme that benefited the respondent the most was seen as being the fairest. To facilitate data analysis, the distribution of reported miles-per-gallon figures was split at the median to form a high-efficiency

Table 13.3. Proportion of Respondents Indicating That Various Allocation Plans Were "Fairest" as a Function of Degree of Scarcity and Own Vehicle's Gasoline Efficiency[a]

Degree of scarcity and own vehicle's gasoline efficiency	Allocation plan		
	Equal allotment per vehicle	Allotment by previous use	First-come–first served
Scarce gasoline			
High-efficiency	.29	.11	.15
	(32)	(12)	(16)
Low-efficiency	.08	.27	.10
	(9)	(30)	(11)
Abundant gasoline			
High-efficiency	.14	.13	.19
	(15)	(14)	(21)
Low-efficiency	.14	.17	.24
	(15)	(19)	(26)

[a]There were 110 respondents each in the scarce and the abundant conditions. A subsequent median split of own vehicle's gasoline efficiency resulted in placing 60 high-efficiency and 50 low-efficiency respondents in the scarce condition, with the figures reversed in the abundant condition. Proportions not totaling 1.0 are due to rounding. The n in each cell is shown in parentheses.

group and a low-efficiency group. As shown in the top two rows of Table 13.3, for scarce gasoline, the majority of the high-efficiency drivers found equal allocations to be fairest, whereas low-efficiency drivers found proportional-to-use allocations as being fairest, $\chi^2(2) = 27.32$, $p < .001$. The figures in the bottom two rows of Table 13.3 demonstrate that there was no one overridingly perceived fair method for distributing abundant supplies of gasoline, although a definite preference for first-come–first-served allocations was expressed, $\chi^2(2) = 4.82$, $p < .10$. Presumably, this trend resulted because this is the usual method of gasoline allocation, and the easiest to implement. Maintenance of prevailing allocation plans may be preferred because they minimize cost (Calabresi & Bobbitt, 1978) and avoid angry reactions (Moore, 1978).

These data reiterate the tendency for self-favoring allocations to be followed, particularly when resources are scarce. When resources are truly plentiful, self-favoring allocations are trivial; so much of the resource is available that benefiting the self is not a concern.[4] These data are reminiscent of the survey results reported in Tables 13.1 and 13.2. In that survey, it will be recalled, the perceived fairness of the allocation

[4] As previously noted, although the male subjects in Messé's investigations (Lane & Messé, 1972; Katz & Messé, 1973) kept the majority of the oversufficient reward for themselves, it would be unwise to assume that these subjects perceived that the monetary reward available comprised an abundant supply in any absolute sense.

scheme was unaffected by plentiful natural resources, and an equal pricing scheme was viewed as fairest. Similarly, in the context of a bargaining game, Komorita and Kravitz (1979) found that equality was most prevalent when resources were abundant.

4.2. High Wages Commanded by Scarce Skills

The principle according to which holders of scarce skills command higher wages has been alternatively called the "canon of supply and demand" (Rescher, 1966) and the "canon of scarcity" (Ryan, 1942), or what economists call the classical theory of price determination. Although the moral questions surrounding this issue are better left to the philosophers, some interesting questions are suggested that warrant attention by social psychologists.

Are the holders of scarce skills given greater recognition if they are legitimately responsible for attaining that skill (as through internally mediated factors such as effort) than if they acquire it through external means (such as genetic endowment)? The evidence provided by Leventhal and Michaels (1971) suggests that the cause underlying performance *is* taken into account, and that persons who are internally responsible for outcomes are seen as more deserving than persons who attain the same success by virtue of factors beyond their control. Accordingly, it is not surprising that society's holders of scarce skills, such as the most accomplished surgeons and athletes, command such astronomical salaries. No doubt, these persons' accomplishments are attributable in varying degrees to both internal and external factors. However, it seems clear that society recognizes holders of scarce skills by rewarding them. Being able to do something that others cannot match is perceived as a determinant of deserving.

It must also be recognized that the process of rewarding holders of scarce skills can be viewed as an instrumental act in itself. For example, industrial supervisors often give generous pay raises to employees who have scarce skills in order to retain their difficult-to-replace services (Lewicki, Bowen, Hall, & Hall, 1975). Some insight into this process is derived from Landau and Leventhal's 1976 study, in which the subjects played the role of industrial supervisors giving pay raises that would help retain employees who received counteroffers. It was found that subjects following the policy of attempting to retain productive employees discriminated highly between productive and unproductive workers, greatly favoring the productive ones. Some subjects were told that it was the company's policy to attempt to dissuade turnover among productive employees, which may have suggested to them that good workers were scarce and difficult to replace. Thus, high reward was offered as an inducement to stay. This evidence suggests that scarce skills may command high wages because persons needing these services want to ensure their continued availability.

Other evidence is available that suggests that persons possessing special capacities that predispose them to offer their special talents to persons in need are more likely to render assistance than those who do not possess such special skills (e.g., Clark & Word, 1974; Schwartz & Clausen, 1970). Persons possessing scarce resources may feel morally obligated to put these resources to use, thereby making their possession a special liability. Such is often the case among some physicians who are morally obligated to fulfill their patients' needs even if doing so creates an inconvenience for them (such as by having to provide emergency services to their patients outside of regular office hours).

4.3. Other Important Considerations

4.3.1. Protestant Ethic Endorsement

Among the several individual difference factors studied in conjunction with interpersonal justice (see Major & Deaux, 1981), one particular personality variable, the Protestant ethic (PE) endorsement, stands out as having a particularly important relationship to the issues considered in this chapter.

For example, Greenberg (1979b) found that high endorsers of the Protestant ethic were more likely than low endorsers to perceive allocations based on internally mediated factors as being fair. In addition, PE endorsement has been found to be positively correlated with beliefs such as the following: welfare recipients are lazy, poverty is the fault of the poor, and assistance to the poor is inappropriate; it is also negatively correlated with favoring a minimum guaranteed annual income (MacDonald, 1972). Accordingly, it would not be surprising to find that these attitudinal differences are responsible for different allocation practices regarding the distribution of scarce resources. High PEs may be expected to be more likely than low PEs to reward the holders of scarce skills for their abilities by giving them a higher wage. In addition, high PEs would be more likely than low PEs to eschew an egalitarian approach to allocating scarce resources, ignoring the basic utility floor concept. They appear to believe that hard work makes a person more deserving and that welfare economics is patently unfair. In other words, many of the beliefs underlying the allocation of scarce resources as discussed in this chapter may be moderated by endorsement of the Protestant ethic, although direct empirical evidence is lacking. Evidence showing that locus of control, a correlate of PE endorsement (Mirels & Garrett, 1971), significantly interacts with the size of the available reward in influencing allocation behavior (Lewis & Leventhal, 1976, Study 2) provides some encouragement for now.

4.3.2. Group versus Individual Decisions

An additional factor that may affect allocation decisions is whether the decision-making unit is a group or an individual. Recent evidence has made it clear that the allocation decisions of groups and individuals are often different; they are not equally subject to social pressures from recipients (Greenberg, 1979a). Archival research needs to be conducted comparing decisions on how to allocate scarce resources made by individuals and by groups. The practice of having committees composed of various community leaders make allocation decisions as to who should receive scarce kidneys (Calabresi & Bobbitt, 1978) may be questionable in light of the well-documented tendency for groups to make polarized decisions (Lamm & Myers, 1978), even on issues of justice (Greenberg, 1979a). Research needs to be conducted that will uncover the allocation biases of groups and individuals in such critical situations.

Calabresi and Bobbitt's (1978) account of the Seattle God Committee offers some interesting insight into the issues involved. The committee was formed in 1961 in order to select suitable candidates for hemodialysis, therefore relieving hospital administrators and doctors from having to make these delicate decisions. As in the case of the individual decisions that preceded the committee's organization, the committee's decisions were also charged with showing favoritism. This external conflict, along with a number of internal stresses, contributed to the committee's demise in 1967 (Calabresi & Bobbitt, 1978). Post hoc archival analyses comparing the decisions made by groups of this sort and individual decision-makers would be an interesting source of information for testing and generating hypotheses about group and individual allocations of scarce resources.

4.3.3. Reward Value

Another potentially important consideration is reward value, that is, the subjective importance of the units being allocated to the recipient (cf. Greenberg, 1978). In one of the few studies germane to this issue, Burgess and Nielsen (1974) have observed that persons will engage in a disadvantageous exchange with another in order to attain some scarce and valuable resource from them rather than engage in no exchange at all. This evidence suggests that scarce resources may be valued so much that people will tolerate inequitable distributions of them in order to attain them.

This finding raises some questions about the fundamental nature of justice under conditions of scarcity. Both crucial and less essential commodities may be in short supply, although it cannot be assumed that their allocation would be identical, or that it would be governed by the same factors. Are the considerations the same in dividing a nation's scarce food supply among its peoples as in dividing its scarce supply of toys among its children? The inherent value of the units in question makes these situations very different. An equal division of toys may be seen as fairest because the disappointment is shared equally. However, should an equal distribution of food result in a situation in which nobody has enough to sustain life, such an allocation may be seen as inappropriate. Probably, what would be seen as fairest under these conditions would be implementation of the utilitarian principle of allocating the bare minimum to as many people as possible in order to keep them alive, sacrificing as few as possible. These speculations are, of course, subject to empirical test and should be the concern of future investigators. It has already been established that justice can be subverted by other motives as a function of reward value (Greenberg, 1978). These findings may be qualified, however, by future research introducing the scarcity–abundance variable.

4.4. Some Practical Considerations[5]

Some very interesting applied research questions are suggested that social scientists may wish to consider. With the cutbacks in supplies of gasoline, what policies do service station owners apply in selling their available supply? Do "full-service" stations promulgate policies that favor their repair service customers? Are regular customers rewarded for their loyalty by being given preferential treatment? The answers to these and related questions may well depend on such variables as the estimated duration of the scarcity, the nature of the clientele, the location of the business, external policies, and numerous other factors.

An interesting comparison with the procedures used to allocate (and conserve) home heating fuels can be made. A crucial difference may have to do with storage capacity: one can buy only one containerful of oil or coal at a time, but natural gas or electricity can be delivered continuously to one's home. Users of oil or coal heat are more vulnerable to interruptions in service than are users of continuously delivered resources, such as electricity or natural gas. Might there be differences in the allocation and conservation practices of users of each type of fuel

---

[5] Many of the ideas contained in this section were derived in collaboration with Gerald Leventhal.

because of method of delivery in conjunction with scarcity–abundance? This is just one of the many relevant questions that warrant future research consideration. The answers would not only yield valuable information about our personal conceptions of justice but could also contribute to our understanding and shaping of public policies on these matters. There is reason to be encouraged about the eventual consultation of social psychologists on matters of resource allocation in view of their past involvement on policy-making boards examining other crucial social issues, such as organ and tissue transplants (Saks, 1978).

5. CONCLUSION

We have identified and discussed several major themes and reached conclusions regarding the preeminence of need and efficiency under scarcity, and the tendency for abundance to lead to conditions of felt scarcity. Related issues, such as the self-serving bias in perceiving justice and the tendency for scarce skills to command high wages, have also been examined. Together, the foregoing analyses of these themes clearly showed that justice is moderated by scarcity–abundance in the philosophical sense of what justice morally requires, in the sociological sense of what actually occurs in society, and in the psychological sense of how justice is perceived. Our analysis was decidedly enriched by the insights afforded by these varied perspectives.

ACKNOWLEDGMENTS

I am indebted to Gregory J. Agamy for conducting some of the data analyses, and to Richard Mena for gathering reference materials. Ronald L. Cohen is also acknowledged for his valuable comments on an earlier draft of this chapter.

REFERENCES

Adams, J. S. Inequity in social exchange. In L. Berkowitz (Ed.), *Advances in experimental social psychology*, Vol. 2. New York: Academic Press, 1965.
Barraclough, G. The haves and the have nots. *New York Review of Books*, May 13, 1976, pp. 31–41.
Bartell, T. Political orientations and public response to the energy crisis. *Social Science Quarterly*, 1976, *57*, 430–436.
Becker, G. S. A theory of the allocation of time. *Economic Journal*, 1965, *75*, 493–517.
Berelson, B. (Ed.). *Population policy in developed countries*. New York: McGraw-Hill, 1974.

Berger, J., Zelditch, M., Anderson, B., & Cohen, B. P. Structural aspects of distributive justice: A status-value formulation. In J. Berger, M. Zelditch, & B. Anderson (Eds.), *Sociological theories in progress*, Vol. 2. Boston: Houghton Mifflin, 1972.

Blau, P. *Exchange and power in social life.* New York: Wiley, 1964.

Borgstrom, G. Overfed, wasteful, we live in a luxury oasis and risk a third world war. *Detroit Free Press Magazine*, September 21, 1975, pp. 26–32.

Bowie, N. E. *Towards a new theory of distributive justice.* Amherst: University of Massachusetts, 1971.

Bowie, N. E., & Simon, R. L. *The individual and the political order.* Englewood Cliffs, N.J.: Prentice-Hall, 1977.

Brickman, P., & Campbell, D. T. Hedonic relativism and planning the good society. In M. H. Appley (Ed.), *Adaptation level theory.* New York: Academic Press, 1971.

Brown, L. Rich countries and poor in a finite, interdependent world. *Daedalus*, 1973, *102*, 153–164.

Burgess, R. L., & Nielsen, J. M. An experimental analysis of some structural determinants of equitable and inequitable exchange relations. *American Sociological Review*, 1974, *39*, 427–443.

Cahn, E. *The moral decision.* Bloomington: Indiana University Press, 1955.

Calabresi, G., & Bobbitt, P. *Tragic choices.* New York: W. W. Norton, 1978.

Clark, R. D., III, & Word, L. E. Where is the apathetic bystander? Situational characteristics of the emergency. *Journal of Personality and Social Psychology*, 1974, *29*, 279–288.

Cook, K. S., & Parcel, T. L. Equity theory: Directions for future research. *Sociological Inquiry*, 1977, *47*, 75–88.

Coon, R. C., Lane, I. M., & Lichtman, R. J. Sufficiency of reward and allocation behavior. *Human Development*, 1974, *17*, 301–313.

Crosby, F. A model of egoistical relative deprivation. *Psychological Review*, 1976, *83*, 85–113.

Deutsch, M. Equity, equality, and need: What determines which value will be used as the basis of distributive justice? *Journal of Social Issues*, 1975, *31*, 137–149.

Eckhoff, T. *Justice: Its determinants in social interaction.* Rotterdam: Rotterdam University Press, 1974.

Ehrlich, P. R. *The population bomb.* New York: Ballantine, 1968.

Ehrlich, P. R., & Ehrlich, A. H. *The end of affluence.* New York: Ballantine, 1974.

Ferber, S. Energy research: Coming in from the cold. *APA Monitor*, March 1977, pp. 4–5.

Foster, G. Peasant society and the image of the limited good. *American Anthropologist*, 1965, *67*, 296–311.

Fraker, S., & Cook, W. J. San Francisco: Going dry. *Newsweek*, April 18, 1977, p. 32.

Fuel cutbacks discounted. *New Orleans Times-Picayune*, March 4, 1979, Sec. 1, p. 3.

Greenberg, J. Effects of reward value and retaliative power on allocation decisions: Justice, generosity, or greed? *Journal of Personality and Social Psychology*, 1978, *36*, 367–379.

Greenberg, J. Group vs individual allocation decisions: Is there a polarization effect? *Journal of Experimental Social Psychology*, 1979, *15*, 504–512. (a)

Greenberg, J. Protestant ethic endorsement and the fairness of equity inputs. *Journal of Research in Personality*, 1979, *13*, 81–90. (b)

Greenberg, J. Scarcity and the distributive fairness of natural resource allocations. Paper presented at the meeting of the Southwestern Psychological Association, New Orleans, March 1979. (c)

Gudeman, S. Anthropological economics: The question of distribution. In B. J. Siegel, A. R. Beals, & S. A. Tyler (Eds.), *Annual review of anthropology*, Vol. 7. Palo Alto, Calif.: Annual Reviews, Inc., 1978.

Halévy, E. *The growth of philosophic radicalism* (M. Morriss, trans.). London: Faber and Gwyer, 1928.

Hardin, G. Lifeboat ethics: The case against helping the poor. *Psychology Today*, September 1974, pp. 38, 40–43, 123–124, 126.

Hillis, J. W., & Wortman, C. B. Some determinants of public acceptance of randomized control group experimental designs. *Sociometry*, 1976, *39*, 91–96.

Homans, G. C. *Social behavior: Its elementary forms*. New York: Harcourt, Brace, and World, 1961.

Hospers, J. *Human conduct*. New York: Harcourt, Brace, and World, 1961.

Hume, D. *A treatise of human nature*. In T. H. Green & T. H. Grose (Eds.), *David Hume: The philosophical works*, Vol. 2. London: Scientia Verlag Allen, 1964. (Originally published 1740.)

Kahn, H., Brown, W., & Martel, L. *The next 200 years*. New York: William Morrow, 1976.

Karsh, B., & Cole, R. E. Industrialization and the convergence hypothesis: Some aspects of contemporary Japan. *Journal of Social Issues*, 1968, *24*, 45–64.

Karuza, J., & Leventhal, G. S. *Justice judgments: Role demands and perception of fairness*. Paper presented at the meeting of the American Psychological Association, Washington, September 1976.

Katz, A. Process design for selection of hemodialysis and organ transplant recipients. *Buffalo Law Review*, 1973, *22*, 373, 402–403.

Katz, M. G., & Messé, L. A. *A sex difference in the distribution of rewards*. Paper presented at the meeting of the Midwestern Psychological Association, Chicago, May 1973.

Komorita, S. S., & Kravitz, D. A. The effects of alternatives in bargaining. *Journal of Experimental Social Psychology*, 1979, *15*, 147–157.

Labich, K., LaBrecque, R., & Camper, D. Women vs. veterans. *Newsweek*, March 5, 1979, p. 59.

Lamm, H., & Myers, D. G. Group-induced polarization of attitudes and behavior. In L. Berkowitz (Ed.), *Advances in experimental social psychology*, Vol. 11. New York: Academic Press, 1978.

Landau, S. B., & Leventhal, G. S. A simulation study of administrators' behavior toward employees who receive job offers. *Journal of Applied Social Psychology*, 1976, *6*, 291–306.

Lane, I. M., & Messé, L. A. Distribution of insufficient, sufficient, and oversufficient rewards: A clarification of equity theory. *Journal of Personality and Social Psychology*, 1972, *21*, 228–233.

Latané, B. *Journal of Personality and Social Psychology*: Problem, perspective, prospect. *Personality and Social Psychology Bulletin*, 1979, *5*, 19–31.

Lenski, G. E. *Power and privilege: A theory of social stratification*. New York: McGraw-Hill, 1966.

Lerner, M. J. Social psychology of justice and interpersonal attraction. In T. Huston (Ed.), *Foundations of interpersonal attraction*. New York: Academic Press, 1974.

Leventhal, G. S. The distribution of rewards and resources in groups and organizations. In L. Berkowitz & E. Walster (Eds.), *Advances in experimental social psychology*, Vol. 9. New York: Academic Press, 1976. (a)

Leventhal, G. S. Fairness in social relationships. In J. W. Thibaut, J. T. Spence, & R. C. Carson (Eds.), *Contemporary topics in social psychology*. Morristown, N.J.: General Learning Press, 1976. (b)

Leventhal, G. S., Karuza, J., & Fry, W. R. Beyond fairness: A theory of allocation preferences. In G. Mikula (Ed.), *Justice and social interaction*. New York: Springer-Verlag, 1980.

Leventhal, G. S., & Michaels, J. W. Locus of cause and equity motivation as determinants of reward allocation. *Journal of Personality and Social Psychology*, 1971, *17*, 229–235.

Leventhal, G. S., & Weiss, T. *Perceived need and the response to inequitable distributions of reward*. Unpublished manuscript, Wayne State University, 1975.

Leventhal, G. S., Weiss, T., & Buttrick, R. Attribution of value, equity, and the prevention of waste in reward allocation. *Journal of Personality and Social Psychology*, 1973, 27, 276–286.

Lewicki, R. J., Bowen, D. D., Hall, D. T., & Hall, F. S. *Experiences in management and organizational behavior: Instructor's manual.* Chicago: St. Clair, 1975.

Lewis, S. A., & Leventhal, G. S. *Some determinants of professors' and lawyers' preference for merit pay: A test of the justice judgment model.* Unpublished manuscript, Wayne State University, 1976.

MacDonald, A. P. More on the Protestant ethic. *Journal of Consulting and Clinical Psychology*, 1972, 39, 116–122.

Major, B., & Deaux, K. Individual differences in equity behavior. In J. Greenberg & R. L. Cohen (Eds.), *Equity and justice in social behavior.* New York: Academic Press, 1981.

Mill, J. S. *Utilitarianism.* New York: Bobbs-Merrill, 1957.

Mirels, H. L., & Garrett, J. B. The Protestant ethic as a personality variable. *Journal of Consulting and Clinical Psychology*, 1971, 36, 40–44.

Moore, B., Jr. *Injustice: The social bases of obedience and revolt.* White Plains, N.Y.: M. E. Sharpe, 1978.

Morrison, D. E. Growth, environment, equity and scarcity. *Social Science Quarterly*, 1976, 57, 292–306.

Parke, R. J., & Glick, P. C. Prospective changes in marriage and the family. *Journal of Marriage and the Family*, 1967, 29, 249–256.

Pauly, D., & Walcott, J. Plans for the pinch. *Newsweek*, January 19, 1979, p. 25.

Pérez, C. A. The president of Venezuela responds to the president of the United States. *New York Times*, September 18, 1974, p. 19.

Rainwater, L. *What money buys: Inequality and the social meanings of income.* New York: Basic Books, 1974.

Rescher, N. *Distributive justice.* Indianapolis: Bobbs-Merrill, 1966.

Ross, W. D. *Foundations of ethics.* London: Oxford University Press, 1939.

Roszak, B., & Roszak, T. (Eds.). *Masculine–feminine: Readings in sexual mythology and the liberation of women.* New York: Harper & Row, 1969.

Runciman, W. G. *Relative deprivation and social justice.* London: Routledge & Kegan Paul, 1966.

Ryan, J. A. *Distributive justice.* New York: Macmillan, 1942.

Saks, M. J. Social psychological contributions to a legislative subcommittee on organ and tissue transplants. *American Psychologist*, 1978, 33, 680–690.

Scanzoni, J. *Sexual bargaining: Power politics in the American marriage.* Englewood Cliffs, N.J.: Prentice-Hall, 1972.

Scarce medical resources. *Columbia Law Review*, 1969, 69, 620, 639–654.

Schwartz, B. *Queuing and waiting.* Chicago: University of Chicago Press, 1975.

Schwartz, S. H., & Clausen, G. T. Responsibility, norms, and helping in an emergency. *Journal of Personality and Social Psychology*, 1970, 16, 299–310.

Singer, P. Famine, affluence, and morality. *Philosophy and Public Affairs*, 1972, 1, 222–243.

Singer, P. *Democracy and disobedience.* New York: Oxford University Press, 1974.

Smith, H. *The Russians.* New York: Quadrangle, 1976.

Smith, J. N. (Ed.). *Environmental quality and social justice in America.* Washington: The Conservation Foundation, 1974.

Thomas, R. D., & Baker, K. G. A growth and developmental model for water resource policy. *Social Science Quarterly*, 1976, 57, 445–454.

Törnblom, K. J. Distributive justice: Typology and propositions. *Human Relations*, 1977, 30, 1–24.

Tyler, G. *Scarcity.* New York: Quadrangle, 1976.

U.S. National Advisory Commission on Selective Service. *In pursuit of equity: Who serves when not all serve.* Washington: U.S. Government Printing Office, 1967.

Veblen, T. *The theory of the leisure class.* New York: Macmillan, 1899.

Wallich, H. C. Gasoline rationing. *Newsweek,* December 24, 1973, p. 112.

Walster, E., Berscheid, E., & Walster, G. W. New directions in equity research. *Journal of Personality and Social Psychology,* 1973, 25, 151–176.

14
The Allocation and Acquisition of Resources in Times of Scarcity

ANDRE deCARUFEL

If current projections are correct, societies like our own, which are based on highly complex technology and rapid rates of energy consumption, may be forced to endure some degree of scarcity in resources that are central to our way of life. These projected conditions of scarcity imply that at least some individuals in these societies will experience a sense of deprivation in these areas. The focus of the first section of this paper is on the origin and forms of this scarcity-based deprivation and on an examination of its potential behavioral consequences. Subsequently, suggestions are offered concerning the potential effects of scarcity on the person who occupies the position of resource allocator. In these sections, no attempt is made to provide exhaustive reviews of the literature. Several excellent sourcebooks in these areas have recently been published (e.g., Berkowitz & Walster, 1976; Crosby, 1976; Walster, Walster, & Berscheid, 1978; Lerner, 1977). Instead, the focus is pointing to gaps in this literature, on making suggestions as to future research directions, and on commenting on existing research traditions.

ANDRE deCARUFEL ● Faculty of Administration, University of Ottawa, Ontario, Canada K1N9B5.

1. RESOURCE SCARCITY AND THE EXPERIENCE OF DEPRIVATION

Psychological theories of deprivation (Stouffer, Suchman, Devinney, Star, & Williams, 1949; Crosby, 1976) postulate that individuals experience deprivation if their outcome level is less than they feel they deserve. Thus, deprivation is relative rather than absolute. Individuals may use any one of several standards to decide what they are entitled to. Equity theory (Adams, 1965; Walster *et al.*, 1978) specifies that if individuals perceive that the ratio of their net gains to their contributions is equal to that of a comparison other, they feel fairly treated. Other standards commonly employed to determine deservingness include the individual's own needs, social obligations, and negotiated contracts. Lerner (1977) has proposed a model that specifies the conditions under which one or another of these diverse standards may be used as the criterion of fair allocations. For our purposes, it is sufficient to note that an individual's experience of deprivation at a given outcome level is intimately bound up with considerations of fairness, justice, and deserving. Individuals may experience very low outcome levels and yet not necessarily feel deprived if they do not perceive their outcome level as being unfairly low. Conversely, individuals may receive very high outcome levels and yet experience acute deprivation. The case of highly paid professional athletes who demand to have their $200,000-a-year contracts renegotiated provides a compelling example of this process. In fact, recent experiments (Folger, 1977; deCarufel, 1979) have shown that even those who experience *increases* in their outcome level may be dissatisfied and may feel unfairly treated if the improvements still do not give them the outcome level they believe they deserve.

How does scarcity relate to the experience of relative deprivation? Scarce resources may be a major factor that will cause an individual to receive less that what the he or she feels is deserved. This may come about in several ways. In the simplest case, relatively static comparisons are being made between an individual's outcome level and a standard of deservingness. The effect of scarcity here is on the distribution of the resources and the possibility that not everyone will be able to receive a fair share. The more complex cases involve dynamic comparisons of changes in outcome levels and changes in the standards of fairness in some cases as well. It is important to realize that the scarcity of resources is being embedded in a pattern of resource consumption that in the case of Western societies has been of ever-increasing affluence and comfort. Thus, a scarcity of important resources would not necessarily entail a low level of outcomes for certain individuals but would involve *lower* outcomes than many are used to. This decrease may take the form of an actual deterioration of outcome level or a slowing down of the rate of increased comfort. Each of these patterns is examined here in turn, starting with the static case, moving to the slowed rates of improvement, and ending with cases where actual deterioration takes place. However, in *all* of these cases, subjective deterioration has taken place relative to the standards of outcomes expected and deserved by the individuals in question.

1.1. Unfairly Distributed Scarce Resources

Scarce resources have the potential to create difficult outcome allocation problems. When resources are scarce, the likelihood increases that individuals will be required to make sacrifices. Low outcomes by themselves do not necessarily create the experience of deprivation, but unfairly low outcomes do. The effect of scarcity may be to make unfair allocations more likely because scarcity may arouse in individuals and groups a sense of self-interest. Self-interest may be an important factor in biasing the distribution of these resources away from ones that would be considered fair on all sides. This self-interest does not necessarily result in outright exploitation. It may simply involve people's taking special care to ensure that they themselves will not be exploited. The summation of these individual efforts in the service of protecting one's interest may create an uneven and possibly an unfair distribution of the hardship. A study by Miller (1977) examined how the self-interest–justice-motive conflict is resolved in different circumstances. Subjects were recruited to participate on a concept formation task. At the conclusion of this study, the subjects were given an opportunity to donate some of their time to a worthy cause. The results indicated that the opportunity to help a victim enhanced the number of hours volunteered when the subject was able also to satisfy his self-interest. However, when self-interest was at stake, the subjects were *less* likely to help the victim. Thus, Miller's study indicates the importance of self-interest and suggests that when these interests are threatened, justice for others becomes a secondary consideration. To the extent that individuals are faced with threats to their own self-interest by a scarcity of valued resources, some members of the group, and probably those of relatively low social power, are likely to receive a less-than-adequate share, leading to the experience of relative deprivation.

This possibility is echoed at the intergroup level by Campbell (1965) in his discussion of "realistic group conflict theory." Campbell suggested that group conflicts may occur when the groups have incompatible goals and are in competition for scarce resources. To the extent that groups of individuals each pursue their self-interest in a condition of scarce resources where not everyone's desires can be realized, the possibility of ethnocentrism and group conflict increases. The relevant issue for our purposes here is the effect of in-group–out-group conflict over resource allocation.

Towson, Lerner, and deCarufel (1979) examined precisely this issue. Male and female 5th- and 6th-grader "supervisors" watched a videotape of a boy and a girl stacking shoe boxes. Half of the subjects saw the boy stack twice as many boxes as the girl and half saw the girl stack twice as many as the boy. The subjects watched the videotape under one of two instructional sets. In the low-competition conditions, the subjects were told that the "workers" on the tape were competing as individuals to see who was better at the job, while those in the high-competition condition were told that the individuals on the screen were representatives of boys' and girls' teams engaged in an ongoing competition. It was expected that in the high-competition conditions, the subjects' own sex would cause them to identify with the same-sexed worker and alter their subsequent pay allocations in the direction of group loyalty. After watching the tape, the subjects indicated how a sum of money should be allocated between the two workers. The results indicated that in the low-competition conditions, both boys and girls allocated the pay in proportion to the worker's performance level on the shoe-box-stacking task. However, in the high-competition cases both groups deviated from these equitable allocations. The girls demonstrated favoritism to their in-group by giving the winning female more than she deserved, while the boys discriminated against the out-group by giving the winning female less than she deserved. The results indicate that although justice considerations are central to allocation decisions, in-group loyalties may override justice rules when competition becomes intense. What is even more striking about these results is that they occurred when the self-interest of the subject was based solely on group identification rather than on actual group membership.

Thus, evidence from both the individual and the group level discussed so far suggests that under conditions of competition of the sort that scarce resources might engender, self-interest and in-group loyalties may override justice rules in the allocation of rewards. This effect could lead to a situation in which at least some individuals would experience relative deprivation as a result of receiving unfairly low outcomes. The experience of injustice is implicated in attempts to restore justice either by changing the actual outcomes or by psychological means (cf. Walster *et al.*, 1978; Crosby, 1976). We shall examine the responses of the group members to the experience of deprivation shortly.

Let us now turn to a consideration of how scarce resources may affect the experience of deprivation when examined *over time*, that is, when resources are becoming scarce in a context where they had been abundant in the past.

1.2. Progressive Deprivation

Gurr (1970) used the term *progressive deprivation* to refer to the case of relative deprivation where one's outcomes are steadily increasing but one's feelings of entitlement are rising at an even faster rate. The source of the experience of the deprivation is in the ever-widening gap between experienced and deserved outcomes. This is an explanation for the paradoxical case where an individual is dissatisfied even when he experiences improvement. It may seem a bit unusual to mention the evaluation of improvement in a chapter on the effects of scarcity. Recall, however, that the experience of deprivation is relative and subjective. People in our society have become accustomed to ever-increasing levels of outcomes; in Brickman and Campbell's (1971) terms, we are on a "hedonic treadmill." Scarcity of important resources may be one important mechanism by which the rate of improvement is restrained. Dissatisfaction and a sense of injustice result from the slowed rate of improvement relative to constantly rising expectations. This principle is particularly important because it is a likely pattern for the near future since important resources may become very scarce over an extended period of time. It also provides an explanation for discontent in the face of improvement that is on firmer theoretical ground than trait attributions of "greed" or "ingratitude."

Evidence of the principle of progressive deprivation comes from several experiments designed to examine the experience of improvement in the case of initially disadvantaged parties. In a study by deCarufel and Schopler (1979), subjects were assigned the role of worker in an industrial simulation. They worked on a clerical task and were paid by a (fictitious) allocator over a series of 10 pay periods. The subjects were always unfairly paid initially to create the situation of disadvantage. At this point, the subjects had an opportunity to threaten the allocator or to appeal to the allocator's sense of fairness, or they had no opportunity to try to influence subsequent allocations. During the remaining pay periods, the subjects experienced an improvement that gave them equality with the allocator from that point on, equality plus some compensation for the initial period of discrimination, or no improvement at all. In all cases, the *total* amount allocated (first part plus second part) remained in favor of the allocator. This latter arrangement demonstrated that programs designed to help the disadvantaged are rarely able to compensate the individuals for their whole history of disadvantage. The results indicated that satisfaction and perceived fairness were higher with the improvement than with no improvement when it was granted arbitrarily by the allocator or was forced by a threat from the disadvantaged party. In the case of appeals to fairness, deCarufel and Schopler (1979) found that subjects were no more satisfied with the equality-plus-compensation than they were with no improvement at all! The explanation of these results was based on the idea that the improvement in the arbitrary and threat cases was evaluated against the past low outcomes and was thus quite satisfying. However, following the appeal to fairness, the partial improvement as a response by the allocator served first to legitimize the subjects' right to higher pay and subsequently to fall short of the standard of overall fairness. This study, then, demonstrated a sense of deprivation with improving outcomes and suggested that procedural matters such as how demands for improvement are made is an important mediating factor. A subsequent study by deCarufel (1979) confirmed the relative deprivation analysis by independently manipulating the standard of fairness and the magnitude of the improvement itself and finding experienced deprivation where the standard was not matched.

Another interesting aspect of the deCarufel (1979) study was that it included a manipulation of the resources available to the allocator with which to provide the improvement. The results indicated that the subjects were relatively satisfied with even inadequate improvement if the allocator also suffered losses. However, satisfaction and perceptions of fairness were considerably lower when the allocator continued to prosper. This finding confirms some of the speculations offered earlier about feelings of deprivation in the face of unevenly distributed loss. In this case, the "loss" was a slower rate of improvement. The implication it raises is that if those who have relatively low power will probably suffer the most loss, then disadvantaged groups in society whose lot has started to improve and whose expectations may be rapidly rising could be the hardest hit. The possibility of strongly dissaffected groups in the society is an alarming one. Pettigrew (1967) has related the black American protests of the 1960s to the fact that although the lot of blacks had improved, the lot of whites had improved at an even faster rate. This relation between protests and riots and feelings of deprivation is examined in a later section. It suffices for now to suggest that scarcity of important resources may curtail the progress made by disadvantaged groups in our society and that this may be the most explosive area of society in which a slowing of the rate of improvement may be felt.

1.3.   Aspirational Deprivation

A closely related type of deprivation discussed by Gurr (1970) is called *aspirational deprivation*. This is the case in which the individual's expectations are rising but the outcome level is not. We may speculate that this might occur at a later stage in the degree of scarcity, where the outcomes to be allocated are no longer expanding but exist at a relatively constant level. It would take some time for the rising expectations generated by a long period of affluence to level, and the result would be an experience of deprivation. The deCarufel and Schopler (1979) study discussed earlier indicated that when the disadvantaged individuals asked for more and received no improvement, they felt unfairly treated. Their appeal, based as it was on an undeniable case of unfair pay, was instrumental in creating the expectation that some improvement would be forthcoming. Continued discrimination in the form of identical outcomes in the subsequent pay periods produced strong feelings of discontent akin to aspirational deprivation.

Another situation in which scarce resources could produce aspirational deprivation is one where gains in salary and benefits received by workers would be immediately chewed up by higher prices for the scarce goods and by higher taxes to provide a minimum living standard for those put out of work by industries no longer able to obtain resources to continue production. In this way, improvements are balanced by increased costs. The likely result would be that expectations would be raised when the increased salaries were first received. This expectation would be replaced by feelings of deprivation when the individual realized that he or she was only keeping pace with the increased cost of living. If progressive deprivation seems to be quite likely in the disadvantaged sectors of society, aspirational deprivation may be more common among those who are salaried workers and whose salary increases are automatically passed on by companies in the form of higher prices or layoffs.

This completes our look at the possible effects of scarcity on the discrepancy between outcome level and rising expectations. In neither of these cases does the hardship result in actual deprivation; rather, it serves to retard the individual's expected progress toward affluence and comfort. We now turn to those cases where there is an actual as well as a subjective deterioration in outcome level.

1.4. J-Curve Hypothesis

Davies (1962) has presented a model of responses to an outcome pattern in which there is a steady rise in the outcome level followed by a sharp drop-off. The expectations of the individuals would continue to rise, and a large discrepancy between experienced and expected outcomes would occur at the point of drop-off. This J-curve hypothesis has been related to outcome patterns predicting historical revolutions and rebellions. Sears and McConahay (1973) analyzed the Watts riot and suggested that the rioters were most likely to have been "new urban blacks"—relatively successful and upwardly mobile young blacks. In addition to the subjective status-deprivation as compared with whites, a major cause of the riot was that the usual grievance mechanisms available to whites were blocked for the blacks. These blocked mechanisms may have served a function similar to the sharp downturn in outcome level itself. The result was a riot, which, according to the authors, was a symbolic act designed to express grievances. Thus, there is evidence that continually rising outcomes that are blocked at a certain level may result in frustration and collective violence. The effects of scarcity may also serve the function of creating a sharp downturn in outcome levels. If, for example, our supply of foreign oil is shut off over a relatively short period, the standard of living in developed countries, based as they are on high energy consumption, would take a sharp downturn. The J-curve hypothesis suggests strongly that this may be an instigation to violent action.

The J-curve hypothesis was examined experimentally by Ross, Thibaut, and Evenbeck (1971). In that study, young boys competed in a rope-pulling contest. Some learned that they were quite competent at the task and others learned that they were not very competent. Experimental confederates acted as "managers" who distributed points in such a way that the subjects received an ever-increasing share over time, and then suddenly, the managers began giving them fewer and fewer points. The results indicated that the "competent" boys were more indignant over the unfair treatment than were the "incompetent" boys and were also more likely to try to retaliate against the manager.

Thus, the J-curve hypothesis has been implicated in perceptions of unfairness and violent activity in historical, survey, and experimental studies. A sudden shortage of essential resources may create these problems in widespread sectors of the population.

1.5. Decremental Deprivation

Gurr (1970) described decremental deprivation as the situation where an individual's expectations remain relatively constant but the experienced outcome level actually deteriorates. Relative deprivation is experienced as the level of experienced outcomes falls further and further below the individual's expectations. We may speculate that this pattern might occur after the individual has had enough experience with scarcity to level off his expectations of increasing prosperity as resources continue to deteriorate. It may also occur in specific sectors of the population whose lives have been untouched by the general level of increasing prosperity but who may be expected to be touched by the scarcity. A study by Ross and McMillen (1973) demonstrated that subjects whose outcomes declined over time felt less satisfied than those who received consistently high or consistently low outcomes, confirming the prediction of discontent with experienced deterioration over time.

The economist A. O. Hirschman (1970) has developed a model of responses to outcome deterioration. He suggested that individuals adopt one of two general strategies to deal with deterioration. One is to "exit" or leave the relationship. This may take many forms, such as buying another brand of toothpaste if a favorite brand deteriorates, or leaving a political party, organization, or state. The other option, "voice," is to the tendency to protest the deterioration with the hope of restoring the relationship to its former level of satisfaction. These options have been described by Tajfel (1976) as "social mobility" and "social change." In the analysis presented by Hirschman, the voice option picks up the theme of protest, violence, etc., that we have seen in other sections.

This section has examined how scarcity of desired resources can result in the experience of deprivation either by making likely the possibility of unfair distribution of hardship or by making it difficult for individuals to attain outcome levels that keep up with their expectations. The results of experienced deprivation have been linked to numerous effects, most prominently among them the possibility of retaliation or violence against society. This aspect is examined in the next section.

2. COLLECTIVE RESPONSES TO RESOURCE SCARCITY

Following the practice of most analyses of relative deprivation (e.g., Crosby, 1976), a variety of sources of evidence have been cited as support for the basic propositions. These have included data on individual feelings of discontent (deCarufel, 1979) and desire to retaliate (Ross et al, 1971), as well as historical and survey analyses of revolutions (Davies, 1962; Sears & McConahay, 1973). Normally, it is considered good to have such "converging" evidence from a variety of paradigms and subject populations. However, this practice has led to glossing over a critical distinction: the unit of analysis. More specifically, the problem is that social psychologists in this area have treated individual and collective actions as equivalent and have explained collective actions, such as riots and revolutions, in terms of the individual psychological processes of the participants. It is almost as if a riot were nothing more than a collection of individuals, all of whom are experiencing relative deprivation and each independently deciding to "go wild" at the same time and in the same place.

The purpose of this section is to present an account of responses to outcome scarcity within the general confines of a relative deprivation approach. The major focus here is on the factors that are necessary to convert individual relative deprivation into collective action.

An attempt is made here to provide a psychological approach to the development and consequences of scarcity-based collective action. However, it will be necessary to consider the interaction of micro- and macro-level variables. The interested reader may wish to consult other works on collective behavior that amplify the "sociological" approach (e.g., Evans, 1975; Smelser, 1962).

2.1. Individual Deprivation and Collective Deprivation

It is important to realize, as Tajfel (1976) has pointed out, that collective or group actions must be explained by processes that occur at the group level. This has not been the case in relative deprivation accounts of collective actions, such as riots and revolutions. These collective actions have been explained by reference to the psychological state of the individuals involved. What is needed is a comparable concept at the group level.

The concept of *fraternal deprivation* (Runciman, 1966) refers to the perception that one's group is deprived relative to the level of outcomes they feel they deserve. This standard of deserving maybe established by legal or constitutional standards of equal opportunity, or by the outcomes of a salient comparison group. To the extent that *groups* believe that they are being unfairly treated, an important precondition is established for group action to eliminate the deprivation. With specific reference to the case of scarcity, if groups feel that the attendant hardship is not being distributed fairly among groups (e.g., a government intervention such as gas rationing may be criticized by salesmen as unfairly burdening their source of livelihood), or that a group's progress is being impeded (e.g., union workers whose large wage settlements are used as a justification by management to lay off a certain number of them), they experience fraternal deprivation.

How is fraternal deprivation aroused? A number of factors seem to be important. One, of course, is the realization by a group member that a state of deprivation exists relative to the outcome level perceived to be deserved for a member of this group. This perception of deprivation is not sufficient for fraternal deprivation to be experienced. A major aspect is the belief that other members of one's group are also deprived. This perception of a "common fate" is crucial to the belief that collective deprivation exists. It is also an important aspect in the development of group cohesiveness, which is a precondition for group action (to be discussed below). A second crucial determinant is the realization not only that deprivation is widespread among members of one's group, but that this deprivation exists because of a *shared group characteristic*. For example, before individual women would take collective action on behalf of women as a group, there must be a belief that the reason for the perceived deprivation is that they are women. It would seem unlikely that women would take collective action if the individuals involved attributed their deprivation to idiosyncratic causes, or if they were unable to reach a consensus about the reasons for their perceived deprivation. In the case of ghetto rioters, Caplan and Paige (1968) and Murphy and Watson (1970) demonstrated a relationship between riot participation and the belief that racial discrimination existed. Similarly, Sears and McConahay (1973) implicated positive black identity and political socialization as factors important in the profiles of participants in the Watts riot. Finally, deCarufel and Carere (1978) demonstrated the importance of the cause of the deprivation in mediating responses to ameliorative gestures by harm doers. Subjects felt more satisfied and more fairly treated when they received compensation after they had been discriminated against than when they had been disadvantaged because of "bad luck."

Thus, the first step is for individuals to realize that they are deprived. This involves a process of *comparison* with a relevant standard of deserving. The next stage involves the belief in a collective deprivation involving similar others. Finally, these collective individual deprivations must be explained as being the result of a shared group characteristic. This is a process of *attribution*. This group-based, or fraternal, deprivation is the first major step necessary to explain collective responses to deprivation because it bases the deprivation at a collective rather than at an individual level.

2.2. Fraternal Deprivation and the Development of Group Cohesiveness

The experience of fraternal deprivation by individuals is not sufficient to impel group activity. In contrast to the egoistical relative deprivation, where it can be posited that the experience of deprivation of an individual motivates that person to do something to eliminate the aversive state, collective action requires processes of mobilization, organization, and leadership of the individuals involved. In other words, group action requires that the theorist consider group variables. These include the formation of groups, the development of group structure, and processes of group dynamics such as leadership and social influence. Without these, one is not explaining group activities; one is simply adding up individual processes and ignoring the group dynamics of collective action.

The process of group formation may take several routes. The individuals may be informally organized around a common interest or a distinguishing characteristic (e.g., homosexuals, war veterans), or they already have a formal organization (e.g., a political party or a religion), or they may be in an "involuntary" group (e.g., ghetto dwellers, the handicapped). Out of this contact with similar others, the individuals may be exposed to the belief that the group is deprived for group-related characteristics. This perception of "consensus" is an important source of "informational influence" (Deutsch & Gerard, 1955), which helps to establish for the individual the beginnings of a group-based explanation for their experienced state of deprivation. This explanation serves to organize the experiences of the individuals involved into a relatively coherent "social reality" (Festinger, 1950). The function of this social reality is to smooth over inconsistencies in individual accounts of their deprivation (e.g., blacks from Alabama and Michigan may both learn to explain their experiences in terms of race) and to bring forward new information that is favorable to the evolving explanation. The state of the collectivity is really still a set of individuals collecting information to refine and reinforce perceptions of their group-based disadvantage.

These individuals must become organized before group action can occur. The individuals may begin to differentiate into different aspects of group activity based on their individual abilities and interests and the distribution of talent in the group. The most important differentiation of function, however, is the emergence of a leader. The leader (e.g., Gandhi, Charles Manson, Adolf Hitler) serves to organize and direct the activities of the group. The leader is often a charismatic figure who inspires the group to act, convinces them that their cause is just, and suggests how the state of deprivation can be alleviated. Notice that the collection of individuals has now started to acquire some of the trappings of "groupness": they have a set of common beliefs, they have begun the process of division of labor, and they have begun to acquire a group-based stance of action.

In order to organize the group members still further and to direct their activities, the leader may begin to demand and receive obedience. This is an important process that sets the stage for the development of group norms and the use of "normative influence" (Deutsch & Gerard, 1955) to secure compliance. It also provides an important element of discipline to the group's activities and demands that group members behave in an appropriate fashion over time, even when group members are not present. In many groups of this sort (e.g., the Moonies), penalties for noncompliance or for leaving the group are very severe. This pressure to stay in the group and to conform to its norms may be due to threats made by the group members themselves (e.g., the threats of retaliation against those who tried to leave Jonestown) or to beliefs instilled in the members about reactions from those outside the group (e.g., Patty Hearst's testimony to the effect that she was afraid of the reception she would receive from the authorities if she left the SLA).

The final stage of group organization and influence occurs when the beliefs of the group and the norms of conduct become more-or-less crystallized into a group ideology. The development of an ideology is central to the formation of numerous "revolutionary" groups, in the realm of politics (e.g., Nazis, Communists), religions (e.g., Black Muslims), or terrorist cells (e.g., IRA, PLO, JDL). The functions of the group ideology are several and crucial to organized collective action. They are (a) to sharpen and further define the social reality of the group; (b) to specify the form of the actions needed to rectify the state of deprivation (e.g., violent or nonviolent, political or spiritual); (c) to select targets for the revolutionary acts to be directed against (e.g., the police, the government, big business, trade unions?); (d) to establish standards of conduct for group members; (e) to serve as a platform to propagandize for support and new membership; (f) to establish a goal against which the success or failure of group action is measured; and (g) to justify group actions.

The result is a collection of individuals who have been organized into a group, and who have a relatively clear idea of what the trouble is, what needs to be done and to whom, and how these acts are to be accomplished. This may seem like rather a lot of steps before group action can take place. It is important to realize, however, that group action does require more coordination than does individual action. It might also be noted that a variety of important aspects have been left out of the accounts of responses to individual deprivation as well (e.g., the attribution of a cause, the selection of a target), but group or collective actions require more explanation still. All of these activities can occur on a continuum of formality. It is not always necessary to have a full-blown ideology like the Marxists do to justify group actions; more informal social realities will do. However, these elements of group formation, organization, and direction appear to be necessary at some level to explain the actions of even informal or relatively transient groups.

Now that the group has the potential to act, it is necessary to examine the conditions under which collective action will take place.

2.3.   Enacting Collective Activity

In describing the actual occurrence of collective action, several important distinctions need to be made. One is between planned and unplanned action, the other between instrumental and symbolic action.

Planned action means actions that are largely premeditated by the group, whereas *unplanned action* deals with spontaneous activity. Planned actions may take the form of political intrigue, converting followers to the new religion, or planning and executing a guerrilla operation. These actions usually have a specific objective in mind. Thus, important considerations in predicting the occurrence of group action include the value of the goal itself and the probability of success in attaining it by group action. These components combine into some sort of "expectancy-value" model to predict the type of group action chosen and the timing of the action itself. The important thing to keep in mind is that these expectancies and values are group-generated, or they are developed by the leader and communicated to group members. The value of the goal may change over time, with the increasing scarcity of valuable resources causing the value of the resources to increase. Similarly, the probability of success may fluctuate, depending on the political and economic climate, the strength of the group, and so on. Planned group actions will probably not occur unless these conditions are favorable, even with a strong sense of deprivation and the existence of an organized group.

Unplanned action, on the other hand, is just that—relatively spontaneous, stimulus-triggered group action. Resentment may be smoldering under the surface, but action does not occur until some critical precipitating event takes place. In the case of the Watts riot, Milgram and Toch (1969) described the precipitating event as involving the alleged use of force by the police in arresting a young black on a drunk-driving charge. According to witnesses, this act seemed to infuriate the crowd of bystanders and touch off the riot. The precipitating event may take a number of different forms. One is what we might call the *last-straw effect*. In this case, the group may be experiencing some degree of deprivation but is managing to contain its desires to retaliate until a final, sometimes trivial, event pushes it beyond the threshold for enduring deprivation. The result is that the group explodes into activity designed to eliminate the deprivation.

A second way in which a precipitating event might cause collective action would be when there is an abrupt change in the status quo. In Davies's (1962) theory of revolution, he described an abrupt downturn in a steadily rising outcome level as a cause of revolutionary activity. The experiment by Ross *et al.* (1971), described earlier, provides data to support this contention. This change in the status quo need not necessarily be negative. Crosby (1976) quoted de Tocqueville as saying, "evils which are patiently endured when they seem inevitable become intolerable when once the idea of escape from them is suggested" (p. 85). This quote suggests the possibility that sudden improvement or sudden hope precipitates negative affect and possible action to eliminate the cause of the deprivation. Experiments by Folger (1977) and by deCarufel and Schopler (1979) provide evidence in support of this point. In the case of scarce resources in particular, one might suggest that both a sudden cutoff of the resource and, following a period of deprivation, a small degree of availability of the once-scarce resource would be sufficient to trigger collective action.

A third way in which unplanned collective action could arise is if some element in the social environment serves to disinhibit the groups from restraining their actions. This disinhibition may occur as a result of a provocation, as in the Watts riot example cited above. The members of the group may be willing to endure the deprivation, but a provocation—perhaps in the form of a slight, an insult, or a direct attack—breaks down any inhibitions that the group members may have had against retaliation. Disinhibition may also occur when the group members observe a model who engages in retaliatory activity and is at least moderately successful. The group members may follow the lead of the model and engage in similar activities. There may also be environmental causes as well that serve to disinhibit the group members. These may be factors like deindividuation (Zimbardo, 1969), a breakdown in the surveillance process (e.g., a police strike or a power failure), or a change in social norms.

In all of these instances, a precipitating event triggers a response that may be spontaneous with regard to the time and the place of the event. However, this precipitating event serves only to determine the specifics of the outburst. As Sears and McConahay (1973) have shown in the case of the Watts riot, many other preconditions of a sort similar to those posited here were necessary before this type of response was observed. Thus, the differences between planned and unplanned collective responses are important with respect to triggering mechanisms and mode of expression, but each requires some set of preconditions based on fraternal deprivation and group formation before they can occur. A major difference, though, between a planned and an unplanned collective action involves the composition of the group engaged in the activity. In a planned action, the group is much more homogeneous with respect to feelings of deprivation and beliefs about the justness of the action. In a spontaneous action, many individuals who may not be experiencing relative deprivation at all may join in. The consequences for them of participating are explored in the next section.

Before going on, however, it is necessary to consider the implications of the distinction between collective action taken for symbolic or instrumental reasons. Instrumental acts are those undertaken actually to eliminate the source of the deprivation, while symbolic acts are designed more to achieve other goals, such as drawing attention to the problem, saving face, or demonstrating solidarity. This distinction is important in understanding the forms that a collective action takes and the targets that are chosen. One could expect that instrumental acts would be directed at the source of the deprivation in ways that are less costly and have the highest probability of success for the group. Symbolic acts, on the other hand, may be directed toward targets that may not be the source of the deprivation but that may be able to satisfy some of the alternative aims suggested above. It may also be that symbolic acts are sometimes *designed to fail*—to illustrate lack of power, to provide martyrs, or to try to gain the support that is needed. The significance of this distinction, then, may be felt in important ways that shape the emergence and the cause of group action. Sears and McConahay (1973), in fact, interpreted the Watts riot as symbolic activity—not necessarily meant to eliminate the social conditions of the ghetto but to serve as a substitute method of grievance ("the functional equivalent hypothesis"). In the Walster *et al.* (1978) discussion of responses to inequity, two types of responses are posited: restoring actual equity and restoring psychological equity. Symbolic acts represent a sort of compromise between these two extremes. They are not merely justification tactics; neither do they directly seek to restore actual equity. Possibly, they represent a third category of responses. It would seem to be a valuable avenue to explore.

2.4. The Aftermath of Collective Action

After the collective action has taken place, other important steps come into play. It is important to analyze behavior *sequences*. Often, research in social psychology studies only a single act and does not examine the impact of this behavior on the individual's subsequent behavior. To study behavior sequences of this sort, it is not necessary to use longitudinal studies. Sometimes, the same effects can be measured in multistage experiments carried out within the same experimental session.

Once the group has completed a collective action, several consequences are likely to ensue. If the action was planned, some evaluation relative to the group's goals would take place. This would affect the group's level of aspiration (Zander, 1971) for future activities and also possibly alter the mode of achieving desired ends. A successful exercise of collective action might lead to a higher aspiration level to achieve more radical change. An unsuccessful one might lower the level of aspiration or cause an alteration in the means. For example, an unsuccessful attempt to negotiate with the individuals who have control over a desired scarce resource might cause the group to lower its sights and ask for less, or it might lead the group to adopt a different approach (such as violence).

A second major consequence is that the members of the group might become more committed to their cause. Possibly, this could occur as a function of self-justification (cf. Festinger, 1957) or as the result of a need for positive evaluations of group action (Ferguson & Kelley, 1964), which would make the group more attractive. This increased group cohesion is particularly marked if the collective action was strenuous or dangerous, or if it was directed against a strongly disliked out-group. This aspect of group cohesiveness as a result of action is also important in the case of unplanned actions when the group members may not have possessed a strong attraction to the group previously, or when they may have participated in the group activity not because of feelings of deprivation but for other reasons. However, as a consequence of the group activity, they might begin to experience the sense of deprivation in order to justify their actions. In other words, an individual who went along with the group may now be placed in the situation of having to justify to himself his part in the collective action. The pressure of the group to provide positive evaluations of their action and to reinforce one another's belief that what they did was right would be important in bringing individuals more strongly into the group. The consequence of this increased cohesion might be to increase the probability of future occurrences of collective action.

In sum, the focus of this section has been to examine collective action as one possible consequence of the experience of deprivation brought about by scarcity of resources. The major thrust of this section has been to show how group activity must be explained—at least, in part—by processes that occur at the group level. Precisely to make this point, this section has examined several of these processes, including group formation, social influence, collective action, and the evaluation of group action. What is urgently needed is a research effort in this domain to examine group-based explanations of group phenomena, and to integrate individual and group processes in order to explain these potentially dangerous occurrences.

To this point, the experience of scarcity has been the sole theme. Let us now turn to the role of the allocator and the possible contributing and inhibiting effects that the allocator may have in this collective process.

3. THE EFFECTS OF SCARCITY ON RESOURCE ALLOCATION

We have seen how scarce resources may cause the recipient of allocations to experience a sense of deprivation. This sense of deprivation does not necessarily rely on actual deprivation; all that is required is that the scarcity cause the individual to receive less than he feels is deserved. We have also seen that this experience of deprivation may motivate individual actions and may be a contributor to collective actions to redress the experienced deprivation.

The focus now turns to the effects of scarce resources on the allocator of resources. The role of the allocator is a central one because it is this person who is often perceived as the cause of the deprivation, and it is this person who will be the target of action designed to redress the deprivation. In this discussion, we make the assumption that the outcomes of the allocator and of the recipient are interdependent. This interdependence may occur when the allocator is a co-recipient, dividing up resources between himself and others (e.g., management and union representatives bargaining over the division of the company's revenues) or when the allocator is linked to the recipients (e.g., as an elected representative, or because of sharing with the recipients a group characteristic such as race or sex). This section concerns itself with some of the consequences for the allocator who is not completely indifferent to the resulting distribution of resources between himself and the other recipients.

3.1. Self-Interest

Since the allocator is not indifferent to the resource distribution itself, and since the issue of scarcity implies that not everyone will receive the amount that they would like, the possibility of a conflict between self-interest and the justice motive looms as a strong possibility. This conflict was discussed earlier from the perspective of its effects on recipients. The studies of Miller (1977) and Towson, Lerner, and deCarufel (1979) indicate that self-interest biases allocations away from considerations of justice. Let us now look at the conditions under which the allocator's behavior is influenced.

An important variable would seem to be the power of the allocator over the recipients. Power is a neglected variable in social-psychological theorizing (Cartwright, 1959; Schopler, 1965), particularly at the group level. However, one only has to think for a moment about the tactics used in union–management bargaining, or the way many nations settle conflicts, to realize the importance of this concept in the allocation of resources. This is not to say that power is the only operative variable, nor would I suggest that power is always used to "exploit" the less powerful, but it is important to give power its due.

There may be several reasons that allocations are led in the direction of self-interest as a result of power apart from "exploitation." For one, there is often little contact between the powerful and the less powerful.

The result is that the powerful may not recognize and appreciate the inputs or needs of these other groups. Thus, they may not be acutely aware of the existence of injustice. Lerner (1977) has suggested that individuals try to avoid threats to their belief in a just world, and as a consequence, the powerful may be psychologically unaware of the effects that their allocations have on the less fortunate. The other point is that the lack of contact and the psychological screening out of the injustice may make it difficult for the powerful to perceive the link between their relative affluence and the suffering of others. Finally, even if they did admit the existence of injustice and their role in it, it is quite likely that the consequences to themselves of restoring justice would be so damaging that they would find it far less costly to try to rationalize away the injustice or their role in it (see below). Thus, power may be used by the allocator to allocate in his self-interest if the recipients cannot prevent him from doing so. In addition, the segregation in lifestyle between the powerful and the less powerful may reinforce this principle because the injustice is not perceived, or because the powerful fail to see how they are responsible for the injustice or why they should sacrifice to right it.

A second important consideration in determining the strength of self-interest is the importance of the scarce resource. Resources may differ in how central they are to the activities of the allocator. For example, an oil shortage would motivate considerably more self-interest, because it is central to so many aspects of our lifestyle, than would a shortage of hockey sticks. Resources may also be central because of their symbolic value. A piece of land or access to a certain resource may not be of central instrumental value, but it may be sought nevertheless because of its significance to the holder. A second dimension of the importance is the allocator's perception of whether the scarcity is likely to be severe. To the extent that a shortage of a desirable resource is perceived as serious (e.g., widespread and possibly permanent), one would expect self-interest considerations and competition to become intense.

Naturally, the allocator takes into account some longer-term considerations as well in the allocation of resources. If the allocator feels that it is in his best long-term interests to minimize the expression of self-interest in the short run, then he will do so. An example of this is the moderate OPEC states that try to convince the other, more extreme states that it is not in their best long-term interests to ruin the Western economies with huge price hikes. In a sense, self-interest is still operative here, but the effects on the recipients are radically different.

Thus, it is important to examine the factors that affect the allocator's distribution of scarce resources. Certainly, self-interest plays a part, but the form of that self-interest may work to the benefit or to the detriment of the recipients.

3.2. Justification of the Injustice

A second consequence concerns the tendency for the allocator to eliminate the injustice psychologically, that is, by justifying it. This is a strategy that may be used as an alternative to restoring justice in the allocations themselves. In their formulation of equity theory, Walster *et al.* (1978) suggested several forms that these justification techniques may take, such as blaming the victims for their own suffering, minimizing the perceived suffering endured by the victims, or denying responsibility for any part in causing the suffering. The result, of course, is that the victim does not receive a redistribution of resources, and the allocator may be more prone to allocate unfairly in the future if he feels that the recipients do not deserve fair treatment. Similarly, Lerner (1977) and Lerner and Simmons (1966) suggested that threats to an individual's belief in a just world may cause him to isolate himself from the victims' suffering or to blame the victims themselves.

Berscheid and Walster (1967) demonstrated that individuals were less likely to try to compensate a victim when insufficient resources were available than when sufficient resources were available. The implications in the case of scarcity of resources are clear. If an allocator's self-interest conflicts with the justice motive and insufficient resources are available for the individual to provide compensation to the victim, then compensation is less likely to occur. In order to reduce the distress engendered by participating in an unjust situation, the allocator may justify the injustice. Justifications are probably quite easy to find when considerations of self-interest are involved and if there already exist precedents of injustice in the system. The consequences for the recipients are clearly negative. They do not receive compensation, and because of the justification by the allocator, they may be even less likely to receive it in the future—in fact, they may become targets of further injustice. It is always possible, of course, that a third party may step in to provide the compensation for the victim that is not forthcoming from the harmdoer. It has been suggested in legal circles (e.g., Fry, 1956) that the state should routinely assume responsibility for compensating victims—in this case, of violent crimes. However, a recently completed study by the author (deCarufel, in press) has indicated that individuals are less satisfied and feel less fairly treated when the compensation comes from a third party rather than from the harmdoer himself. Thus, the consequences of justification of the inequity by the allocator can be very marked for the recipients and may contribute to the experience of deprivation and the possible attendant consequences discussed earlier.

3.3. Substitute Rewards

The choices for the allocator that we have outlined thus far generally have negative consequences for the recipient. The scarcity of resources provokes self-interest and increases the likelihood of justification of the injustice. However, the allocator may be able to restore justice to the relationship by allocating substitute rewards for those that are unavailable. The allocator may have at his disposal resources that are similar in function that he may substitute to make up the existing injustice. Naturally, this strategy would depend to some degree on the availability of desirable substitutes but the allocator might be able to find these in some circumstances.

A second possibility would be for the allocator to try to cultivate in the recipients an interest in resources that are different but that may be desirable, more plentiful, and less costly to allocate. An excellent example of just such a process comes from recent work in the area of worker motivation called *work design* (Hackman, 1977). Organizations have tried for many years to solve the problem of worker motivation (Herzberg, 1968). Many of these interventions have been designed to change the extrinsic rewards of the organization, such as increasing pay and benefits and cutting the length of the work week. Two things quickly became apparent: (a) these interventions did not produce the desired effects; and (b) it was a very expensive system to maintain for reasons that are well known to students of adaptation-level theory (e.g., Brickman & Campbell, 1971). The fact that these extrinsic rewards were in relatively short supply led to a revolution in the area of work motivation. The alternative was to try to design work so that it was more interesting, challenging, and rewarding to the employees. Thus, the object here was to substitute intrinsic rewards for the extrinsic ones that were in short supply. This is not to say that such efforts are certain to work (e.g., Dowling, 1977; Winpisinger, 1977), but they represent an interesting attempt to substitute rewards, and to the extent that these substitute rewards exist, they may represent an alternative to the injustice of self-interest and justification techniques.

3.4. Attention to Procedural Matters

To this point, the discussion has centered on the issue of distributive justice—the perceived fairness of the actual distribution of rewards. Another important matter is *procedural justice* (Thibaut & Walker, 1975), the fairness of the methods or procedures used to generate these outcome distributions. Procedural justice combines with distribution justice to influence the individual's feelings of satisfaction and perceptions of fairness. In the context of legal decision-making, Walker, LaTour, Lind, and Thibaut (1974) found that subjects who were judged guilty of a crime of which they knew they were innocent were considerably less satisfied and felt less fairly treated when the trial procedure was one that they felt was also unfair. These procedural elements are also directly implicated in the experience of deprivation with progressively increasing outcomes (Folger, 1977; deCarufel & Schopler, 1979).

The philosopher John Rawls (1971) has presented a method for arriving at fair distributions. It results from an agreement produced in a properly defined initial situation that is itself fair. An important aspect of this original position is the "veil of ignorance": prior to the decision as to how resources are to be allocated, the parties involved do not know whether they will be advantaged or disadvantaged by the result. In other words, a choice of allocation procedure is agreed upon as fair in advance, before the parties know what the result will be. A study in the context of legal decision-making by Thibaut, Walker, LaTour, and Houlden (1974) used this procedure and found that subjects "behind the veil" preferred systems that favored the disadvantaged, while preserving for all of the parties access to information and to mechanisms of control. The implications of this procedure in the case of scarce resources are that the parties would decide on a procedure for allocation before the situation reached a stage where self-interest is very pronounced. This procedure would certainly be difficult to implement in practice, because in many cases, the parties are not behind the veil at all and have quite a good idea of what the likely results of various procedures would be. Nevertheless, it is likely that there are situations in which this is not the case, such as in periods of relatively rapid change, or when new resources (see above) are coming into play.

Finally, I would like to finish off this section with a hypothesis. Basically the hypothesis states that allocators become more attuned to the fairness of the procedures for allocating resources when they believe that the recipients will not be satisfied with the distribution than when they anticipate that the recipients will be satisfied. This attitude stems from the allocator's desire to avoid censure by the recipients and from a general tendency to avoid transmitting negative information (Tessor & Rosen, 1975). Thus, the allocator, given the choice, tries to present the procedures involved as being scrupulously fair. In this way, the allocator can distance himself from the allocations and can avoid retaliation or demands for more from the recipients. One expects that the allocator would exhibit this tendency particularly when he himself is a co-recipient, so as to avoid the impression of impropriety, or when the recipients control a powerful retaliatory potential. This hypothesized tendency has important implications for the presentation of outcome distributions. This aspect of "packaging" is a neglected aspect of the allocation process, but it is crucial to an understanding of how individuals experience this process. This hypothesis concerning the attention to procedure is currently being tested in the author's lab, along with related aspects of the presentation of outcome distributions to potential recipients. It might also be added that if it is true that procedural fairness is of more consideration when negative rather than positive effects are anticipated, then allocators of positive consequences may not be producing optimum satisfaction in their recipients if they ignore these procedural elements. This principle has implications for the introduction of ameliorative social programs, changes in the organizational environment, and so on. If the allocator expects that the allocations themselves will be sufficient to produce high satisfaction, procedural factors may be ignored to the detriment of the recipients' full appreciation of the benefit. Thus, the elements of procedural justice may have important consequences in the introduction of both positive and negative outcome patterns.

REFERENCES

Adams, J. S. Inequity in social exchange. In L. Berkowitz (Ed.), *Advances in experimental social psychology*, Vol. 2. New York: Academic Press, 1965.

Berkowitz, L., & Walster, E. (Eds.). Equity theory: Towards a general theory of social interaction. *Advances in experimental social psychology*, Vol. 9. New York: Academic Press, 1976.

Berscheid, E., & Walster, E. When does a harmdoer compensate a victim? *Journal of Personality and Social Psychology*, 1967, *6*, 435–441.

Brickman, P., & Campbell, D. T. Hedonic relativism and planning the good society. In M. H. Appley (Ed.), *Adaptation-level theory: A symposium*. New York: Academic Press, 1971.

Campbell, D. T. Ethnocentrism and other altruistic motives. In D. Levine (Ed.), *Nebraska Symposium on Motivation*, Vol. 13. Lincoln: University of Nebraska Press, 1965.

Caplan, N., & Paige, J. M. A study of ghetto rioters. *Scientific American*, 1968, *219*(2), 15–22.

Cartwright, D. Power: A neglected variable in social psychology. In D. Cartwright (Ed.), *Studies in social power*. Ann Arbor, Mich: Institute for Social Research, 1959.

Crosby, F. A model of egoistical relative deprivation. *Psychological Review*, 1976, *83*, 85–113.

Davies, J. C. Toward a theory of revolution. *American Sociological Review*, 1962, *27*, 5–19.

deCarufel, A. Factors affecting the evaluation of improvement: The role of normative standards and allocator resources. *Journal of Personality and Social Psychology*, 1979, *37*, 847–857.

deCarufel, A. Victim's satisfaction with compensation: Effects of initial disadvantage and third party intervention. *Journal of Applied Social Psychology*, in press.

deCarufel, A., & Carere, R. *The evaluation of outcome improvement by disadvantaged parties.* Paper presented at the annual meeting of the *Canadian Psychological Association*, Ottawa, June 1978.

deCarufel, A., & Schopler, J. Evaluation of outcome improvement resulting from threats and appeals. *Journal of Personality and Social Psychology*, 1979, *37*, 662–673.

Deutsch, M., & Gerard, H. B. A study of normative and informative influence upon individual judgment. *Journal of Abnormal and Social Psychology*, 1955, *51*, 629–636.

Dowling, W. F. Job redesign on the assembly line: Farewell to blue-collar blues? In J. R. Hackman, E. E. Lawler, & L. W. Porter (Eds.), *Perspectives on behavior in organizations*. New York: McGraw-Hill, 1977.

Evans, R. R. *Readings in collective behavior* (2nd ed.). Chicago: Rand McNally, 1975.

Ferguson, C. K., & Kelley, H. H. Significant factors in overevaluation of own-group's product. *Journal of Abnormal and Social Psychology*, 1964, *69*, 223–228.

Festinger, L. Informal social communication. *Psychological Review*, 1950, *57*, 271–282.

Festinger, L. *A theory of cognitive dissonance*. Evanston, Ill.: Row Peterson, 1957.

Folger, R. Distributive and procedural justice: Combined impact of "voice" and improvement on experienced inequity. *Journal of Personality and Social Psychology*, 1977, *35*, 108–119.

Fry, M. Justice for victims. *Journal of Public Law*, 1956, *8*, 155–253.

Gurr, T. R. *Why men rebel*. Princeton, N.J.: Princeton University Press, 1970.

Hackman, J. R. Designing work for individuals and for groups. In J. R. Hackman, E. E. Lawler, & L. W. Porter (Eds.), *Perspectives on behavior in organization*. New York: McGraw-Hill, 1977.

Herzberg, F. One more time: How do you motivate employees? *Harvard Business Review*, January–February 1968.

Hirschman, A. O. *Exit, voice, and loyalty: Responses to decline in firms, organizations, and states*. Cambridge, Mass.: Harvard University Press, 1970.

Lerner, M. S. The justice motive: Some hypotheses as to its origins and forms. *Journal of Personality*, 1977, *45*, 1–52.

Lerner, M. S., & Simmons, C. H. Observer's reaction to the innocent victim: Compassion or rejection? *Journal of Personality and Social Psychology*, 1966, *4*, 203–210.

Milgram, S., & Toch, H. Collective behavior: Crowds and social movements. In G. Lindzey & E. Aronson (Eds.), *The handbook of social psychology*, Vol. 4. Reading, Mass.: Addison-Wesley, 1969.

Miller, D. T. Personal deserving versus justice for others: An exploration of the justice motive. *Journal of Experimental Social Psychology*, 1977, *13*, 1–13.

Murphy, R. J., & Watson, J. W. The structure of discontent: The relationship between social structure, grievance, and riot support. In N. Cohen (Ed.), *The Los Angeles riots: A socio-psychological study*. New York: Praeger, 1970.

Pettigrew, T. F. Social evaluation theory: Convergences and applications. In D. Levine (Ed.), *Nebraska Symposium on Motivation*, Vol. 15. Lincoln: University of Nebraska Press, 1967.

Rawls, J. *A theory of justice*. Cambridge, Mass.: Belknap Press, 1971.

Ross, M., & McMillen, M. I. External referent and past outcomes as determinants of social discontent. *Journal of Experimental Social Psychology*, 1973, *9*, 437–449.

Ross, M., Thibaut, J., & Evenbeck, S. Some determinants of the intensity of social protest. *Journal of Experimental Social Psychology*, 1971, *7*, 401–418.

Runciman, W. G. *Relative deprivation and social justice: A study of attitudes to social inequality in twentieth century England*. Berkeley: University of California Press, 1966.

Schopler, J. Social power. In L. Berkowitz (Ed.), *Advances in experimental social psychology*, Vol. 2. New York: Academic Press, 1965.

Sears, D. O., & McConahay, J. S. *The politics of violence* Boston: Houghton Mifflin, 1973.

Smelser, N. *Theory of collective behavior*. New York: Free Press, 1962.

Stouffer, S. A. Suchman, E. A., Devinney, L. C., Star, S. A., & Williams, R. M. *The American soldier: Adjustment during army life*, Vol. 1. Princeton, N.J.: Princeton University Press, 1949.

Tajfel, H. Exit, voice and intergroup relations. In L. H. Strickland, F. E. Aboud, & K. J. Gergen (Eds.), *Social psychology in transition*. New York: Plenum Press, 1976.

Tessor, A., & Rosen, S. The reluctance to transmit bad news. In L. Berkowitz (Ed.), *Advances in experimental social psychology*, Vol. 7. New York: Academic Press, 1975.

Thibaut, J., & Walker, L. *Procedural justice*. Hillsdale, N.J.: Erlbaum, 1975.

Thibaut, J., Walker, L., LaTour, S., & Houlden, P. Procedural justice as fairness. *Stanford Law Review*, 1974, 26, 1271–1289.

Towson, S., Lerner, M. J., & deCarufel, A. *Justice rules or ingroup loyalties: The effects of competition on childrens' allocation behavior*. Paper presented at the annual meeting of the *Canadian Psychological Association*, Quebec City, June 1979.

Walker, L., LaTour, S., Lind, E. A., & Thibaut, S. Reactions of participants and observers to modes of adjudication. *Journal of Applied Social Psychology*, 1974, 4, 295–310.

Walster, E., Walster, G. W., & Berscheid, E. *Equity: Theory and research*. Boston: Allyn & Bacon, 1978.

Winpisinger, W. W. Job enrichment: A union view. In K. P. Magnusen (Ed.), *Organizational design, development and behavior*. Glenview, Ill.: Scott, Foresman, 1977.

Zander, A. *Motives and goals in groups*. New York: Academic Press, 1971.

Zimbardo, P. The human choice: Individuation, reason, and order versus deindividuation, impulse, and chaos. In W. Arnold & D. Levine (Eds.), *Nebraska Symposium on Motivation*, Vol. 17. Lincoln: University of Nebraska Press, 1969.

15

Justice in "The Crunch"

MORTON DEUTSCH

1. INTRODUCTION

Tallulah Bankhead is reported to have said "I've been poor and I've been rich and, believe me, rich is better." Most of us would agree with Tallulah. Nevertheless, many individuals, groups, and societies during the course of their existence face the necessity of coping with an economic "crunch," that is, with a diminution of their resources. Often, they cope badly and feel a loss in their self-esteem, their unity, and their sense of purpose as well as in their standard of living. However, it is not always the case that economic loss makes one poorer psychologically and socially. Thus, it is reasonable to ask: What conditions lead to the stimulation of latent paranoia, to the emergence of suspicious relations with others, to the breakdown of civility, to the decrease of individual and group morale, and to individual and social disruption in the face of an "economic crunch"? What conditions foster the effective mobilization of self and community to deal with adversity? These are the basic questions to which this paper is addressed.

My approach to these questions is guided by two interrelated assumptions:

1. The ability to cope is impaired to the extent that the economic loss is experienced as a threat to one's self-esteem. One of the important

MORTON DEUTSCH • Department of Psychology, Teachers College, Columbia University, New York, New York 10027. Work on this paper has been supported by the National Science Foundation, Grant # BNS 77–16017.

factors contributing to this experience is the sense that one's loss is personally unjust, that one has been deprived unfairly.

2. The ability to cope is impaired to the extent that the economic loss weakens the cooperative bonds that exist among individuals, within groups, organizations, or societies. The bonds of solidarity that help individuals and collectives to cope with adversity are weakened by the sense that the loss is not being justly shared or distributed.

It should be emphasized that the maintenance of self-esteem and the cooperativeness of the group bear reciprocal relationships with coping with a crunch. Thus, the ability of a group, institution, or society to cope with economic adversity is a function of its cooperativeness and self-esteem, and, in turn, its cooperativeness and self-esteem are influenced by the success of its coping.

2. CONDITIONS THAT AFFECT COPING

The above two assumptions lead me to place considerable stress on the importance of justice in coping with hardship. However, it would be amiss if we did not recognize that coping is affected by many factors and that the sense of justice in the face of adversity both affects and is affected by these other factors.

Below, I list and discuss a variety of conditions that affect coping with economic loss. These are (a) the individual's and the group's situation prior to adversity; (b) the salience of economic value in the individual's and the group's eyes; (c) the nature of the crunch; (d) the causal attribution of the crunch; (e) the distribution of the loss; (f) the constructiveness versus the destructiveness of the conflict; (g) the ability to be creative; (h) the potential for mobilization; and (i) participation in decision making. My discussion is, for the most part, speculative rather than based on well-established research findings.

2.1. The Individual's and the Group's Situation prior to the Adversity

Here I refer to the prior economic condition as well as to the preexisting psychological and social state. The economic conditions include factors such as one's level of economic expectations, one's existing economic reserves for coping with adversity, and one's prior experiences and skills in dealing with hardship. Uninterrupted prior affluence may be a poor basis for coping with adversity: it leads to a high level of economic expectations and, thus, a sense of severe deprivation when difficulty is experienced; it does not promote the development of the skills for coping with adversity; and it might lead to an optimism regarding future affluence that does not encourage the accumulation of reserves for hard times. Prior poverty might, on the other hand, be good training for dealing with current hardship, unless the new adversity were to push one below the level necessary for physical, psychological, or organizational survival, or unless the prior poverty had left one so debilitated that one had no capacity to deal with further stress. The best basis for coping might well be a moderate degree of prior affluence, providing that this were associated with prudent expectations developed under the influence of past experiences of adversity as well as comfort and it were combined with unused resources that could be mobilized to deal with the current hardship.

As noted above, one's vulnerability to an economic crunch is determined not only by one's prior economic condition but also by one's preexisting psychological and social state. In general, it could be expected that the worse off an individual is in problem-solving skills and other personal resources, then the lower his self esteem is, and thus the less able he would be to cope with new adversity; similarly, the more poorly a group, institution, or society normally functions, the less skills and resources it has developed for promoting cooperative relations; and the more rigidly it is organized, the less capacity it would have to cope with additional hardship.

2.2. The Salience of Economic Value in the Individual's and the Group's Eyes

Material affluence and comfort is often intimately linked to an individual's self-esteem or to a group's functioning. A person might define his psychological worth in terms of his economic worth, in terms of his possessions and material standard of living. Similarly, a group or a society might have economic production as a highly central value and might place primary stress on the production and distribution of consumer goods in its mode of functioning. If individuals or groups define themselves in terms of economic values, then they have increased vulnerability to economic adversity. It seems likely that much of the difficulty of coping with an economic crunch in the relatively affluent Western world would come from the social and psychological meanings of economic loss rather than from the associated physical hardships.

2.3. The Nature of the Economic Crunch

Clearly, the ability to cope with a situation is, in part, a function of its nature. An economic crunch has such characteristics as magnitude, onset, expected duration, types of hardship, predictability, and per-ceived cause. An economic loss that is of minor magnitude, that has a gradual onset and a short duration, that was expected and is perceived to be of natural causes, and that creates tolerable hardships will cause little upset. In fact, it may not be sufficiently disruptive to motivate coping behavior. On the other hand, the nature of the economic crunch might be such as to induce a large discrepancy between what a person obtains and what he feels entitled to in the way of economic well-being—or, in other words, a strong sense of *relative deprivation*. This is apt to occur when the individual has been led to expect a high or increasing standard of living and he experiences instead a sudden, unpredicted, sharp decrease that he believes is not likely to be of short duration. Such circumstances are very apt to arouse an intense feeling of injustice.

2.4. The Causal Attribution of the Crunch

There are many factors that might influence whether an economic crunch is perceived as justifiable. Perhaps the most important is the social distribution of the economic loss, which is discussed below. An-other key factor is the causal attribution or the subjective explanation for the occurrence of the crunch. The explanation might give rise to the view that someone or some group is responsible for it and can be blamed for the resulting hardships, or it might support the contrary belief, that it is no one's fault—the crunch has resulted from a natural disaster or an unavoidable, uncontrollable concurrence of mishaps. The latter view is much less likely to support a sense of injustice about the hardships that one is experiencing than the former view.

It is evident that whom one blames could affect how and how well one copes with the crunch. If the blame is "intropunitive," so that it is directed against oneself or one's group, internal turmoil and immobili-zation could result. If it is "extropunitive," so that the hardship is ex-ternalized, problems that are internal to an individual or to a group can be projected onto an external adversary or a disliked out-group (espe-cially if these are judged to be weaker than oneself or one's group). Thus, increased internal cohesion and mobilization of resources may result, and the individual or the group might function better than they would otherwise.

2.5. The Distribution of the Loss

Crucial to whether one views one's adversity as being fair or unfair is how one evaluates the social distribution of the hardships resulting from the economic crunch. In discussions of distributive justice, a number of values underlying distributive justice have been repeatedly identified (Rescher, 1966). These are that all people should be treated:

1. So that they have equal "inputs"
2. So that they have equal "outputs"
3. According to their needs
4. According to their ability or potential
5. According to their efforts and sacrifices
6. According to their performance or according to their improvement in performance
7. According to the social value of their contributions
8. According to the requirements of the common good
9. So that none fall below a certain minimum

It is evident that these different values may conflict with one another: the most needy may not be the most able; those who work the hardest may not accomplish the most; giving everybody equal "inputs" may not result in their having equal "outputs"; treating everyone equally may not maximize the common good.

Under conditions of an economic crunch, how should the adversity be distributed? What distribution would be most likely to foster effective social cooperation to promote individual well-being? Rescher (1966) has suggested that in an economy of scarcity, the just rule is "The number of individuals whose share of utility falls below the "minimal" level is to be made as small as possible" (p. 97). Acceptance of a minimal rule implies the basic equality of human life: all people are entitled to at least the minimal conditions necessary for a humane life in a given society. How can the minimum be defined in a way that achieves social consensus? Rawls (1971) has suggested an interesting procedure that seems adaptable to the decision of how to define minimality: principles of justice are chosen behind a veil of ignorance. If decisions about the minimal share were made behind a veil of ignorance, each responsible individual in that society would be asked to define the minimal level that he would require for a humane life and, in making his definition, he would also be asserting that this would be the minimum for all others in that society. In making his definition, he would not know whether he would be sacrificed if there were not enough for everyone to have the minimum; he would know only that each person had the same chance to be sacrificed.

It seems unlikely that the amount available for distribution will be precisely the amount necessary to meet the requirements of a minimal distribution: it will be either too little or too much. If it is too little, it seems likely that Rescher's rule ("the number of individuals whose share of utility falls below the minimal level is to be made as small as possible") will be invoked rather than dividing the total equally so that all individuals are below the minimum. An "exception" might occur if an equal division would paradoxically make the "lower" minimum more humane. This might be the case if the lower minimum does not lead to physical debilitation or death but, if equally shared, leads to a higher sense of trust and cooperativeness with the full members of the community. If the lower minimum were physically debilitating, and Rescher's rule were applied, then some procedure for selecting those who would not receive the minimum (i.e., those who would be "sacrificed") must be developed. Two procedures that maintain the equal value of human life have been described for such circumstances: a lottery in which everyone has an equal risk of being sacrificed and a self-sacrificing or volunteer system. In both systems, those who are sacrificed would be honored for their sacrifice.

However, it seems likely that "randomization" of sacrifice and volunteering of self-sacrifice would not be adequate, by themselves, to ensure the survival and the perpetuation of the group in a catastrophic crunch, a crunch that would permit only a small percentage to survive. To achieve these ends in such a crunch, it would be necessary to make certain that those selected to receive the minimum include the group members who are most willing and able to produce the goods and services required for the survival of the group and also the members who are most willing and able to bear and rear children so as to ensure the group's perpetuation. The introduction of "high-priority" categories of group members does, of course, impair the equality principle and could lead to intragroup strife about the selection of members to compose the to-be-favored categories. The equality principle would be least impaired if either of two conditions prevailed: if all or many members of the group have the characteristics that would place them in high-priority categories or if only a very small percentage fall into such categories. In any case, if people were classified into various categories, the number required from each category could be selected by lottery designed to equalize chances within categories and also to take into account the differential priorities of the various categories.

From the above discussion, it can be concluded that where there is not enough for all to exist above the minimal level, scarcity should lead to the application of the equality value as the basis for distributive justice; and this value would be modified to give priority to those who can contribute most to the group's survival and perpetuation when the scarcity is catastrophic. Suppose, however, that there is more than enough to provide everyone with a humane minimum but not enough to give everyone what he has been used to: How is the extra to be distributed? One principle might be to distribute the extras so that everyone's income before the crunch is reduced by equal numbers of JNDs (just noticeable differences). Thus, if Person A had a yearly income of $1,000,000, he might just notice a drop in his standard of living if his income were reduced by $500,000; if Person B earned $500,000, he might just notice a loss of $25,000; if Person C's income was $10,000, he might just notice a drop of $500. In a moderately severe economic crunch, where each individual would have to lose many JNDs of income, I believe it can be shown that there would be a convergence of incomes of people at initially quite different levels of affluence. Thus, for all but relatively minor or catastrophic economic crunches, it seems that equality should be the dominating value for distributive systems.

Yet, it is evident that equality will result in an inequality of absolute economic loss; the previously affluent will lose more than those who were previously poor. As a consequence, they are more apt to experience the change to a system of equality as unjust and to resist it. Is there any way that they can be psychologically compensated for their larger economic losses without undermining the value of equality? I am not sure that there is a positive answer to this question. One can think of ways of easing the pain of the transition, by substituting honors, social appreciation, and such for income and wealth. But the fundamental problem for the formerly rich and the formerly poor alike in adjusting to a distributive system based on equality is to disengage their conceptions of themselves from a system of unequal distribution based on notions of relative individual worth. If they could take esteem for themselves and enjoy the human warmth of being part of a mutually respecting, friendly community of equals, their loss of relative position in a mutually suspicious, cold community of competitive unequals might not be a source of regret.

2.6. The Constructiveness versus the Destructiveness of the Conflict

In the economic crunch, the number and magnitude of conflicts among members of a society are bound to increase, partly because of the increased scarcity of many resources, and partly because of the social and individual changes required to adjust to the crunch. Changes will challenge vested interests, habits, loyalties, and commitments.

Conflict can take a destructive or a constructive course. Its course will be very much influenced by whether it occurs in a cooperative or a competitive context, and the course that a conflict takes will, in turn, influence whether the conflicting parties develop cooperative or competitive relations with one another. Destructive conflict has the characteristics of a competitive process and tends to elicit competitive relations, while constructive conflict has the characteristics of a cooperative process and tends to elicit cooperative relations. I have summarized the differences in these two processes as follows (Deutsch, 1973).

2.6.1. Differences between Constructive and Destructive Conflict

2.6.1a. Communication. A *cooperative process* is characterized by open and honest communication of relevant information between the participants. Each is interested in informing, and being informed by, the other.

A *competitive process* is characterized by either lack of communication or misleading communication. It also gives rise to espionage or other techniques of obtaining information about the other that the other is unwilling to communicate. In addition to obtaining such information, each party is interested in providing discouraging or misleading information to the other.

2.6.1b. Perception. A *cooperative process* tends to increase sensitivity to similarities and common interests while minimizing the salience of differences. It stimulates a convergence and a conformity of beliefs and values.

A *competitive process* tends to increase sensitivity to differences and threats while minimizing the awareness of similarities. It stimulates the sense of complete oppositeness: "You are bad; I am good." It seems likely that competition produces a stronger bias toward misperceiving the other's neutral or conciliatory actions as malevolently motivated than the bias induced by cooperation to see the other's actions as benevolently intended.

2.6.1c. Attitudes toward One Another. A *cooperative process* leads to a trusting, friendly attitude, and it increases the willingness to respond helpfully to the other's needs and requests.

A *competitive process* leads to a suspicious, hostile attitude, and it increases the readiness to exploit the other's needs and to respond negatively to the other's requests.

2.6.1d. Task Orientation. A *cooperative process* enables the participants to approach the mutually acknowledged problem in a way that utilizes their special talents and enables them to substitute for one another in their joint work, so that duplication of effort is reduced. The enhancement of mutual power and resources becomes an objective. It leads to the defining of conflicting interests as a mutual problem to be solved by collaborative effort. It facilitates the recognition of the legitimacy of each other's interests and of the necessity of searching for a solution that is responsive to the needs of all. It tends to limit rather than expand the scope of conflicting interests. Attempts to influence the other tend to be limited to processes of persuasion.

A *competitive process* stimulates the view that the solution of a conflict can only be one that is imposed by one side on the other. The enhancement of one's own power and the minimization of the legitimacy of the other side's interests in the situation become objectives. It fosters the expansion of the scope of the issues in conflict, so that the conflict becomes a matter of general priniciple and is no longer confined to a particular issue at a given time and place. The escalation of the conflict increases its motivational significance to the participants and intensifies their emotional involvement in it; these factors, in turn, may make a limited defeat less acceptable or more humiliating than mutual disaster might be. Duplication of effort, so that the competitors become mirror images of one another, is more likely than division of effort. Coercive processes tend to be employed in the attempt to influence the other.

2.6.2. Factors That Shape the Course of a Conflict

What factors influence whether a conflict will take a constructive–cooperative or a destructive–competitive course? What will enable a group to react cooperatively rather than competitively in the face of economic adversity? In the *Resolution of Conflict* (Deutsch, 1973), I have presented a detailed answer to such questions. Here, I will state a general principle, which provides a basis for deriving the more specific answers. This principle, which I have labeled "Deutsch's crude law of social relations," is that the *characteristic processes and effects elicited by a given type of social relationship tend also to elicit that type of social relationship.* Thus, the strategy of power and the tactics of coercion, threat, and deception result from and also result in a competitive relationship. Similarly, the strategy of mutual problem-solving and the tactics of persuasion, openness, and mutual enhancement elicit and also are elicited by a cooperative orientation.

Among the many implications that can be drawn from Deutsch's crude law, a few are listed below:

1. Egalitarian and need-oriented systems of distributing the benefits and costs of group membership are more apt to foster cooperation than a competitive, meritocratic system.

2. The opportunity for direct, full, open, and honest communication among group members and between group leaders and group members encourages cooperation; infrequent communication, evasiveness, and lack of open, direct communication gives rise to rumors that stimulate suspicious, paranoid thinking and competition.

3. Increasing the salience of common interests and of similarities in values among group members stimulates cooperation; emphasizing the divergence of interests and values elicits competition.

4. Encouraging more frequent, friendly, informal interactions

among group members strengthens cooperativeness; restricting contacts to formal, distant impersonal relations is more apt to evoke competition.

5. Fostering member participation in group problem-solving aids cooperation; restricting problem solving to a few members encourages competition.

The issue of cooperation within a group is closely related to the issue of cooperation among different generations. Heilbroner (1975) raised the horrendous possibility that humanity may remain indifferent to the dangers of the future, diminution of resources being one of them. The question is: On what considerations should we make sacrifices now to ease the difficulties of future generations? There is only one possible answer to this question: it lies in our capacity to form a collective bond of identity with those future generations. Indeed, it is the absence of such a bond with the future generations that casts doubts on the ability of contemporary society to take now the measures needed to mitigate the problems of the future. In contemporary society, where economic productivity is a primary goal, individuals have only a limited motivation to form such bonds. In a society where competitive rather than cooperative relations are predominant, the conception of one's community is narrow, and thus, the scope of his responsibility for future generations is narrow. It is probable that in a more cooperative society that is concerned more with the general welfare and indulges less in selfish calculations, such an identificatory sense could be strengthened.

2.7. The Ability to Be Creative

Effective cooperation and the ability to confront conflict constructively are important ingredients of creative solutions to problems. It is evident that an economic crunch faces individuals, groups, institutions, and societies with the necessity of creatively developing new ways of relating to their changed realities. The creative process can be described as consisting of several overlapping phases: (a) an initial period that leads to the experiencing and recognition of a problem that is sufficiently arousing to motivate efforts to solve it; (b) a period of concentrated effort to solve the problem through routine, readily available, or habitual actions; (c) an experience of frustration, tension, and discomfort that follows the failure of customary processes to solve the problem and leads to a temporary withdrawal from the problem; (d) the perception of the problem from a different perspective and its reformulation in a way that permits new orientations to a solution to emerge; (e) the appearance of a tentative solution in a moment of insight, often accompanied by a sense of exhilaration; (f) the elaboration of the solution and the testing of it against reality; and finally, (g) the communication of the solution to the relevant audiences.

There are three key psychological elements in this process: (a) the arousal of an appropriate level of motivation to solve the problem; (b) the development of the conditions that permit the reformulation of the problem once an impasse has been reached; and (c) the concurrent availability of diverse ideas that can be flexibly combined into novel and varied patterns. Each of these key elements is subject to influence from social conditions and the personalities of the problem solvers.

Thus, consider the arousal of an optimal level of motivation, a level sufficient to sustain problem-solving efforts despite frustrations and impasses and yet not so intense that it overwhelms or prevents distancing from the problem. Optimal motivation presupposes an alert readiness to be dissatisfied with things as they are and a freedom to confront one's environment without excessive fear, combined with a confidence in one's capacities to persist in the face of obstacles. The intensity of motivation that is optimal varies with the effectiveness with which it can be controlled: the more effective the controls, the more intense the motivation can be without having disruptive consequences.

Although acute dissatisfaction with things as they are and the motivation to recognize and work at problems are necessary for creative solutions, these things are not sufficient. The circumstances conducive to the creative breaking-through of impasses are varied, but they have in common that they provide the individual with an environment in which he does not feel threatened and in which he does not feel under pressure. He is relaxed but alert. Threat induces defensiveness and reduces both the tolerance of ambiguity and the openness to the new and unfamiliar; excessive tension leads to a primitivization and a stereotyping of thought processes. As Rokeach (1960) has pointed out, threat and excessive tension lead to the closed rather than the open mind. To entertain novel ideas that may at first seem wild and implausible, to question initial assumptions of the framework within which the problem or conflict occurs, the individual needs the freedom or courage to express himself without fear of censure. In addition, he needs to become sufficiently detached from his original viewpoints to be able to see the conflict from new perspectives.

Although an unpressured and unthreatening environment facilitates the restructuring of a problem or a conflict and, by so doing, makes it more amenable to solution, the ability to reformulate a problem and to develop solutions is, in turn, dependent on the availability of cognitive resources. Ideas *are* important to the creative resolution of conflict, and any factors that broaden the range of ideas and alternatives available to the participants in a conflict will be useful. Intelligence, exposure to diverse experiences, an interest in ideas, a preference for the novel and complex, a receptivity to metaphors and analogies, the capacity to make remote associations, independence of judgment, and the ability to play with ideas are some of the personal factors that characterize creative problem-solvers. The availability of ideas is also dependent on such social conditions as the opportunity to communicate with and be exposed to other people who may have relevant and unfamiliar ideas (i.e., experts, impartial outsiders, people facing similar or analogous situations); a social atmosphere that values innovation and originality and encourages the exchange of ideas; and a social tradition that fosters the optimistic view that, with effort and time, constructive solutions can be discovered or invented to overcome problems that initially seem intractable.

It can be shown that a cooperative process produces many of the characteristics that are conducive to creative problem-solving: openness, lack of defensiveness, and full utilization of available resources. However, in itself, cooperation does not ensure that problem-solving efforts will be successful. Such other factors as the imaginativeness, the experience, and the flexibility of the parties involved are also determinative.

2.8. The Potential for Mobilization

Individuals, groups, institutions, and societies rarely function at their capacity. They have resources and assets that are used inefficiently or not at all. "Mobilization" (see Etzioni, 1968) is the process by which individuals or social units increase the number of effective resources or assets that they have available to bring to bear on the problems confronting them. Mobilization can increase the availability of any of a variety of types of resources: (a) *motivational resources,* such as energy, drive, commitment, dedication, and determination; (b) *cognitive resources,* such as attentiveness, consciousness, information, memory, and intellectual and other more specific skills; (c) *economic resources,* such as manpower, tools, land, and capital; (d) *organizational resources,* such as leadership, division of labor and specialization of function, communication and coordination, and planning and evaluation; and (e) *social resources,* such as cohesion, trust, loyalty, and solidarity.

As Etzioni (1968) has pointed out, "major societal changes are propelled by small changes in the absolute level of mobilization because they constitute sharp increases in the *relative* level of energy available" (p. 398). Typically, the level of political mobilization is low in modern societies (e.g., only about a third of American adults know the name of their Congressman; even fewer actively participate in political organizations). Similarly, the level of economic mobilization is commonly low in modern societies (e.g., it has been established that by a more efficient and less wasteful use of energy, energy consumption in the United States could be reduced by more than one-third without a change in the standard of living).

What determines whether an individual or a group can mobilize itself to deal with a continuing crisis such as an economic crunch? Being prepared to mobilize, knowing where one's unused resources are and how and when they can be used more effectively, and having practiced or rehearsed mobilizing one's resources will facilitate an effective mobilization when it is required. Also, having confidence in one's resources and in the possibility of amplifying them significantly through a process of mobilization will increase the likelihood of an effective mobilization.

Mobilization has costs in terms of discipline, self-denial, postponement of pleasures, and the like. These costs are not likely to be borne if one has little trust that others will cooperate and assume their responsibilities in the process of mobilization. If one knows that many are cheating and obtaining more rations than they are entitled to in a rationing program, it is difficult to withstand the temptation to cheat when it becomes possible for one to do so. Also, one will be less likely to conserve energy by driving below the 55-mph speed limit if one sees others disregarding it. The felt inequality of disciplining oneself and making sacrifices while others are perceived to be "taking advantage" of the situation undermines the cooperativeness necessary to an effective group mobilization. In contrast, the sense that the duties and obligations involved in the mobilization are being fairly shared enhances one's commitment to it.

2.9. Participation in Decision Making

It seems reasonably well established that people who participate in making decisions that affect their lives are more likely to accept the decisions and to feel that they are just than if they have had no part in the decision-making process. Thus, a participatory decision-making process in response to an economic crunch is more likely to give rise to policies that are considered fair and is more apt to foster a cooperative process in relation to the issues involved. However, it is very difficult or impossible to have a meaningful participatory decision-making process as the size of the group expands beyond a certain point. This difficulty is enhanced in emergency situations that require quick decisions. These problems often lead to more centralization in decision making as organizations and communities grow in size; more centralization, in turn, results in an increase in alienation and a decrease in the sense of cooperativeness.

How can one handle this dilemma for participatory decision making and cooperation that is created by size? No one really knows. There is, however, widespread agreement that "small is beautiful." But it is not yet clear how in a large world, with world-scale problems, the small units can be nested together into increasingly larger units so as to avoid the problems associated with remote, centralized decision making. Nevertheless, there seems to be an emerging consensus among utopian thinkers that as much decision making as possible should be placed in the hands of small, cooperative local units that are democratically controlled by the immediate members of the local unit—by the inhabitants of the local community, by the people working in a given institution— and that cooperative, regional federations of local units should coordinate activities of the local units to enable the economies of large-scale purchasing, production, and distribution. Similarly, cooperative national federations of regional units would coordinate the regional units. Ideally, the control of resources and of decision making would remain in the small, local units and would not move upward. There are, unfortunately, many factors conducive to the flow of power and decision making to centralized units. How to inhibit this flow while maintaining effective coordination and cooperation among smaller units is a problem that needs our most creative work. Social injustice is fostered by the accumulation of power and the control of resources in remote, central decision-making bodies, and it is also fostered by the inequalities of local communities, by their out-group prejudices, and by their ethnocentric neglect of larger, worldwide concerns.

3. CONCLUSION

In conclusion, I have advanced the thesis that the ability of an individual or a group to cope with an economic crunch is heavily dependent on the maintenance of self-esteem and group pride as well as on the strengthening of cooperative bonds. These are likely to be impaired if one thinks that the loss one experiences during an economic adversity is not being justly shared or distributed. In a situation of economic scarcity, it seems likely that the sense of justice is most likely to be satisfied by a distributive system that seeks to provide all or as many as possible with at least a humane, minimal level of goods. Such a distributive system is essentially a socially egalitarian system leavened by particularistic responses of local units to individual needs and situations. The advantage of such a distributive system during a period of economic hardship is that it not only promotes the sense of justice but also strengthens the ties of solidarity among the members of a community and enables them to function cohesively and productively in difficult circumstances. The resulting experience of participating in a cohesive, productive, solidary group enhances one's feeling of personal and social well-being; this may more than compensate for the economic losses that one has during a period of economic crunch. The kibbutzim in Israel have provided many notable examples of just such a process.

ACKNOWLEDGMENT

I am grateful for Shula Shichman's help.

REFERENCES

Deutsch, M. *The resolution of conflict: Constructive and destructive processes.* New Haven, Conn.: Yale University Press, 1973.
Deutsch, M. Awakening the sense of injustice. In M. Lerner and M. Ross (Eds.), *The quest for justice: Myth, reality, ideal.* Toronto/Montreal: Holt Rinehart and Winston of Canada, 1974.
Etzioni, A. *The active society.* New York: Free Press, 1968.
Heilbroner, R. L. *An inquiry into human prospect.* New York: W. W. Norton, 1975.
Rawls, J. *A theory of justice.* Cambridge, Mass.: Harvard University Press, 1971.
Rescher, N. *Distributive justice.* New York: Bobbs-Merrill, Co., 1966.
Rokeach, M. *The open and closed mind.* New York: Basic Books, 1960.

16

The Relationship of Economic Growth to Inequality in the Distribution of Income

THOMAS D. COOK and BARBARA PEARLMAN

1. INTRODUCTION

It is by now commonplace in Western Europe and the United States to note how difficult it will be in the future to stimulate the gains in economic growth to which we have become accustomed. Productivity is not increasing at its former rate; the cost of many important raw materials is escalating; externalities like pollution have to be dealt with; and expectations about transfer payments to the poor, the retired, the disabled, and other "victims"—payments that do not contribute in the most direct fashion possible to economic growth—are by now institutionalized.

It is also commonplace to note that one reaction to these changes is to adopt economic policies designed to give an even higher priority to increasing economic growth. Thus, in many countries, a tilt toward the economic right is taking place that is associated not only with more conservative politicians' being elected but also with plans for tax cuts aimed at stimulating incentives and growth, with attempts to relax regulatory controls, and with campaigns to break any legal or social forces that index many transfer payments to increases in the cost of living.

THOMAS D. COOK and BARBARA PEARLMAN ● Department of Psychology, Northwestern University, Evanston, Illinois 60201.

The concern of some persons is that economic strategies for increasing growth will exacerbate inequalities of condition. This is because many of the strategies, it is currently believed, that stimulate growth require a decrease in tax burdens. While such cuts apply to all taxpayers, the savings are greatest in both percentage and absolute dollar terms for the more affluent. Other strategies that are advocated decrease government spending, but this is more likely to involve a decrease in transfer payments to the poor and disabled than the closing of tax loopholes. Finally, other advocated strategies involve decreasing the amount of regulation. But this often entails a failure to improve living and working conditions for the kinds of people who work in more hazardous occupations or live near polluting plants. Once again, such persons are likely to be among the less affluent in society. The fear, then, is that the promotion of economic growth complements other social mechanisms whereby "those that have, get" and "the rich get richer."[1]

For many persons, this concern would not be warranted if we had valid economic theories and practices that simultaneously promoted a high level of economic growth and also a decrease in the variance of the income distribution. Then, we could have both growth and less income inequality. Others see a decrease in the income variance as too strict a criterion and would instead suggest the advisability of theories and practices that promote growth without making existing income differences more acute. In this chapter, we shall explore two theories that purport to stimulate economic growth while also reducing the variance in the income distribution or leaving it unchanged relative to a higher growth-induced mean.

Not everyone is convinced that public policy should be concerned with equality of condition. To some, the global goal of economic policy should be a widely shared growth. All sectors of the public will then do better in absolute terms, even if some do better than others and the ones that do better are those that initially have more anyway. Other com-

---

[1] "The rich get richer" reflects a belief in the proposition that in an environment of freedom, the persons who flourish will be those who already possess the most resources. Evidence to support the general proposition is not restricted to the economic domain. In education, we know that children from more affluent homes who also happen to know more than poorer children of the same age watch "Sesame Street" more often as early as age 3 and thus benefit more from it; we also know that adolescents from more affluent homes are more likely to attend and graduate from universities. In health care, we know that even when health care is free—as in England—persons from more affluent backgrounds, who are on the average healthier than less affluent persons, stay in the hospital longer and receive more expensive medication for the same symptoms. In all these cases, resources that create growth, whether it be in wealth, knowledge, or health, also exacerbate existing differences between social groups that differ in the aggregate in wealth, knowledge, and health.

mentators take a strict laissez-faire approach and leave to policy only the enforcement of contracts and the keeping of public order. To these persons, the level and distribution of resources are matters best left to the "invisible hand of classical economics" rather than to the more visible hands of politicians and bureaucrats. A more cynical position is to argue that individuals compare themselves with individuals like themselves rather than with persons of different social strata. This being the case, individuals are not likely to notice increased differentials and so will not do anything about them. Thus, the cynical argument continues: Why should we care whether growth exacerbates income differentials if nothing follows from the increased differences?

Two of the above reasons for ignoring issues of income distribution are ideological, and we have no wish here to enter into these particular debates. But the last argument is based on social-science hypotheses concerning social comparison. One of the major shortcomings of the social comparison literature is that much more is known about the consequences of social comparison than about its determinants. When people compare themselves to individuals with different backgrounds is still largely unknown, despite some theoretical and empirical work (e.g., Festinger, 1964; supplementary number of *Journal of Experimental Social Psychology*, 1966). Moreover, little is known about social forces (including the media) that highlight to individuals comparisons that they might not normally make. Also, the psychological literature has focused largely on egoistic relative deprivation rather than fraternal deprivation (Runciman, 1966; Cook, Crosby, & Hennigan, 1977). Yet political scientists assume that one reaction to increased differentials might be an increase in fraternal solidarity and a corresponding willingness to confront more powerful others in the formal and the informal political arena. For all of these reasons, we would be loath to accept the assumption that social comparisons rarely occur across social strata or the assumption that nothing of social importance follows from any particular comparisons that occur when income differentials are increased. Indeed, one of the working assumptions of this chapter is that undesirable social and political consequences can sometimes follow from the increased income differentials induced by pursuing growth-oriented economic policies.

This chapter reflects the following concerns: (a) that economic growth rates have slowed down in the United States and Western Europe; (b) that political pressures have exerted themselves that assign economic growth a predominant priority, even though (c) at a superficial level it would seem that such growth policies may increase income differentials. However, (d) some theories and practices exist that proponents claim will maximize growth without increasing inequalities of condition, and that might perhaps even reduce them. Such theories and practices deserve special scrutiny, for if they are valid they reduce the need to make sensitive, political choices between options that, in making growth the major priority, may increase inequality or that, in making inequality the major priority, may reduce the rate of growth over what it might otherwise have been.

The empirically minded scholar might object to this formulation of a choice, for current evidence suggests that while growth and equality may be negatively related in developing nations, they are not so related in developed nations. Indeed, we shall later see that in developed nations growth has been the norm for decades; yet, despite some fluctuations, indicators of the degree of inequality have remained roughly constant since World War II. We would only point here to the fact that the relationship we have presumed between growth and inequality is theoretical in the sense of being the basic relationship that would operate, all other things being equal. However, many actions can be taken to reduce the inequalities that the pursuit of economic growth may cause or to increase economic growth over the lower than desired rates that the pursuit of equality may cause. In the last part of this essay we discuss some political events that may be designed to prevent inequalities from increasing when growth policies are implemented or that may be designed to stimulate growth after egalitarian policies have been implemented. Our discussion highlights a number of ironies that follow from the absence of a viable, single theory that predicts both equality and growth rates of a socially acceptable level. One irony depicts governments following each other into power in order to redress the imbalance between growth and equality perpetrated by their predecessors; while a second depicts Federal agencies with egalitarian or growth mandates simultaneously and independently designing policies to counteract the unintended side effects of the other's policies. But more about these ironies later.

2. THE TRICKLE-DOWN THEORY

2.1. Overview

The trickle-down theory aims, by stressing growth in the short run, to stimulate equality in the longer run. The theory is portrayed in Figure 16.1 in terms of six propositions that are related in temporal sequence: (1) business can be stimulated by providing direct benefits to entrepreneurs and innovators; (2) such stimulation will increase the growth of the enterprise; (3) the profits of this growth will be invested or reinvested; (4) the investment will create new jobs; (5) these jobs will help fulfill the

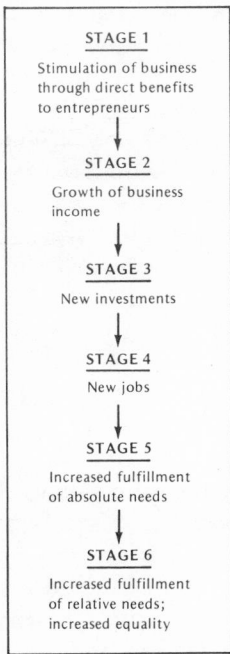

Figure 16.1. The trickle-down theory.

absolute needs of poorer persons who come to hold the jobs; and (6) through earnings, savings, and new opportunities in an open society— including education, vocational training, etc.—income differentials may eventually be reduced.

The intellectual origins of Stages 2–5 are in the classical liberal theory associated with Adam Smith. The first stage is not, having its origins in theories that stress how production can be stimulated by externally provided incentives such as tax credits or guaranteed loans. The final stage reflects what may be an incidental side effect of increased growth, especially if the growth has resulted in a better-trained, and eventually a better-educated, labor force.

Our suspicion is that the trickle-down theory reflects a historical parochialism on the part of Western scholars, for it may well be based on the experience of Western nations, where innovation and investment facilitated growth, and growth was followed by more jobs; universal education; a mass market of consumers, many of whose basic needs were met (e.g., for food, clothing, and housing); and a democratic society where equality was understood in terms of achieving a greater equality of rights, educational opportunities, and perhaps income. But parochial or not, after World War II, the presumed success of the trickle-down theory in Western nations led many agencies that give aid to third-world countries to prefer this theory of development over others. After all, it promised growth; the satisfaction of unmet basic needs; and perhaps more equality at a later date. The most obvious risk associated with the theory was that economic growth might be realized at the cost of an initial increase in inequality that would not be reduced later. Note from Figure 16.1 that at Stage 1, resources are diverted to the persons who it is thought will use them most profitably, even though they, as individuals or social groups, might need them least.

2.2. Assessing the Effectiveness of the Trickle-Down Theory

We cannot assess the efficacy of the theory by the conventional hypothetico-deductive method of science, which involves collecting data that provide validating and falsifying tests of the basic and unique postulates of the theory. No unambiguous deduction or controlled tests can be based on a theory as poorly specified as trickle-down. For instance, no time lines are mentioned in the theory at all. If we assume that the theory implies "long" (i.e., multigenerational) time intervals, then this specification precludes new data collection and throws a heavy reliance on fallible historical records and secondary sources. Moreover, a multitude of forces affect growth and the distribution of income over a long time period, making it difficult to assess whether any observed changes are due to pursuing a trickle-down policy. These difficulties mean that we shall be more than normally inductive and judgmental in reviewing this particular literature. For reasons of space, we shall not discuss each stage. Rather, we shall concentrate only on the stages that we consider most significant and problematic.

2.2.1. Stage 3

Major difficulties with the trickle-down theory arise in Stage 3, since new investments do not necessarily result from increases in business income. Indeed, several factors weaken the relationship between investment and income.

Consider, first, the significance of limitations in natural resources, that is, oil. For many businesses a rapid increase in the price of such resources diverts what would have been profit into higher prices for raw materials. Although some of the profits that accrue to the persons owning the raw materials might be reinvested in the country from which the profits were derived, there is no guarantee that all or most of the profits will take this route. Consider, also, externalities. For instance, in the United States concern about the environment has led to mandating costly antipollution regulations and to instituting severe penalties for noncompliance. Precautions designed to protect future generations inevitably arouse political controversy because they divert capital from short-term investment and growth. Another weakness with Stage 3 concerns where the profits are invested. When only a portion of the profits made in a country are reinvested in that country, growth is reduced over what it might have been. And in some cases, tax laws provide multinationals with an incentive to expatriate profits.

An important final point needs to be made about Stage 3. When a disproportionately large share of national income is earned by a small segment of the population—as happens in Stage 2 of the trickle-down theory—purchasing power is initially restricted to that small population segment. Such persons may choose to consume rather than invest, and their consumption may be directed toward expensive goods that hardly benefit the local economy, especially if they are imported. This has happened in some developing nations, as well as in more economically advanced nations, when lower growth rates have been associated with increases in inflation. In this situation, all have a motivation to consume rather than to invest.

2.2.2.   Stage 4

One might expect new jobs to be generated once the earlier stages specified in the trickle-down theory have occurred. In underdeveloped countries, however, many of the poor live in rural, agricultural areas where they remain somewhat unconnected to the rapid growth and increased jobs that are primarily located in and around newer urban areas. Consider, too, that in developed nations, technological advances typically have a labor-saving bias that limits the increase in jobs. Indeed, the bias may even reduce the number of jobs relative to population size. This problem is exacerbated if rural workers migrate to the cities and add to the increasing excess supply of labor. Thus, although trickle-down strategies may initially generate new jobs, it is not at all certain that these strategies will continue to generate jobs or will keep up with population growth. The empirical issue is to determine which groups are most likely to suffer from these social forces: the unskilled, the skilled blue collar worker, low-level executives, or others.

2.2.3.   Stage 5

Do the processes listed in the trickle-down theory reduce absolute poverty? Some evidence suggests that, as expected during the initial periods of economic growth, absolute poverty may increase. Let us first consider developing nations. In an exploratory study of the relative importance of social, political, and economic factors in accounting for differences in the income distributions among 43 non-communist developing nations between 1950 and 1963, Adelman and Morris (1973) concluded that in countries with basically a subsistence economy, the position of the poorest 60% typically worsened, absolutely and relatively, following an initial growth spurt. They also found that, on the average, it took at least one generation for this poorest segment to recover the absolute income loss. Even in countries with higher levels of development, Adelman and Morris found that the poorest segments typically benefited from economic growth only when additional widespread efforts were deliberately made to improve the human resource base (pp. 178–181).

Consider the case of Brazil, where trickle-down policies were aggressively pursued in the 1960s and 1970s. These were probably of great benefit in restoring growth to the economy, and the GNP per capita grew by 2.5% annually during this time. However, there seems to have been as much as a 20% drop in the per capita income of the poorest 40% of the population, and between 1964 and 1970, the real purchasing power of the minimum wage declined by over 25%. Thus, the price of stimulating economic growth was paid largely by the lower-income segments of the population (Adelman & Morris, 1973; Looney, 1975).

If the United States can be said to have followed any one theory of development more than others over its history, it would probably be the trickle-down theory. In the United States, the theory is associated with reducing poverty as defined by the official absolute standard. Census Bureau statistics reflect that in 1959 about 39.5 million individuals lived below the (retrospectively applied) official poverty line of $2,973. By 1973, this number had declined to about 23 million persons, who were then living below the revised standard of $4,540. (Of course, these numbers depend on the standard being used, and the numbers of poor can be decreased or increased almost at will by varying the standard.) Nonetheless, the number of poor persons may have decreased between 1959 and 1973, and it has almost certainly decreased if we adopt a larger historical perspective.

However, it is not at all clear that any decrease has been due to trickling down. Instead, it may have been due to explicit national policies designed to lift the income floor so as to prevent inequalities in income from increasing. Or it may be that some poor persons can benefit from trickling down while others cannot. If so, the most likely beneficiaries are persons whose motivational structure most resembles the prevalent one in the economic domain, that is, those who are hard-working, self-disciplined, and willing to take reasonable risks to achieve greater material success. Others, whose social and psychic attachment to the labor force is relatively weak, are called the *backwash poor*, and they may be less likely to benefit from general national economic growth (Aaron, 1967; Lampman, 1965; Galloway, 1967). Noting the difficulty of reaching and stimulating the backwash poor—as opposed to the poor with more traditional economic motives, who can be "creamed off" by trickle-down strategies—Anderson (1964) concluded that "elimination of poverty through trickling down is likely to be slower and more uncertain in the future than it has been in the past" (p. 512). Parker (1973) wrote that the expectation held in the 1950s, when the trickle down was *the* dominant theory of economic development, was that the backwash poor would make up a small residual group "handily cared for by the public dole" (p. 31). However, 23 million can hardly be considered a small residual group, and of these, 9.5 million were members of female-headed households, which are somewhat less likely than male-headed families to be connected to the labor force. One can perhaps also see evidence of the backwash thesis in relation to certain racial and ethnic groups. For instance, discrimination may have resulted in fewer opportunities for the mass of blacks and Hispanics to benefit from economic growth, even though some blacks and Hispanics have been "creamed off" in a highly visible fashion. This may be why 10% of all whites were officially poor in 1972, 25.6% of all known Hispanics, and 32% of blacks (Kohler, 1973, p. 8). It can be assumed that nearly every society has a large segment of the backwash poor who—for whatever reasons—do not share in the benefits of trickle-down policies.

2.2.4. Stage 6

In Stage 6, income differentials are supposed to narrow. The most famous exposition of this hypothesis is by Kuznets (1966), who postulated a U-shaped relationship between growth and equality. Equality is supposed to be greatest in countries with the least and the most economic growth. Analyzing the relationship between economic growth and income in developing nations is difficult and depends on using very fallible economic indicators. Therefore, all the statements that follow have to be treated gingerly.

In support of part of Kuznets's hypothesis, the evidence does seem to suggest that income inequalities increase in developing countries during the initial stages of growth. This is particularly true where growth has occurred rapidly in subsistence agricultural economies. But it is undoubtedly also true in developed nations whose government, or laissez-faire economic philosophy, allows entrepreneurs and other persons with capital the chance to invest for large after-tax returns.

According to Kuznets, as development continues, we have the "long secular swing" from greater to lesser inequality. Recent studies indicate that if there is a secular swing, it may be very long indeed—so long, in fact, that one might doubt whether an eventual upswing will take place, at least in developing nations. Thus, in their 1950–1963 study of developing nations, Adelman, Morris, and Robinson (1977) wrote:

> In the absence of domestic policy action aimed specifically at redirecting the benefits of growth a nation must attain a level of development corresponding to that which exists among the socioeconomically most highly developed of the underdeveloped countries (Argentina, Chile, Taiwan, Israel) before the income distribution tends to become as even as it is in countries that have undergone virtually no economic development (e.g., Dahomey, Chad, Niger). (p. 194)

In other words, the authors can sometimes discover the swing, but it may only be to the same level of equality from which the nation started!

In a recent paper, Henry Levin (1978) wrote, "by the mid-seventies it became apparent that the historical data of Kuznets and the predictions of neo-classical economic theory had not been achieved for most of the developing countries of the world" (p. 4). He went on to note that even in the relatively developed of the developing nations (e.g., South Korea and Brazil) there has been little improvement in the distribution of income. From 1960 to 1970, the relative income share received by Brazil's poorest 40% of the population declined from 10% to 8%, while the share of the richest 5% of the population grew from 29% to 38% (Adelman & Morris, 1973; Looney, 1975). In Mexico, the annual GNP growth rate during 1940–1970 ws about 6%, uncorrected for population growth. Most of the relative income benefits resulting from this growth went to the middle-income groups. Between 1950 and 1969, the income share of the poorest 60% of the population declined from 24.6% to 20.0%, the share of the richest 5% declined from 40.0% to 36.0%, and the share of those in the 60th to the 90th percentiles increased from 35.4% to 41.0% (Looney, 1975, Loehr, 1977). Similar increases in relative income inequality have been reported for other countries experiencing rapid economic growth (e.g., Argentina and Puerto Rico, according to Weiskoff's 1970 data; prerevolutionary Iran, according to Looney, 1975).

The next question is: Is there evidence that the high rates of growth in developed countries have led to a decrease in the inequality of income? A crucial issue here is how we define *income*. For the United States, we shall use the official census definition, recognizing that it does not include the cash value of in-kind transfers, of government services at all levels (including police, fire, roads, airports, etc.), and legal loopholes in the tax system. Such features render fallible the income figures we shall use, although there is no indication that sources of "unearned income" are greater for the poor than for the more affluent. It may be that they are; but this is not known yet.

Consider the distribution of family personal income between 1929 and 1972 in the United States. When historical and more current Census Bureau figures are used, the income share of the poorest 40% of U.S. families rose from about 13% in 1929 to about 17% in 1947, while the share of the richest 20% declined from 54% to 43% during the same period. This trend, which encompasses the Great Depression and the war years that saw relatively little economic development in terms of broad-ranged industrial growth, compares favorably with the distribution from 1947 to 1972. During the more recent period, the income share of the poorest 40% remained at about 17%, while that of the richest 20% declined to about 41%. Figures that we shall later cite for Sweden show even greater stability over time.

The U.S. example suggests that while income differentials can eventually reverse in highly developed economies, the rate of reversal is slow. Additionally, the argument can be made that the obtained decreases are due to deliberate political interventions in the economic system in order to improve the human resource base and that the reduced inequalities have not resulted from economic growth alone. Of course, one could argue that growth provided a political climate in which redistribution was possible and that, in this sense, growth was necessary but not sufficient for the obtained diminution of inequality.

2.3. Summary

W. H. Locke Anderson (1964) noted of the trickle-down theory that

> The dispersion of income expands along with its central tendency . . . thus bringing widening absolute differentials. . . . This accords with our retributive standards of justice; people are rewarded in proportion to their relative degrees of social worth or productivity. It incidentally guarantees the persistence of poverty without policies designed to redistribute income or earning power. (p. 513)

Anderson's comments reflect that by the mid-1960s, and certainly by the early 1970s, the scholarly consensus had shifted away from a dependence on the trickle-down strategy as a means of meeting the relative (or even the absolute) needs of the poor in developed nations. Galloway (1965) quoted Walter Heller (then Chairmen of the Council of Economic Advisors) in his 1964 testimony before the legislative subcommittee hearings on the Economic Opportunity Act, "Clearly, we cannot rely on the general progress of the economy—or on job-creating programs alone—to erase poverty in America" (p. 122). In a 1965 paper, Richard Titmus noted that in both the United States and Great Britain, there had been a shift away from a "naive" belief in the trickle-down strategy.

The general disavowal of the trickle-down strategy is also evident in reference to developing countries. Adelman *et al.* (1977) wrote:

> After two decades of concern with the problem of raising per-capita GNP in low-income developing countries, the development community of the 1970's has shifted its focus to the challenge of increasing the equity of the distribution of income. The shift proved to be dramatically needed, as the empirical studies of the distribution of benefits from economic growth showed that the expected trickle-down was not taking place. (p. 191)

3. THE BUBBLE-UP THEORY

3.1. Overview

A second theory that appears to reconcile the conflict between economic growth and equality of income is referred to as the *bubble-up theory*. It is based on initially directing resources to the needy—as education or job training, government-financed jobs and housing, access to credit, etc.—in order to increase the productivity and purchasing power of the poor. They will then be "bubbled up" into the economic mainstream, where their newly stimulated productivity, creativity, and purchasing power will increase economic growth.

Just as tenets of classical capitalism buttress the trickle-down theory, so tenets of Keynesian thinking buttress the bubble-up theory. The role of demand was stressed by Keynes as a cause of growth. The principal determinants of aggregate demand are government spending, consumption, investment, and exports. Of these, the government can most readily increase its own spending. However, not all sections of the economy consume their income to the same degree; poorer persons probably consume a greater percentage of their total income. Since the time lag from investment to growth is somewhat longer than the time lag from consumption to growth, the shot in the arm that follows from tax cuts to a wide range of individuals is usually more rapid than the shot that follows when the government provides investment incentive.

A society where most of the income goes to the rich is a society where widespread demand can be stimulated but not satisfied because the money to buy goods is not available. Consequently, Keynes saw a causal connection between the degree of inequality in a social system and the ability to consume in sufficient amounts to affect aggregate demand and growth in a significant manner. Too much inequality means a small domestic market. However, if most of the inequality were removed and most persons still consumed all of their income, the only source of indigenous investment would be through retained profits rather than through individual investment and reinvestment. Thus, some modern thinkers stress the need to create as much equality of income as is required to create and maintain a mass society of consumers, but not so much equality that investment and reinvestment are threatened. However, this last point is not presented with any saliency by advocates of bubble-up, who stress the role of demand rather than of savings or retained profits.

By stimulating the education of children and adults, government can theoretically help individuals to increase their own "human capital," that is, their ability to create more of a product out of themselves and so enhance their market value. Any person can enhance his or her own "human capital," but in being able to steer certain individuals to an education, governments can alter the distribution of educational and vocational opportunities and can thus stimulate bubbling up. Much of the pressure for school integration and job-training programs in the 1960s was obviously predicated on a bubbling-up theory. So, too, is the new thrust of the past five years or so in international agencies that have abandoned the trickle-down theory and are operating on the assumption that education, on-the-job training, and community development projects aimed at helping the poorest of the poor with "appropriate technologies" can achieve greater equality initially and higher growth rates subsequently.

The bubble-up theory has two related aspects, as we have seen. The first emphasizes the creation and maintenance of a mass-consuming society by the appropriate manipulation of taxes, subsidies, and government spending; the second emphasizes the role of mass education and vocational training in creating a more literate society, which will unleash talents that currently have few means of economic expression. The first aspect is more relevant to short-term considerations and the second to longer-term considerations. But whichever of these aspects we stress, the stages in the underlying theory are similar. They are laid out in Figure 16.2.

The trickle-down and bubble-up theories have a great deal in common. Consider the stages of the latter. They can be interpreted as (1) resources are disproportionately assigned to the poor and the deprived; (2) this effort releases a previously untapped source of creativity and talent, which (3) stimulates productivity and creates a mass-purchasing

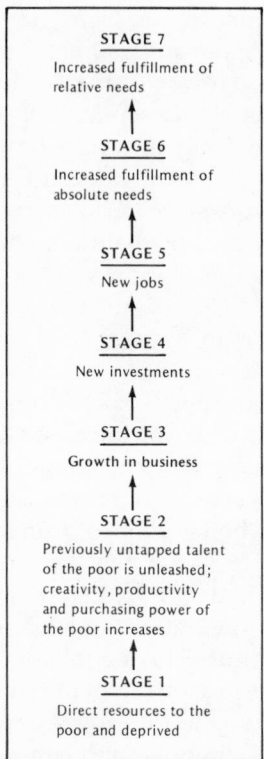

Figure 16.2. The bubble-up theory.

power, which (4) stimulates business in the aggregate; (5) much of the profits of the business are then invested where (6) they create new jobs; (7) increase the fulfillment of absolute needs; and (8) narrow income gaps. The last five stages are the same in both theories. The major differences are at the beginning.

Given its meritocratic origins, trickle-down starts by creating and rewarding growth in the abstract and more "meritorious" individuals in particular. Bubble-up, on the other hand, emphasizes egalitarian goals from the start, and resources are devoted to the needy rather than to the "meritorious." Anderson (1964) and Chenery, Ahlvwalia, Bell, Duly, and Jolly (1974) noted that policies based on a bubble-up strategy are almost exclusively redistributive in the short run, since they require that the earnings of the more productive sectors and individuals of the economy be spent to build up the human and physical capital of the unproductive. In the long run, however, economic growth is supposed to occur from which all are supposed to benefit.

3.2. Evaluation of the Effectiveness of the Bubble-Up Strategy

3.2.1. Stage 1

Perhaps the most significant issue in evaluating the first stage of the bubble-up strategy is that many of the resources initially distributed are likely to reach only a small percentage of the total population. Income distributions are pyramidal in form, and short of a revolution, massive reallocations of resources are needed if many at the bottom are to have access to resources in the amounts from which they can benefit. With lesser amounts of resources, a crucial policy decision has to be made. Devoting the resources to the neediest of the needy means that the targets are the persons least likely to make the initial investment productive. Hence, in many cases, the resources are actually given to a select group of the needy who are sometimes referred to as the *cream*. These are the individuals thought to have the greatest motivation and ability, and IQ testing, vocational testing, and the like are used to select them. However, these will also be the persons most likely to pull themselves out of poverty without the additional resources provided to them by a bubble-up program. Thus, the distributional principle employed in Stage 1 is rarely that of distribution to the needy. Rather, it is distribution to those presumed meritorious among the needy.

Inefficiencies in the initial distribution of resources may be much greater in the first stage of bubble-up than in the first stage of trickle-down. We are referring here not to the financial cost of the resources disbursed but to the cost of the distribution process itself. The massive bureaucracy that develops to administer bubble-up programs may result in a much leakier bucket, in Okun's (1975) term, than is the case in trickle-down programs. If so, this would be because of the larger numbers of recipients involved; and because of the need in bubble-up programs for expensive professionals to provide and monitor services. As implemented thus far, some instances of the bubble-up strategy have a less-than-enviable efficiency record in the initial distribution process.

In considering some of the reasons for the failure (or the limited success) of the War on Poverty, Pilisuk and Pilisuk (1973) and Liebman (1974) particularly noted the practical problems at the first stage. Liebman concluded that adequate resources were not provided; first, because of the need to finance the Vietnam war; second, because the middle class resisted the additional taxation needed to supply adequate levels of resources for a large number of the poor; and third, because the redistribution of some resources would have been too costly to powerful individuals and organizations that benefited from the economic status quo. While what was promised was partly an "absolute increase in the level of goods and services: housing, schools, hospitals, jobs . . . this promise was often interpreted to mean something else—that the percentage distribution would be altered, that some would have more and others less of the same" (p. 17). This suggests that short of revolution, systems of redistribution may be preferred that shift people from one side to the other of a defined need line rather than systems that fundamentally affect the relative standing of some individuals or groups relative to others.

3.2.2. Stage 2

Stage 2 of the bubble-up—which involves the vitalization of hidden talent among the poor—rests on several crucial assumptions. While we are prepared to accept that there may be dormant talent among the poor, we are not prepared to assume that programs exist that can tap this talent to a meaningful degree. Richard Titmus (1965) suggested that the faulty assumption prevailing in the early 1960s in both Great Britain and the United States was that "the poverty war can be painlessly won . . . all we need is activity, efficiency, and programs that can change the attitudes, motivation, and employability of the poor" (p. 130). Nine years later, in 1974, Lance Liebman concurred that this was indeed a faulty assumption when he listed among the reasons for the failure of the War on Poverty that we do not yet know how to do certain things, "like teaching poor lads to read competently."

Lester Thurow (1975) has outlined the prevalent economic theory whereby equalizing educational opportunities is supposed to equalize incomes eventually. Education is supposed to (a) transform low-skill workers into high-skill workers who merit higher wages; (b) reduce the supply of low-skill workers, leading to an increase in their wages; and (c) increase the supply of high-skill workers, causing a decrease in their wages relative to what they would have earned; and it is also supposed to (d) increase overall average earnings because the greater number of more-educated workers with higher skills will raise total productivity and so cause growth. A salient assumption behind this theory is that increasing educational opportunities changes the quality or quantity of schooling. Does this in fact occur? Assuming that enrollment in higher education has any relation to the talent one possesses, we would expect successful bubble-up programs in a country like the United States to increase the proportion of the needy population attending college. However, figures reported by the American Council of Education for the period between 1966 and 1972 reflect increases in the proportion of students enrolled only for those whose annual family income was in the highest category, above $15,000. Those from the neediest families, with incomes below $8,000 in 1972 dollars, showed a steady decline in enrollment from about 30% in 1966 to about 16% in 1972. Enrollment figures for 1976 from the Digest of Education Statistics 1977–1978 indicate a continuation of those trends. In terms of college enrollment, then, there appears to have been little increase in the bubbling up of needy students, though some needy students have doubtless been "creamed" off. It may be, though, that one has to wait several generations for bubbling up to occur.

3.2.3. Stage 5

Little new needs to be said about Stages 3 and 4 that has not been said previously about the trickle-down theory. Stage 5 is more interesting. Assuming that the bubble-up strategy has been successful in the previous stages, new jobs should become available in Stage 5. The issue is: Will they be made available to all, including qualified persons from poor backgrounds? If so, will this mean "creaming" or "raising the floor" of the poor? Blacks have been disproportionately represented among the poor in the United States, and many of the bubble-up policies have aimed at reversing and compensating for previous discrimination. Thus, the economic progress of blacks is of significance in evaluating this stage of the theory. Thurow (1977) reported the U.S. Census Bureau data on employment in jobs providing earnings in the top 5% of the income distribution. In 1960, black males were 9% as likely to hold such jobs as were white males, while in 1973, the relative figure was 19%. Thus, blacks have increased their access to the best-paying jobs. Additionally, Thurow reported that the relative earnings of full-time employed blacks rose from 56% of whites' earnings in 1955 to 66% in 1973. However, as Thurow importantly noted, this rate of relative gain was no faster in the 1960s than it was in the 1950s, suggesting that the bubble-up programs had little apparent effect. Moreover, the ratio was still 66% by 1978, according to the June 1980 Current Population Reports of Consumer Income. Further, the relative unemployment rate of blacks held constant between 1950 and 1975 at twice the rate of whites. The overall picture begins increasingly to support the suggestion that while a small elite group may have bubbled up, the majority is not much better off than prior to the initiation of bubble-up programs.

Consider what would happen if, as a result of the failure of one of the earlier stages of bubble-up, there was no substantial generation of new jobs, but there was increased education. This has probably occurred in many developing nations, and it may be the case with some groups in the United States. In the same paper by Thurow (1975), he noted that the result of a more equal distribution of human capital investments through education may actually result in an expanding wage differential. When those who previously would have received only a grade-school education continue through high school (as a result of bubble-up programs), the new high-school workers will take what would have been the best jobs available to the grade-school workers, including being taxi drivers or garbage collectors. The average wage of grade-school workers will then fall, and those who failed to bubble up by receiving more education may be even more needy than they were prior to the initiation of such programs.

3.2.4. Stage 7

Omitting Stage 6, we come to Stage 7, where relative differences in resources are supposed to be less than was found before the bubble-up strategy began. Since some persons interpret the bubble-up theory as having provided the rationale for redistributive policies in the United States in the 1960s, it is appropriate to ask whether the distribution of family income has narrowed since then. Nearly all commentators agree that it has not, with the sole exception being for college-educated blacks. They have caught up with their white counterparts, though they are also more likely than whites to be in families with two working adults.

A number of methodological difficulties beset conclusions of the above sort that are drawn from U.S. Census Bureau data. First, the definition of *income* is in terms of wages, salaries, rents, dividends, and cash payments from "welfare" sources. No consideration is given to in-kind transfers and noncash services. Some persons believe that poorer Americans receive more of transfers and services, and they cite Medicaid, AFDC, food stamps, and other means-tested programs that are worth money, as well as public services that the poor use more often (e.g., public transportation). Whether these sources of income offset the additional sources available to the more affluent is a moot point. These sources include the way tax laws are written, the ceiling on social-security payments, and better police, fire, and other public services in more affluent districts, as well as subsidies for airports and road construction. It is only within a sensitive multiple-operationalist perspective that we can adequately assess how income (broadly understood) is distributed.

Second, the point can be made that the census figures include many persons who have not been affected by bubble-up programs. These programs have been targeted primarily at the young via educational or job-training programs; and even among this target group, not everyone has been reached, especially in job-training programs. Thus, a strong test of the bubble-up theory would examine changes in the "income" distribution over time for persons in younger age brackets. And such studies cannot be done well until all the children who began schooling in the middle 1960s are of age to participate in the labor force and are beyond the stage of receiving salaries and wages that are depressed because training is still taking place.

3.3. Conclusion

It is very difficult to assess whether the bubble-up theory is successful in outlining a strategy that will increase both growth and equality. In non-Communist Western nations, bubble-up has never been attempted alone; rather, it has been part of a larger macroeconomic package and has assumed different degrees of political importance at different moments in history. In the United States, the major recent period in which it was in favor was the 1960s, and the data on income distribution that are currently available—which are only partially relevant—do not suggest that much equality resulted from that period, though the absolute needs of more people may have been met. In addition, many commentators have stressed that the costs of the bubble-up of the 1960s may have been to retard economic growth by preventing the accumulation of capital for new investments in production.

It is possible to interpret the revolutions in Eastern European countries as attempts to institute mass bubble-up campaigns. A heavy initial investment in public education was begun in order to create a new type of nationally committed, ideologically acceptable, and academically trained citizen; and from the ranks of these citizens, a new class of managers was supposed to emerge whose zeal and risk-taking proclivities would be channeled into state-run enterprises, the profits from which would be reinvested in the enterprise or would be invested elsewhere. If this interpretation is accepted, it is not unreasonable to conclude that in these countries, the bubble-up strategy may indeed have run through most of the stages illustrated in Figure 16-2 and may have led to the impressive degrees of economic growth and equality in income noted in most East European countries since their "revolutions." These changes have had their costs, of course, particularly the loss of some personal, legal, political, and economic freedoms. (It is important to note the loss of economic freedoms, for in the more capitalistic countries it is these very freedoms that permit the negative correlation between growth and equality to appear.)

In some developing countries, bubble-up strategies have recently been attempted to replace trickle-down strategies that did not seem effective. It is still too early to know how these countries, with their heavy emphasis on a human capital approach, will develop and whether they will succeed in creating both higher growth rates and a "reasonable" level of equality. However, we cannot be sanguine about this possibility, given all the problematic social and psychological links in the prevailing theory and given the external forces that can affect the transition from investments in the poor to economic growth and a reduction of inequalities in income.

4. INDEPENDENT TINKERING

4.1. Introduction

At present, the social sciences have no well-tested theory that can be relied upon to produce high levels of growth and to reduce inequality. Operating alone, the capitalist-flavored trickle-down theory is associated with high levels of economic growth, a widespread meeting of absolute needs, and a probable increase in the variance of the income distribution. Bubble-up has more heterogeneous roots in Keynesian economics, socialism, and the human-capital theory of modern economics. When resources are distributed to the poor on a preferential basis, the variance in the income distribution is initially reduced to a trivial degree, growth is slowed down at first, and it is not clear just when, or if, at a later date, a higher growth rate and more equality result. However, it seems clear from considering the underlying theory with its multiple causal links that bubble-up will not *inevitably* result in more growth and equality.

It is our distinct impression that much of the deliberate manipulation of the economy that occurs takes place in order to have a positive effect on either growth or income variance, but not in order to affect both positively by the same manipulation. In this sense, policymakers have to tinker independently with each in order to affect both at about the same time. But this leads to a great irony. Policymakers who try to increase growth will often create the unintended side effects of exacerbated income differentials. At the same time, policymakers from the same administration who are trying to reduce income differentials may inadvertantly reduce growth rates under what they would have been. Each set of policymakers is therefore working to counteract the consequences of the other. Yet, if they did not work on growth and equality at the same time for their different offices, growth would be created in a more laissez-faire economic environment at the cost of equality of condition, and equality might be achieved in a more controlled society but possibly at the cost of growth. Thus, independent tinkering is needed in the absence of a theory that stimulates both growth and greater equality. Two kinds of tinkering can be distinguished.

4.2. Managed Independent Tinkering

All governments in developed countries have some agencies designed to stimulate economic growth and other agencies designed to protect the interests of poorer persons. In the United States, the Departments of Commerce and Defense, and much of the Department of Labor and the Department of Agriculture, presumably help trade and business, while HUD, Health and Welfare, and some offices within Labor and Agriculture strive to protect poorer persons. In the legislative branch, Congressmen make themselves and their staffs available to representatives of business and the poor and minorities, and Congressmen have to balance the interests of one type of group against those of the other. So, too, do members of the highest offices in the executive branch, where preliminary decisions are made about what share of the budget should go to stimulating growth and what share should go to helping the less economically advantaged.[2]

Not all commentators would agree that in the nation's capital we have minimally interacting agencies, some of which help business stimulate growth, while others help guard against the income gaps that widen because of growth. Okun (1975), for instance, has argued that the "welfare" agencies help both to stimulate growth and to reduce inequality. He stresses the importance of removing racist and sexist barriers to accomplishment that operate within any particular level of qualification, with particular concern for the effects of discrimination on qualified female or black workers who, excluded from productive jobs, are forced virtually to function as if one of their hands had been tied behind their backs. This state of affairs creates inefficiency as well as inequality, and Okun estimates that eliminating discriminatory barriers has added at least $10 billion a year to the gross national product and

---

[2] It should not be thought that all, or most, of the institutional concern about the poor is aimed at reducing income differentials. Indeed, our guess is that the goals of the relevant agencies and offices are not stated explicitly and that vague phrases such as "to help the poor and needy" are used. Such phrases are capable of many interpretations: (a) meeting absolute needs, (b) preventing income gaps from widening, or (c) narrowing income differentials. Our speculation is that welfare in the United States is justified in either of the first two ways but not in the third. It is as though we have institutionalized a concern for propping up the floor of the poor—a floor that, because of economic growth that is not equally shared, continually tends to drop relative to other incomes.

has led to a bonus of pride and justice. In a similar vein, Okun refers to the gains in efficiency and equality that he thinks would occur once the poor were allowed the same access to bank loans as richer persons.

One problem with Okun's analysis is that in the social structure of our day, persons of different races are likely to differ in when they finished schooling, especially at more advanced levels of education. Indeed, it is possible that minority members in the aggregate are hurt more in the job market by receiving less schooling of inferior quality than they are by the barriers to achievement that operate against them when they compete for jobs with whites of comparable education. Another problem is that relatively few of the "welfare" agencies' funds go to detecting and eliminating discrimination. Most are spent on direct and indirect payments for unemployment, medical care, housing, food, child care, and job training.

Economic growth is a goal of the private secto as well as of some governmental and political offices. Scandinavian countries—in particular, Sweden—have frequently been invoked as examples that the private sector can create high levels of growth even when a sharply progressive tax structure is in place that contains few legal loopholes and provides considerable sums of money for the poor that can be used to reduce income differentials. The importance of the Swedish example lies in the assumption that sharply progressive taxes need not undermine individual and corporate motivations to work and produce, so that the private sector can handle growth and the government can handle redistribution. We shall now examine whether this argument holds by assessing whether any gains in equality have been made in Sweden since World War II. To do this, we shall use income data that are subject to all the limitations of definition mentioned earlier. However, it is worth noting that many more governmental services are available in Sweden on a universal (rather than a means-tested) basis than is the case in the United States. This probably decreases the extent to which income data underestimate the income of the poor relative to the more affluent.

With these limitations in mind, we shall now consider the pretax money-income distribution in Sweden. (Posttax data will be considered later.) Recall that the gini-coefficient, a measure of inequality in distribution, ranges from 0 for perfect equality to 1 for complete inequality of income. Scase (1976) offered evidence that Sweden's gini-coefficient rose from .38 to .40 between 1954 and 1963. (In the United States during that period, the gini remained at about .40). Roberti (1978) gave the 1972 Swedish gini as .39. For purposes of comparison, Roberti provided the coefficients for 15 other developed nations and derived an average gini of .38 for the 16 countries. While international comparisons are suspect because of the varying definitions and means of computing income, these figures suggest that Sweden's pretax income distribution is not significantly more egalitarian than is true for other nations.

Gini-coefficients do not make possible a clear view of how income is actually distributed among income classes. Decile (or some other percentage) shares provide a better view of the relative inequality in income distribution. In an Organization for Economic Cooperation and Development (OECD) study, Sawyer (1976) defined pretax income as including wages and salaries, entrepreneurial and property income, and current monetary transfers. The study looked at the income distributions for 10 OECD countries. While in 1972 the top decile in Sweden received a smaller share of national income than was true for the average of the 10 nations (24.4% in Sweden, 27.3% average), the difference in Sweden seems to have been fairly evenly distributed among the 9 remaining decile groups, so that the share of income received by 7 of the remaining groups was greater than the OECD average. For the fourth decile, the share in Sweden was 6.1%, while the average was 6.3%. The bottom decile groups in Sweden received 2% of pretax income, the same as the OECD average.

In a second paper, Roberti (1978) provided pretax income-distribution data for 10 developed nations. Relative to the other nations, Roberti found that Sweden's richest decile group received an exceptionally low percentage of income. However, consistent with Sawyer's figures, the income shares received by the remaining groups were generally within average range. Roberti also provided pretax distribution trends for the decile groups. Comparing the 1971–1972 period with 1948–1950, he found Sweden's bottom four deciles receiving a smaller income share in the more recent period. Of the remaining six deciles, all but the top group received about the same or a greater percentage of income. Consistent with the previous data, the income share received by the top decile group was lower in the more recent period. This analysis suggests that while Sweden's income redistribution policies may have reduced the share received by the richest 10% of earners, the redistribution has benefited middle-income groups more than the poorest.

We recognize that pretax money distributions may be expected to differ from posttax money distributions, with the latter showing more clearly the effects of redistributive programs in reducing income inequality. However, several writers have suggested that Sweden's taxation system has little effect on altering the income distribution. Although the national income tax is progressive, local taxes are proportional, and regressive indirect taxes have been increasing. Schnitzer and Nordyke (1971) noted that the general sales tax (now the value-added tax) accounted for 9.1% of Sweden's national tax revenues in the 1960–1961 fiscal year, compared with 19.7% in the 1967–1968 period (p. 347). Jenkins (1968) believes that because the government was reluctant to push up direct income taxes any higher, by the 1960s subsequent tax increases usually involved regressively higher indirect levies like the value-added tax. Scase (1976) found the Swedish tax structure less progressive than that of the United Kingdom or the Netherlands, for example. He concluded that "the overall system of Swedish taxation does little to alter the distribution of pre-tax incomes and it is reasonable to assume that the structures of post-tax incomes are similar for both Britain and Sweden" (p. 289).

While the data we have found are again limited by the exclusion of in-kind transfers and public expenditures, posttax figures provided by Sawyer's 1976 OECD paper do suggest little dramatic change in Sweden's income distribution. The posttax incomes of families includes wages and salaries, entrepreneurial and property income, and current money transfers, with direct taxes and social-security contributions deducted. The greatest posttax gains (relative to the pretax distribution) are made by the third through the sixth decile groups, with gains in income share received ranging from .67% to 1.1% for these four groups. The bottom group gains only .2%, and the next to bottom group gains .4%. Again, it appears that much of Sweden's income redistribution occurs from the highest to the middle income groups. Relative to 11 other OECD countries, the richest decile in Sweden receives less of an income share than the average of the richest deciles in other nations. However, all of the other deciles in Sweden receive greater income shares. Significantly, the smallest deviation from the average of the other countries occurs for the poorest decile in Sweden, which receives 2.2% of the income compared with 2.1% for the OECD average. Thus, posttax income transfers mirror the pretax picture. Redistribution is from the persons with the highest incomes to persons in the middle of the income distribution, but not the bottom.

Although a definitive income study remains to be done, it is not apparent from the best evidence available to us that Sweden's income distribution is significantly more equal than in other developed nations. It is also not clear that Sweden has been becoming more egalitarian since World War II, though there may have been some income transferred from the richest 10% to the middle but not the lowest deciles. The point is that even in the nation most cited as achieving both increased growth and reduced inequality through independent tinkering with each, there is no evidence that inequality has been reduced or is lower than elsewhere. As in the United States, the phenomenon that needs explaining is the stability in the income structure, rather than any changes in it.

4.3. A "Systems" Approach with "Mindless" Tinkering

It is striking in the United States and Western Europe how often governments with slightly-right-of-center leanings (who stress growth) are voted out of power in favor of governments with slightly-left-of-center leanings (who stress equality more than their conservative friends). However, these left-of-center "rascals" are in their turn voted out, and more conservative "rascals" are voted in. A cyclical pattern of alternation results. At one time, growth is stressed; then, egalitarian concerns become more salient; these then fade from the scene to be replaced by concerns of growth, after which egalitarianism becomes more popular again, and so on.

One of many explanations for this cyclical alternation is that it controls for too violent swings either toward growth and away from equality or toward egalitarianism and away from growth. At the very least, it is a mechanism that prevents income differentials from opening too wide, for to judge from the postwar stability in measured income variance in most of the democratic countries, such cyclical alternation has not greatly influenced differentials. Our guess is that at times of great growth, income differentials widen, which eventually cause feelings of individual and/or fraternal relative deprivation, which, in their turn, lead to verbal or behavioral expressions of the need to catch up. But as catching up takes place and traditional income differentials or growth are threatened, then calls arise to take off the brakes, to increase savings, and to allow a freer rein to investors.

Our further suspicion is that the balance between growth and equality is asymmetrical. That is, increases, but not decreases, in growth rates will be tolerated, whereas no significant increase or decrease in the equality of income will be tolerated. Also, time lags may differ insofar as more time may elapse between the period when growth accelerates and concern is voiced about income differences widening and the period when income differentials narrow and concerns are voiced about the need for more growth.

Consider cyclical alternation as it characterizes the policy emphases of U.S. presidential administrations during the past 50 years. The Roosevelt years were characterized by a concern for bubble-up strategies, with massive investment in the poor taking place to stimulate their welfare, productivity, and entry into the economic mainstream. The Truman and Eisenhower years were more growth-oriented, as was the Kennedy administration. The latter emphasized growth and the merit principle, applying Keynesian economics with a tax cut to increase the rate of growth and to reduce unemployment. Investment tax credits for research and development were intended to foster further growth and economic efficiency. Johnson's Great Society programs, on the other hand, emphasized need to a larger extent, requiring some sacrifices by the majority to provide resources to the needy. It was a time when bubble-up policies found great favor. The Nixon, Ford, and Carter administrations again have stressed merit, productivity, and growth. The recent reverse discrimination suits (i.e., *Bakke* and *Weber*) demonstrate the process of the shifting policy emphasis.

In his autobiographical *Vantage Point*, Lyndon Johnson (1971) reflected on this alternation from a politician's perspective:

> Traditionally, legislation that helps the American people keep abreast of changing times has come only after the problems reach crisis proportion. The pattern of social reform in America has been like a vast pendulum, swinging over the years from creative activity to almost total inaction, and then back again. (p. 70)

A similar point was made by Ginzberg and Solow (1974), who wrote that as long as our political and economic systems function reasonably well, the majority of Americans, including those in Congress, prefer the "quiet life." In describing the cyclical process, they referred to large-scale programs with direct or unintended egalitarian goals and suggested:

> A first lesson is that the public will accept large-scale programs of social intervention only at long intervals. One measure of time is the generation that elapsed between the New Freedom of Woodrow Wilson and the New Deal of Franklin Roosevelt, and again between the New Deal of Franklin Roosevelt, and again between the New Deal and the Great Society . . . but only special circumstances—like the breakdown of the economy in the Great Depression or the rise of the Civil Rights movement and the political awakening of the black population—with an assist from strong leadership can set the stage for major social reform. At other times, a piecemeal approach is the only kind possible. (p. 212)

The utility of a "systems" approach to the problems posed by the relationship between growth and equality is not yet clear. Indeed, much descriptive research still needs to be done to discover whether cyclical alternations exist in the priority accorded to growth and equality and—if they do—to describe the amplitude of the waves.[3] Another need will be to discover the mechanisms whereby deviations from the stable equilibrium levels of equality and sustained growth are detected and to detail the repertoire of responses to deviations. Much anecdotal information suggests that the countervailing consequences of growth and equality

[3] In this respect it is interesting to note the discussion of growth and redistribution in the United States since 1950 that is contained in Carnoy and Shearer (1980, pp. 282–284). Their analysis indicates that in times of higher growth the degree of inequality increased. But fine-grained work over longer periods is required.

are brought into stability not at any one time, but by varying the political saliency of each one at different times, and by having different agencies simultaneously concerned with promoting each of them. If so, the measured stability of indicators of income inequality over the last decades suggests system-level adjustments are made for very small changes in inequality.

5. SUMMARY AND CONCLUSION

We have suggested in this speculative essay that traditional means of producing the greatest economic growth tend, over the short run, to increase income differentials, while most strategies for reducing such differentials also tend to lower the growth rate relative to what it would have been. For societies that value increasing both growth and equality, this is a disturbing relationship; and it is probably at its most disturbing when external factors place appreciable limits on the rate of growth that can be achieved in a particular society. Then, pressures may arise to push for growth. These may well immediately exacerbate inequalities, the social and psychological consequences of which we are not yet aware.

With these speculations as our backdrop, we examined two theories that detail economic policies that are supposed to achieve both high growth rates and reduced inequality. They are dangerous theories in the sense that each deliberately sets out in the short run to increase either growth (trickle-down) or equality (bubble-up) *at the cost of the other*, although each theory also postulates that the value that is initially placed in jeopardy will later be promoted. More evidence is available to provide an approximate test of the trickle-down than the bubble-up theory. All the indications are that with trickle-down, growth rates increase, that some of the absolute needs of some of the less fortunate are met, but that in the long run, inequalities are not reduced and may well be exacerbated. Economic, sociological, and psychological reasons were advanced for the theory's probable failure to predict well.

Fewer data exist for testing the bubble-up theory. But the evidence that does exist suggests that, short of massive infusions of resources to the needy, neither significant levels of growth nor significant increases in equality are likely to occur. Again, economic, sociological, and psychological reasons were advanced to explain why the theory may be found to predict so poorly.

If the two major theories of the relationship of growth to equality do not predict well, how then can we explain that in the developed West, growth has tended to occur but inequality has not increased? Two speculative reasons were advanced. One is a "social systems" theory, which postulates cyclical alternations in governments. If one government stresses growth policies, then the theory holds that soon afterwards another government will stress egalitarian concerns, perhaps in order to make up the loss in equality caused by its predecessor. But since egalitarian policies tend to threaten personal expectancies about "appropriate" income differentials between individuals and/or between groups, and since these policies also tend to reduce growth over what it would otherwise have been, the more egalitarian party is voted out and the more growth-oriented party is voted in.

A second theory is that the social policy of modern governments is increasingly oriented toward managing the widening inequalities that would be generated by a capitalist or oligopolistic economic system if it operated in a void. Such management takes the form of supporting the standard of living and income of the poor to a degree that prevents the level of measured inequality from increasing to an appreciable degree; but it may not at all be aimed at decreasing inequalities. If governments are interested in achieving the latter, they have not been successful, for one of the most striking regularities in income data is the temporal stability of income differentials within most nations, as income has been traditionally measured. There are, of course, small fluctuations, but no radical shifts.

If greater equality of condition has not been achieved in non-Communist developed nations under conditions of sustained high growth rates, what are the prospects for it when growth cannot be taken for granted and is instead sporadic, low, or even negative? This is a question of great practical importance, and it is one where sociological and psychological research on reactions to scarce resources may help. Our guess is that reexamining such research in the specific context of the problem posed in this essay will lead to significant new researchable questions' being asked that may help clarify such issues as: When do individuals and/or groups compare their current standing to their own past standing rather than to the current standing of others? Under which conditions are decreasing resources associated with cooperation rather than competition, with concern for others rather than self? Under which conditions do individuals lose interest in future-oriented competition and become more oriented to enjoyment of the present? Will individuals who perceive external causes for slower growth (e.g., foreign energy costs; the demographic shift toward an older society supported by fewer people in the work force) be more reconciled to growing inequalities?

These and other questions are predicated on a genuine concern with the real possibility that economic growth works to the detriment of equality. Strategies are available to political actors for reducing the likelihood that the increase in inequality will be perceived—at least, at first. One strategy can be labeled *bread and circuses*, and it involves creating plans to turn attention away from economic ills by such visible acts as building neighborhood swimming pools or creating job corps programs that train young people for jobs that do not exist. A whole array of such techniques is available to anyone who wants to substitute symbolic politics for politics.

A second strategy can be labeled the *narrow definition strategy*. This involves focusing public attention on meeting the most obvious absolute needs of the most destitute or the most vocal among the less privileged. It involves, for example, creating a definition like "poverty line" and then publicizing how many persons have risen out of poverty. To many social psychologists, absolute poverty may not, over the long run, be as meaningful as relative poverty, given what seems to be a human proclivity for self-definition through social comparison. But be that as it may, political actors can gain mileage and time by directing public attention to palliatives, including narrow and largely irrelevant definitions of the social problem at hand. Such strategies are hardly likely to provide permanent solutions to the issues posed by the negative relationship between acts designed to foster growth and acts designed to reduce inequality or to prevent it from increasing. And these issues are likely to be exacerbated when growth is slow, sporadic, nonexistent, or negative.

The claim can readily be made that deemphasizing growth in favor of equality will, over the long run, lead to lesser relative differences but also to a lower level of material welfare than would have prevailed had growth been emphasized. The crucial issues to which this claim gives rise, and to which we have no answers at this time, are: (1) Are persons at the bottom end of the income distribution "better off" if they have less but are not so different from others? (2) Are there any conditions under which the majority would seriously countenance having less, or having not much more, in favor of narrowing differentials? (3) What would be the consequences of income differentials' widening if persons at the bottom end of the distribution were also not gaining absolutely or were not gaining anywhere near as much as they expected? This last question is perhaps of most immediate significance if economic growth is emphasized, and continues to be emphasized, over a period of reduced, null, or negative growth.

Acknowledgments

The authors thank Christopher S. Jencks for his detailed criticism, and regret that it came in time in the production process when only minimal modifications could be made to the text. He has made us aware of additional bibliographic sources that, if consulted in time, would have

modified parts of the analysis, though it is doubtful whether they would
have changed its major thrust.

REFERENCES

Aaron, H. The foundation of the "War on Poverty" re-examined. *The American Economic
    Review*, 1967, *57*, 1229–1240.
Adelman, J., & Morris, L. *Economic growth and social equity in developing countries*. Stanford,
    Calif.: Stanford University Press, 1973.
Adelman, I., Morris, C., & Robinson, S. Policies for equitable growth. In W. Loehr & J.
    Powelson (Eds.), *Economic development, poverty, and income distribution*. Boulder, Colo.:
    Westview, 1977, pp. 191–229.
Anderson, W. Trickling down: The relationship between economic growth and the effect
    of poverty among American families. *The Quarterly Journal of Economics*, 1964, *78*,
    511–524.
Carnoy, M. & Shearer, D. *Economic democracy: The challenge of the 1980's*. White Plains: M.
    E. Sharpe, Inc., 1980.
Chenery, H., Ahlvwalia, M., Bell, C., Duly, J., & Jolly, R. *Redistribution with growth*.
    London: Oxford University Press, 1974.
Cook, T., Crosby, F., & Hennigan, K. The construct validity of relative deprivation. In J.
    M. Suls & R. L. Miller (Eds.), *Social comparison processes: Theoretical and empirical per-
    spectives*. Washington: Hemisphere, 1977.
Festinger, L. A theory of social comparison processes. *Human Relations*, 1964, *1*, 117–140.
Galloway, C. The foundations of the "War on Poverty." *The American Economic Review*,
    1965, *55*, 122–131.
Galloway, L. The foundations of the "War on Poverty": Reply. *The American Economic
    Review*, 1967, *57*, 1241–1243.
Ginzberg, E. and Solow, R. (Eds.), *The Great Society: Lessons for the future*. New York: Basic
    Books, 1974.
Jenkins, D. *Sweden and the price of progress*. New York: Coward-McCann, 1968.
Johnson, L. *The vantage point: Perspectives of the presidency, 1963–1969*. New York: Popular
    Library, 1971.
Journal of Experimental Social Psychology, Special Supplement on social comparison, 1966.
Kohler, H. *Economics and urban problems*. Lexington, Mass.: D. C. Heath, 1973.
Kuznets, S. *Modern economic growth: Rate, structure, and spread*. New Haven, Conn.: Yale
    University Press, 1966.
Lampman, R. The future of the low-income problem. In B. Weisbrod (Ed.), *The economics
    of poverty: An american paradox*. Englewood Cliffs, N.J.: Prentice-Hall, 1965.
Leibman, L. Social intervention in a democracy. *The Public Interest*, Winter 1974, No. 34,
    14–29.
Levin, H. Assessing the equalization potential of education. Unpublished manuscript,
    Stanford University, 1978.
Loehr, W. Economic underdevelopment and income distribution: A survey of the litera-
    ture. In W. Locke & J. Powelson (Eds.), *Economic development, poverty, and income
    distribution*. Boulder, Colo.: Westview, 1977.
Looney, R. *Income distribution policies and economic growth in semi-industrialized countries: A
    comparative study of Iran, Mexico, Brazil, and South Korea*. New York: Praeger, 1975.
Okun, A. *Equality and efficiency: The big trade-off*. Washington: Brookings Institution, 1975.
Parker, R. The myth of middle America. In M. Pilisuk & P. Pilisuk (Eds.), *How we lost the
    war on poverty*. New Brunswick, N.J.: Transaction Books, 1973.

Pilisuk, M., & Pilisuk, P. (Eds.). *How we lost the war on poverty.* New Brunswick, N.J.: Transaction Books, 1973.

Roberti, P. Income inequality in some western countries: Patterns and trends. *International Journal of Social Economics,* 1978, 5, 22–39.

Runciman, W. *Relative deprivation and social justice.* Berkeley: University of California Press, 1966.

Sawyer, M. Income distribution in OECD countries. *OECD Economic Outlook: OECD occasional studies,* 1976, pp. 3–36.

Scase, R. *Readings in the Swedish class structure.* New York: Pergamon, 1976.

Schnitzer, M., & Nordyke, J. *Comparative economic systems.* Cincinnati: Southwestern Publishing Company, 1971.

Thurow, L. *Generating inequality: Mechanisms of distribution in the U.S. economy.* New York: Basic Books, 1975.

Thurow, L. The economic progress of minority groups. In M. Sharpe and B. Schieler (Eds.), *The challenge of economics: Readings from Challenge Magazine,* New York: Random House, 1977.

Titmus, R. Poverty vs. inequality: Diagnosis. *The Nation,* 1965, 200, 130–133.

Weisskoff, R. Income distribution and economic growth in Puerto Rico, Argentina, and Mexico. *Review of Income and Wealth,* December 1970, 16.

Legal Institutions and Their Alternatives

The layman who has had little to do with the legal system is likely to view the law as society's noblest attempt to assure that justice is done. People who have had more intimate contact with the law in either its civil or its criminal forms are often less enthusiastic about the legal process as a means of producing just outcomes. Within the legal community itself, there has long been ongoing debate and deliberation as to what changes in laws, legal forums, and law enforcement would best serve to improve the quality of justice dispensed by the system. Most recently, increased attention has been given to proposals and social experiments involving alternate approaches to the resolution of disputes and the control of deviant behavior. The chapters in this section not only provide an overview of contemporary concerns about how well law serves justice but also reflect some of the areas of contention among social scientists interested in this question.

Vidmar is concerned primarily with the psychology of the dispute process as it can be related to policy questions about the provision of forums for dispute resolution. First, he provides an overview of the differences among types of resolution processes, such as adjudication, arbitration, and mediation. Then, after considering in detail the variety of justice and other motives that impel people to become disputants, he addresses the intriguing question of which forums are best suited to satisfying disputants with different motives. A particular value of Vidmar's analysis is that he makes explicit its assumptions and limitations, thus providing a suggestive and useful model of motive–forum linkages. Equally stimulating is his examination of how various motives and other individual differences come into play at the choice points that comprise dispute development. A complex mix of factors influences whether the person experiences a grievance or not, decides to forego redress or seek a resolution, and so on. Vidmar not only sharpens our understanding of these dynamics but also points the way toward a full bill of research priorities in the area. Among these, he argues, is the need to understand fully the costs and benefits to the individual of becoming or not becoming a disputant so that we can make more informed decisions about whether to proliferate forum options. Vidmar is not convinced that more is better.

One type of "alternate" forum—the neighborhood "court" (with its ancillary, the neighborhood crime-prevention program)—is the focus of Kidder's chapter. He, too, questions the wisdom of moving quickly to provide a network of alternate forums, and his analysis develops the sociological perspective on problems that might be encountered in seeking to develop neighborhood-based servicing of justice concerns. Despite evidence that some societies have functioned well with only minimal reliance on a formal legal system, Kidder argues that there are numerous obstacles to moving from a highly legalized society to one employing simpler, decentralized, and presumably more responsive procedures. Drawing on case material from India, Chile, and China, as well as the United States, Kidder catalogs the mistakes that well-intentioned reformers have made in attempting to give people more access to and control over their dispute forums on a local basis. Prime among these has been the fact that change was introduced from outside by agents who had their own agenda and conceptions of "people's needs" based on inadequate information about the perceptions and social organization of the localities involved.

In conclusion, Kidder offers an intriguing, if somewhat pessimistic, scenario of the probable development of a polarized clientele for legal services in an impending period of scarcity. He offers little encouragement to those who see decentralization or neighborhood self-reliance as positive strategies for addressing justice concerns. Rather, he suggests that such developments in the future may be reactive and may present many of the dangers we associate with tribalism.

Peachey and Lerner hold a contrasting view of the uses of decentralization in developing responses to justice concerns. They first summarize the evidence that suggests that reliance on the legal system can function as a social trap. In both the civil and the criminal areas, research indicates that experience with legal forums often causes intensified feelings of having been dealt with unjustly, particularly for members of less powerful groups in society. In general, numerous studies indicate that familiarity with the law seems to breed distrust of its procedures. A central question, as the authors note, is whether these negative consequences can be dealt with by effecting reforms in how the legal system operates, or whether the very nature and pervasiveness of our legal structures create both distrust of the system and of one another. This is the social trap to which Peachey and Lerner allude—the possibility that having an elaborate formal system of laws "in place" increases the extent to which people feel that all human interactions must be predicated on distrust of others' motives and perhaps a hardening of one's own. The result of such attributions can then be demands for still more law and order, leading to more distrust and creating the social-trap effect. While it is possible that a proliferation of forums for dispute resolution could intensify the trap effect, as Vidmar suggests, it can also be argued, as these *authors do, that what is required is a restructuring of our approaches to conflicts of interest.*

Such conflicts, often at the heart of justice concerns, will not disappear, particularly in a period of scarce resources. But the extent to which people see themselves as related to others competitively and impersonally or, rather, as members of interdependent, cooperative groups whose needs and motives can be understood may be crucial to determining how these conflicts are dealt with. The authors maintain that it is possible for people to choose the more positive orientation and to channel the desire for justice into socially beneficial outcomes.

17

Justice Motives and Other Psychological Factors in the Development and Resolution of Disputes

NEIL VIDMAR

1. OVERVIEW

In this chapter, I want to explore some of the psychological dynamics involved in the development, continuance, and resolution of disputes between private citizens. This topic is directly germane to issues of justice in several ways. Disputes arise when justice motives are activated. Understanding the process of disputes gives us insight into some of the conditions under which justice motives are or are not exercised. Dispute-resolution forums involve problems of procedural justice, that is, a "methodology" for handling conflict with maximum possible fairness.

As a focal point for the discussion, I will address an ongoing policy debate regarding the issue of whether neighborhood justice centers, expanded or modified small claims courts, and similar innovations are desirable alternatives to formal adjudication, the principal means of re-

NEIL VIDMAR • Department of Psychology, University of Western Ontario, London, Canada N6A 5C2. Preparation of this manuscript was supported in part, by a research grant from the Social Sciences and Humanities Research Council of Canada and also, in part, by a research grant from the Russell Sage Foundation.

solving such disputes in our society. A number of complex economic, sociological, and ideological issues are involved in the debate,[1] but my main concern is with the psychology of the dispute process. Scholars have given tacit recognition to the importance of psychological elements (e.g., Danzig & Lowey, 1975; Felstiner, 1974, 1975; Nader, 1975), but, in fact, there has been little theoretical analysis of the motives and other psychological variables that affect disputants and consequently the dispute process. My goal here is only to lay a groundwork for more systematic theoretical and empirical work; the development of an integrated model of the psychodynamics of disputes is a task for the future. Nevertheless, even this rudimentary analysis raises thoughts about the extent to which proposed alternative procedures might provide better justice and the extent to which remedies for grievances might be provided for more people if alternative forums were available.

Some initial observations and interviews from a study of a Canadian small claims court provide a tentative data base for some of my analyses. However, for the most part, they are speculative in that I move quite far from what psychologists normally consider a sound empirical grounding. Moreover, let me acknowledge from the start a basic limitation of this data base and my primary focus. I am concerned here only with noncriminal disputes, and my data are derived mostly from working- and middle-class persons. I believe that the dimensions and dynamics that I outline are basic to all private disputes. Yet, the specific nature of disputes and the conditions under which they must be handled may be somewhat different for persons dwelling, say, in urban, ethnically mixed, low-income neighborhoods, such as the one studied recently by Merry (1979). While my focus can be defended on the grounds that calls for reform have been directed to all segments of society, it is more constructive to argue that its limited scope allows a clearer exposition of the psychological issues than would otherwise be possible. Expansion of the scope is a task for future papers.

The chapter is organized as follows. First, I will provide an overview of minor legal disputes between private citizens and the ongoing debate

---

[1] These other issues, sometimes touched on briefly in my discussion, include administrative efficiency, the likelihood of grafting alien resolution procedures onto our own society, and the need for underprivileged classes of persons such as the poor for relief from the many day-to-day minor grievances that in sum constitute major oppression. These other issues are elaborated in many of the references I cite in developing my psychological themes (e.g., Curran, 1977; Danzig & Lowey, 1975; Galanter, 1975; Johnson, 1978; McGillis & Mullen, 1977; Merry, 1979; Nader, 1975; Nader & Todd, 1978). My failure to come to grips with them in this chapter, therefore, does not reflect lack of awareness— or lack of empathy. Further, although my analysis does lead to a certain degree of skepticism about some of the proposals for alternative procedures, it is by no means an argument for the status quo or an argument against social experiments.

regarding the adequacy of the legal forum for dealing with them. Next, I examine the kinds of motives expressed in such legal disputes and consider their relationship to the question of procedural justice. In the following section, I consider the question of whether and to what degree these motives exist in persons who do not engage in legal disputes. Further, I consider the psychological factors that inhibit or facilitate the development of disputes as they proceed through a number of choice points. This section is relevant to the issues of whether disputes are transformed by the legal process, whether there may be dispute-prone persons, and, at the practical–policy level, whether more people would use alternative forums if they existed. The final section summarizes a number of implications for the policy debate and raises some additional research questions about justice motives and disputes.

2. PRIVATE DISPUTES AND THE DEBATE OVER PROCEDURAL ALTERNATIVES

Disputes are an ineluctable part of any society, though the substance of those disputes is colored by the environmental and cultural context of that society. Focusing on our own North American society, we can gain a picture of what some of those disputes are about by considering typical cases that come before a small claims court: a landlord asserts that a tenant owes rent; a tenant claims that his landlord has not made promised repairs; one person claims that an automobile driver's negligence caused damage to his property or person; a businesswoman accuses another businesswoman of failure to honor the terms of a contract or charges a customer with failure to pay a bill; a buyer claims that a merchant sold a defective refrigerator; a customer claims that his mechanic (plumber, dentist, etc.) provided inadequate service or overcharged; a gardener wants her neighbor to restrict the dog that tore up her flowerbed and to pay damages; a man asserts that deceased Aunt Minnie's heirloom vase (worth less than $200) belongs to him instead of his brother; a college student wants "her" television set back from an estranged former roommate.

Anyone who has observed such cases in a small claims court will be aware that frequently they involve very intense emotional feelings about justice and rights. Some other observations can be drawn from these examples as well. With the exclusion of criminal acts and domestic problems, the examples are fairly representative of the kinds of minor disputes in which citizens could be involved. Indeed, for the most part, the basic conflicts or grievances are ones that any of us, regardless of our social class, could experience—or have experienced. It is also probably obvious that disputes that come before small claims courts constitute only the tip of an iceberg of potential or actual disputes. The questions of what happens to these disputes and what are the consequences for the involved persons, as well as the question of what happens to the disputes when the legal system does process them, are ones about which there are no satisfactory empirical data. The goal of this chapter is to provide a framework for collecting data bearing on aspects of these questions.

Various modes of third-party dispute-resolution procedures have been developed in other societies. Some nonlegal third-party forums also exist in our own society, but for the average person with an unresolved grievance, the main recourse is law (see, e.g., Felstiner, 1975; Galanter, 1974; McGillis & Mullen, 1977; Nader, 1975; Nader & Singer, 1976; Palen, 1979). In recent years, however, a number of legal experts and social scientists have argued that there is a need to develop alternative forums that will provide greater access for more people and also provide better justice than exists at present (see e.g., Danzig, 1973; Danzig & Lowey, 1975; Johnson, 1977; McGillis & Mullen, 1977; Sander, 1977). Further, government and private agencies have begun to fund projects such as neighborhood justice centers or expanded small claims courts to fill this assumed void. While most of the neighborhood justice centers are intended to deal primarily with disputes involving persons with ongoing relationships and to cover domestic and minor criminal matters as well as the kinds of disputes that we have described above (see McGillis & Mullen, 1977), the implicit assumption seems to be that the need for alternatives to adjudication is widespread throughout society. Sander (1977) has captured the essence of the logic underlying this movement in a reference to "missing plaintiffs," that is, people who have perceived grievances for which they cannot or do not seek a remedy but who, it is assumed, would do so if they had the chance.

At the risk of oversimplification, let us distill the arguments against adjudication as a means of resolving minor disputes (see Abel, 1973, 1977; Aubert, 1967; Danzig & Lowey, 1975; Felstiner, 1974, 1975; McGillis & Mullen, 1977; Nader & Singer, 1976; Sander, 1977). Recourse to formal legal procedure is viewed as so threatening, inaccessible, or costly in terms of economics, social relationships, or psychological stress that most people with grievances engage in various forms of problem avoidance: severing the relationship with the other party, or exit; redefining the dispute as a nonproblem; or enduring the conflict, or "lumping it," that is, giving up (see Felstiner, 1974, 1975; Galanter, 1974; Merry, 1979). It is asserted, however, that these alternative behaviors are very costly because they make people cynical and harm their sense of justice. Lack of resolution may ultimately cause the parties to expand the conflict to other areas or to involve other persons, or worse, cause them to pursue illegal remedies. Equally important is the argument that once disputes do get into the legal system, they are transformed in a way that may not get at the underlying cause of the conflict and, moreover, may even unnecessarily harm or destroy interpersonal relationships. Finally, it is argued, procedural alternatives to adjudication, such as mediation or arbitration, that allow the disputants more involvement in the process and that are not bound by the rigid decision rules of adjudication are less likely to be threatening and more likely to produce a creative result that both sides consider fair.

It is important to note that the adjudicatory forum does differ from mediation and arbitration in procedurally important ways. Though a full discussion of all the differences is beyond the purview of this chapter, it is important to indicate what some of the differences are. Steele (1977) has summarized the principal dimensions along which dispute forums may vary: public–private, formal–informal, adjudication–mediation, coercive–voluntary, legalistic–therapeutic, emphasis on value dissensus–emphasis on conflict of interest, zero-sum–compromise, decision-oriented–agreement-oriented, rule-oriented–person-oriented. The extent to which the forums are unofficial as opposed to being appended to the legal system (see Galanter, 1974) is also an important dimension. Nader (1969, 1975) has pointed out that studies of forums, both cross-culturally and within North American societies, indicate that no one style or dimension is necessarily always associated with a particular forum. Nevertheless, we can suggest that the modal adjudicatory forum in North America tends to have certain characteristics. In adjudication the issues are narrowed, precluding discussion of additional points of conflict; the dispute has to be translated into arguments over conflicting values and facts instead of conflicting interests; proceedings are public as opposed to being held in private; the decisions tend to be all-or-none rather than based on compromise; the value basis of the decision, combined with its all-or-none nature, tends to produce a moral stigma for the loser, both socially and personally, in that past behavior is pronounced wrong; the disputants' control over and participation in the proceedings is curtailed; the arbiter is removed and distant; the process involves specialized rituals and language (see e.g., Abel, 1973; Aubert, 1963, 1967; Danzig & Lowey, 1975; Felstiner, 1974, 1975; Fuller, 1971; Nader, 1975; Sarat, 1976; Steele, 1977; Thibaut & Walker, 1975, 1978).

In contrast, mediation and arbitration procedures tend to have more flexibility in dealing with problems on a case-to-case basis, to be held in private, to allow compromise, and to allow more participant control over the proceedings. Thus, critics of adjudication argue that to the extent that disputants have ongoing, multiplex relationships, to the ex-

tent that they are intimidated by the adjudicatory process, to the extent
that hidden motives underlie the dispute, or to the extent that the resolution procedures, or the result, conflict with other important socio-
culturally derived values, some alternative third-party procedure is more
likely to be utilized and to arrive at a just result (see e.g., Danzig &
Lowey, 1975; Nader, 1975; Nader & Singer, 1976, for a review and discussion).

Arguments can, however, be marshaled against these alternative
forums, at least for our society. Felstiner (1974, 1975) has argued that
the kinds of mediation forums proposed by Danzig (1973; Danzig &
Lowey, 1975) probably cannot be grafted onto our society, with its need
for coercive authority, its overwhelming number of secondary versus
primary social relationships, and its lack of a structural organization and
a cultural tradition favorable to such procedures. Moreover, while assuming that people have grievances in need of resolution—that is, assuming that the "missing plaintiffs" exist—Felstiner argued that avoidance in its various forms may not be as maladaptive as proponents of
change have argued that it is. One can go further and actually question
the extent to which missing plaintiffs exist. Perhaps many people do not
perceive problems, or they have learned to confront and solve problems,
either through extant alternative forums or through self-help without
recourse to third parties. As I argue below, some people may even prefer
avoidance over almost any form of social confrontation. Finally, law is
coercive. Mediation, in particular, requires the consent of both disputing
parties, but it is frequently the case that the party who has possession
of an object, who is alleged to have caused the damage or to owe the
debt, has no incentive to consent. Adjudication, on the other hand,
gives the party who is asserting the grievance the leverage to force the
dispute into the settlement arena.

Thus, the debate over alternative forums involves some complex
issues that require analysis at legal, sociological, and psychological levels. Certain kinds of legal differences between disputes, say, torts versus
debts, or cases involving strong versus weak factual evidence—may
suggest advantages of one kind of forum over another. At the sociological level, a number of studies have shown that such factors as a prior
relationship between the disputants, alternative norm systems, and different resolution traditions are likely to cause people to prefer nonlegal
forums; scarcity of the object in dispute and superior economic and legal
resources are likely to cause preference for adjudication (see, e.g., Nader,
1975; Sarat, 1976). But knowledge of these parameters does not tell the
whole story. Of two disputes having essentially the same legal and
sociological characteristics, one may end up in a legal forum while the
other may end up in an alternative forum or be settled prior to trial.

Moreover, in the cases that do end up in the legal forum, some parties may come away relatively satisfied and others may come away relatively dissatisfied. By examining disputes from the perspective of the disputants, including their psychological motives and goals, we can begin to understand why and begin to consider another facet of the question about the effectiveness of resolution procedures.

3. JUSTICE MOTIVES AND RESOLUTION PROCEDURE

A basic premise of our psychological perspective, therefore, is that to determine whether a dispute resolution procedure will be effective, we must ask what goals the parties have in pursuing the dispute; and then, we must consider the extent to which the procedure may help to achieve these goals.[2] Our initial interviews and observations of small claims court plaintiffs reveal a number of motives behind the desire for legal action.

3.1. Justice and Other Motives in Disputes

It is fair to assert that justice motives underlie every dispute, but the particular principles of justice may be quite varied, they may arise or become transformed during the course of the dispute, and they may sometimes be subservient to other motives. Most often, multiple motives are involved, but for purposes of analysis it is useful to consider them separately.

One of the most common core causes of legal disputes in the small claims court which we have under study is an individual's assertion of a right to restitution. The owner of a destroyed flowerbed or a damaged automobile simply wants the property restored to the original state, or at least, to receive recompense. The purchaser of the defective merchandise or service wants an outcome commensurate with costs. A closely related cause is obligation. In the case of an agreement or a contract, the party with the grievance wants the other party to fulfill agreed-upon terms (or provide compensation). These restitution, obli-

---

[2] Thibaut and Walker (1975, 1978) make the important point that procedural justice requires both that the resolution forum find truth as nearly as possible *and* that it provide the participating parties with the perception of fairness. Perceived fairness, within bounds, may be even more important than finding truth when the two goals come into conflict. Considering the goals of the disputing parties as a starting point to judge effectiveness of resolution procedure is, as I view it, quite consistent with this perspective. Starr and Yngvesson (1975) provided an interesting anthropological perspective on this point as well (also see Nader & Todd, 1978).

gation, and contract cases may be seen as involving an equity principle. There are sometimes other cases that involve even more straightforward equity considerations, for example, the relative who took care of Aunt Minnie for several years before her death and now asserts that a family heirloom should belong to him or her.

Disputes, however, sometimes evolve out of "rights" (e.g., Friedman, 1971) rather than equity considerations and thus involve a second type of justice motive. For example, another relative of Aunt Minnie may assert the right to possession of the heirloom. Even though this relative did little to help Aunt Minnie—that is, provided few inputs— he or she claims that it was promised to him or her and may even have a letter documenting the claim. Minority-group members may assert their "right" to access to education or jobs. This assertion is made on grounds of natural justice or a statute or a human rights code rather than equity since the minority individual has not had a chance to make the inputs necessary for equity considerations. This particular example would probably be based on equality of opportunity, but cases involving the principles of parity or need might also arise. This civil rights example would also probably be dealt with in some forum other than the small claims court, such as a human rights commission, but it helps to illustrate the concept of rights, or perceived rights, that sometimes motivates people in private disputes.

Occasionally, the motives involved in disputes may be concerned with what Thibaut and Walker (1975, 1978; see also Deutsch, 1975; Leventhal, 1976) have called "procedural justice." Such instances involve conflict that has gone through some third-party forum, but one of the parties feels that his or her position was not given a fair hearing. Thus, the focus is on the way the dispute was handled, not its actual outcome. A student who appeals a mark to a dean's committee but who feels that the committee was biased, or a faculty member who feels that his salary increment does not reflect his contributions, might be responding to the unfairness of the procedure rather than to the outcome *per se*. Several small claims plaintiffs indicated that this was at least part of the motivation in their pursuit of justice. In one instance, the plaintiff perceived poor treatment by the better business bureau. In another case, the plaintiff pursued a grievance against a large store even though that store eventually offered to settle out of court for two-thirds of the claim. In an interview, she indicated that her refusal to settle was based on the discourtesy and unfairness with which the store's manager had heard her complaint. Another case involved a bank manager who, it was asserted, did not give a fair hearing on a claim that the bank had charged excess interest on a student loan.

The retribution motive, explored in this volume by Hogan and Emler and by Miller and Vidmar (see also Vidmar & Miller, 1980), is also an important justice motive in disputes. Some disputes involve retribution at their genesis, but at least as often, retributively based motives arise during the course of the dispute. As disputes progress and the participants make public charges and countercharges about the other person's bad intentions and violation of moral obligations, it becomes perceived to be necessary that each party vindicate himself, a process on which we elaborate more in the next section. Material compensation may become unimportant except to the degree that it serves as a symbol of vindication. Leff (1970), for example, has noted how "spite" in legal disputes oftentimes acts as an autonomous motive. That is, people sometimes pursue a dispute even though their potential material gain is calculated, even by themselves, to be trivial or nil. What is more, it is not unusual for persons to pursue a dispute even after it becomes abundantly clear that they will incur substantially more costs than can be gained even if the suit is won, a condition that Leff appealingly calls "superspite." The manifestation of superspite is quite clear to anyone who has spent several days observing cases tried in a small claims court, and, of course, the basic psychology and dynamics of spite have long been documented in the research by Deutsch (1973) and others.

Anthropological research in communities characterized by multiplex relationships has led Nader and her associates (see Nader, 1975; Nader & Todd, 1978; Yngvesson, 1976) to suggest that social power, status assertion, and similar motives may underlie a dispute even though that dispute is articulated with respect to material or other concerns. We do not know about the extent to which such motives may underlie disputes in modern Western societies, but they might easily be involved in disputes that arise among relatives, neighbors, or co-workers, that is, persons whose relationships are long-term and/or multiplex (see Felstiner, 1974, 1975, for an excellent discussion on this issue). One quite comical (to us) case that we observed involved a rejected male suitor who, with lawyer in tow, sued his former paramour for the cost of the female undergarments and other toiletries incurred by him during a rather hurriedly arranged vacation trip. Upon their return, she had politely but firmly expressed a wish to end the relationship. Unfortunately, she did so in front of some mutual friends, causing a considerable loss of face to the plaintiff. To everyone, including the judge, it was obvious that the motive for the suit involved bruised male ego rather than the cost of three pairs of nylon panties and a toothbrush. Of course, the plaintiff did not view it this way and failed to see how others viewed it. These status motives bear much similarity to retribution and, indeed, may operate in much the same way.

Finally, some disputes arise out of motives that are less directly related to justice concerns. As I have noted elsewhere (Vidmar, 1979), Sarat's (1976) research and my own preliminary research suggest that small businessmen may sometimes use the small claims court as a convenient mechanism for settling a business disagreement. They may approach litigation with relatively low levels of affect, with an attitude that attaches no moral stigma to winning or losing, and with the expectation that the litigation will not necessarily have a major negative effect on the relationship. Sarat found, for example, that a surprising number of persons indicated that small claims litigation actually improved their relationship, presumably because it clarified roles and obligations between the parties.

Simply controlling another person's behavior in the present or at some future time may be another motive behind a suit. For example, in one case we observed an automobile dealer who pressed a claim involving violation of an agreement against several insurance adjusters with whom he had a long-standing relationship and, because of the nature of their respective businesses, with whom he could not avoid future dealings. Relations between the parties were cordial before, during, and after the trial. Though the dealer took obvious satisfaction from winning, the amount of money gained did not, in the single case at issue, justify his costs of time and legal fees. However, as he articulated afterward, his primary goal was to establish ground rules that would guide his future relationships with not only the particular insurance company defendants but also adjusters from other companies.

As a last example, some persons may utilize courts simply as a mechanism for debt collection (see Galanter, 1974, 1975; Sarat, 1976). These persons are usually "repeat players" who know how to utilize dispute forums to induce recalcitrant debtors to pay up. Their attitude toward the dispute may usually be characterized as impersonal, routine, and quite distant from justice considerations as we ordinarily think of them.

3.2. Motives and Forums

By making several assumptions, we can now ask which type of forum, *ceteris paribus*, might best give expression to the motives. The first assumption is that the parties are forced to meet; that is, participation is coercive. A second assumption is that only a single type of motive dominates the dispute. A third assumption, which we will subsequently modify, is that only the complaining party's (plaintiff's) motives and perspectives are important. A final assumption, also to be qualified later, is that the complaining party is able to make a reasoned decision about what he or she hopes to gain from the dispute.

A dispute pressed on grounds of retribution, or spite, is probably the most clear-cut instance where adjudication will be preferred to other procedural alternatives. A person motivated by spite is concerned with moral vindication. Vindication may entail personal humiliation of the other party, but it is usually also a reaction involving the need to affirm social consensus about the negative character of the offender and his or her act (see Vidmar & Miller, 1980). The adjudication forum is public and tends to provide all-or-none decisions that are perceived to be tinged with moral stigma. Thus, it can probably provide a better outlet for expression of this motive than a forum, such as mediation, that proceeds in private and that tends toward compromise solutions.

Persons pursuing disputes on grounds of behavior control may also find better expression of their needs in a legal rather than some other forum. Because a legal forum's decisions are backed by sanctioning power and because it is public, it can best communicate a behavior control message. This holds regardless of whether the first party wants to control the person involved in the present dispute or persons who might be involved in future disputes.

It is likely that persons with disputes based on grounds of procedural injustice will similarly prefer an adjudicative forum. Since the aggrieved party has already experienced what she or he perceived to be unfair treatment, the regularized process of law at least provides more clear-cut rules for hearing the complaint. In such cases, moreover, it is likely that the motives of retribution and deterrence will have arisen and will co-exist with the desire for procedural justice, further fostering preference for a legal forum.

The choice of forum is not so clear in the case of either rights or hidden causes, such as power and status. If the claimed right is clearly established in law, the party's preference will, of course, be for the legal forum. If, on the other hand, the basis of the perceived right is based on custom, tradition, or justice norms outside the legal norms, then preference should be for a forum that will allow those rights to be expressed. In the instance where social power or status is a motive, the preferable forum will be based on the extent to which a legal decision will accomplish that end. If the desire, for example, is status dominance and a legal decision would be recognized by the participants and other interested parties as symbolic of that dominance, then adjudication is probably preferable. However, if the person is simply seeking power or status equalization, a more informal, private, and less morally tinged forum will be preferred.

Disputes pursued for restitution or other equity motives should be independent of these basic dimensions that distinguish resolution forums. The person is interested only in being paid back in some coin commensurate with incurred costs or inputs. Whichever procedure will provide that end will be chosen. Issues of morality, public versus private hearing, and so forth are just not relevant, with one exception. That exception is the tendency for courts to make all-or-none rulings. To the extent that the person is willing to compromise or is pursuing a minimax strategy, a nonadjudicatory forum may be preferable.

This attempt to match dominant motives with form of resolution procedure has admittedly oversimplified a number of considerations and ignored a number of others. For one, various resolution forums sometimes use mixed procedures. For another, the person pursuing the dispute often has mixed justice motives. We have, moreover, also ignored the motives of the other party and his or her interests and preferences in the dispute. Take, for example, a dispute involving conflict based, respectively, on equity and right, as in our example of the two relatives in conflict over possession of Aunt Minnie's heirloom. Houlden, La Tour, Walker, and Thibaut (1978) created conditions similar to this kind of conflict in a laboratory experiment: one subject had a claim based on equity and one had a claim based on strict legal right. The subject with the legal claim preferred a dispute resolution procedure more similar to adjudication, while the subject having a claim based on equity perferred a procedure allowing more flexibility in the presentation of his/her case. Thus, an adequate prediction model ultimately requires assessment of both parties' motives.

The analysis has also ignored some other factors that might compete with these basic motives. For example, the individual's resources, risk-taking propensities, knowledge, and ability to articulate the grievance may influence procedural preference (see Aubert, 1967; Galanter, 1974; Nader, 1975). Cultural differences in traditions and ascription to legal norms might similarly override the basic motives (see Nader, 1975). Additionally, the analysis has assumed that the person has an adequate perception of his/her own motives. It also assumes that the person has accurate knowledge of the characteristics of the forum, including the relevant norms and criteria of that forum; of any facts or evidence which may conflict with his/her own position; and of the behavior and commitment of the other party. Leff (1970), for one, has persuasively argued that ignorance of such factors is very frequently a crucial element causing disputes to be pursued even though a veridical perception of reality would dictate settlement. It should be added that ignorance could also result in preference of an inappropriate forum. The effects of some of these other factors are elaborated on shortly.

The analysis has also ignored a value consideration, namely, the possibility that the resolution procedure might be capable of transforming motives—and, consequently, the dispute—in such a way that it has a therapeutic effect on the parties (see Aubert, 1963, 1967; Danzig & Lowey 1975). For example, a dispute that grew out of conflicts of interests or equity that eventually turned into spite might be reconverted; the disputants might discover their original common interests and live happily ever after. This hypothesis gains some credence from Sarat's (1976) study of a small claims court where, on arrival in court, the disputing parties were given the option of formal public adjudication or private, informal, but binding arbitration. Persons who had a prior relationship together were more likely to choose arbitration over adjudication, whereas persons who had no prior relationship were more likely to choose adjudication. Though the data were confounded because of this self-selection, the interesting finding was that when queried some time later, over 35% of the litigants who chose arbitration reported that the experience actually improved their relationship; for those persons choosing adjudication, only slightly over 9% reported improvement. Adjudication in comparison with arbitration was much more likely to terminate the relationship or make it more difficult to get along. The important point is that Sarat's findings do raise the possibility that a procedural alternative to adjudication might have a more ameliorative effect on disputants and the motives that cause them to take legal action. In considering this possibility, however, we must also reflect on whether the disputants would understand this possibility in choosing a forum— or desire such a choice.

These qualifications notwithstanding, it is important to reemphasize the points that disputes must be understood from the perspective of the disputing parties (see Nader, 1975) and that within this perspective, it is essential to understand what motives have caused the parties to carry their conflict into a resolution forum before we can begin to estimate what type of resolution procedure may be preferable or more effective. It is also important to stress again that the type of motive or motives will be correlated only imperfectly with legal and sociostructural variables. For instance, in our example of the person whose neighbor's dog destroyed her flowerbed, the legal fact of the damage and the dog's responsibility may be clear-cut; the two parties may have had a long-standing and perhaps even positive relationship. This would suggest that a preferable forum would be nonadjudicatory and ameliorative in its approach. However, if the dispute has evolved into recrimination and the predominant motives of the parties are now spite, it might be expected that the parties will prefer an adjudicatory hearing. Such knowledge, therefore, enhances our predictive ability regarding parties' choice of forum (when available) and probably also their ultimate satisfaction with the resolution procedure.

In a word, there are interactive effects between motives and resolution procedure that have implications for disputant behavior and attitudes. Insofar as the debate about alternative forums is concerned, the primary point to be made is that adjudicatory procedure may be more satisfactory than, say, mediation procedures for many disputes, though perhaps less satisfactory for others.

4. DYNAMICS OF THE DISPUTE PROCESS

In the last section, we concerned ourselves with the motives with which persons enter a resolution forum. We now go back to the beginning and attempt to outline some of the psychological dynamics of disputes prior to, and including engagement of, a forum. The analysis is germane to the issues of (a) the extent to which the legal forum transforms disputes and (b) the extent to which people might utilize alternative independent resolution forums if they were made available. To grasp these issues properly, we need to examine the social-psychological factors operating at various stages of the dispute process; these include justice motives, noneconomic costs, and dyadic reactions. We also need to address a comparative question: Why do some potential disputes escalate and others fail to develop into disputes?

One possible answer to this comparative question is that there are individual differences between persons regarding justice motives. Data show that, at least with respect to criminal violations, there are differences between persons in this regard (see Miller & Vidmar in this volume or Vidmar & Miller, 1980); and there is no reason to expect that this could not be true regarding private disputes. Thus, persons who engage in public disputes may be different than persons who do not engage in them in the extent to which they are prone to have motives evoked, in the type of motive that is evoked, or in the strength of the motive. A more sophisticated view suggests, moreover, that these motives are influenced by factors specific to each situation.

Other psychological considerations besides justice motives also affect the disputes, however. We can best illustrate the interplay of all these factors by viewing disputes as a process involving choice points. The path taken at each choice point is a complex function of the parties' perceptions, attitudes and personality dispositions, and their behaviors. To demonstrate the dynamics of each of these points, we will draw on several of the examples of disputes noted at the beginning of this chapter: someone has a consumer complaint; someone sustains minor automobile damage from another motorist; a neighbor's dog tears up a freshly planted flowerbed.

4.1. Choice Points in the Dispute Process

The first choice point in a dispute involves perceiving grounds for complaint, that is, developing a grievance. Friedman (1971, 1975) has elaborated on the concepts of "rights consciousness" and "claims consciousness," that is, the tendency for people to perceive a grievance and to seek a recourse for that perceived grievance. Friedman argued that the consciousness of rights is a product of socialization, and though most of his analysis is focused on what people do or do not do after a grievance has developed, it applies equally well to the perceptions that give rise to the grievance in the first place. While there are differences between social classes or groups in these initial perceptions—and these are the brunt of Friedman's concern—we may also expect a wide range of individual differences within social categories of persons. For example, consider our gardener. The ruined flowerbed may be attributed to some other dog, and even if the dog is caught in the act, the unfortunate gardener may say, "Well, dogs will be dogs." Alternatively, another gardener in the same situation might immediately ascribe responsibility for the act to the dog's owner. To take another example, one person may perceive a defect in a cheap or moderately priced product or service as simply "getting what you pay for," while another person may feel that the defect is grounds for complaint to the merchant or the manufacturer. A recent consumer complaint study by Best and Andreasen (1977) illustrates such differences nicely. Even though many people perceived some deficiencies in purchased products, only a minority tended to identify themselves as consumer victims or as having serious grounds for complaint. In the case of the automobile damage, the owner may assume that she/he was at least partly responsible for the accident or may insist that it was entirely the other person's fault. In short, we can expect differences between persons at the very inception of a dispute, that is, perception of a grievance.

If a grievance is perceived, the second choice point involves the decision to express the grievance or to engage in some other behavior, such as exit from the relationship, enduring it, or "lumping it" (see Felstiner, 1974; Hirschman, 1970). The gardener may say, "His damned dog! But, after all, he is otherwise a good neighbor; rather than embarrass him, I'll just repair the flowerbed and forget it." Or she/he may simply consider the fact that the neighbor is cantankerous, potentially prone to violence—and big. With the defective product example, we can also suggest a number of reasons for not expressing the grievance: "The merchant isn't at fault"; "I don't want to spend the time"; "What's the use—complaining won't do any good"; "I would rather avoid a confrontation." While most theorists (e.g., Best & Andreasen, 1977; Danzig & Lowey, 1975) have tended to emphasize the "what's the use" reason as an explanation for failure to voice complaints or grievances, they have greatly underestimated the simple dislike of confrontation as a reason for withholding expression of the grievance. I have argued elsewhere (Vidmar, 1978) that for many people, the unpleasantness of confrontation may be far more distasteful—that is, psychologically costly—than avoidance or lumping the injustice. This is especially true when the person can engage in forms of rationalization or ways of "getting even," such as telling other persons about the offender's bad behavior or even assuming that fate or God will eventually extract retribution. While there are no empirical data bearing directly on this hypothesis, it seems very likely that it may explain a lot of nondispute behavior. Other research on justice motives shows people exhibiting considerable ingenuity in rationalizing injustice (e.g., Lerner & Miller, 1978).

Of course, other people do give voice to their grievances, and this involves a third choice point. The person may immediately appeal to a third party, circumventing the person perceived to be responsible for the problem. People go "over the head" of their immediate supervisor, to their minister, or even directly to a lawyer and the court. This is done for a number of reasons, for example, the complaining party may feel that the person against whom s/he has a grievance will not give satisfaction or will retaliate. Sometimes the party feels so aggrieved that s/he moves directly to the third party out of spite, hoping to precipitate a major confrontation.

Most people, however, probably choose at least to attempt to communicate directly with the other party. The outcome of direct negotiations is very dependent on the attitudes and behavior of both the complaining party and the second party. The style of voicing a complaint can fall along a continuum from apologetic–conciliatory to bellicose: "Mr. Jones, you are a good neighbor and I hate to bother you but your dog . . ." versus "Jones, you no good SOB, your vicious mutt . . ." The effectiveness of negotiation styles may vary from person to person and, of course, is dependent on the second party. A bellicose approach may intimidate some persons but cause others to become belligerent. Alternatively, a conciliatory approach to a second party who is inclined toward belligerence may be viewed by him/her as a sign of nonresolve or weakness. While two-person game research in social psychology suggests that a conciliatory approach ultimately tends to get the best results for both parties, research on real-life disputes indicates that threat may be a very successful negotiating technique, especially when the complaining party is experienced in disputes and has knowledge and resources (see Galanter, 1974, 1975; Merry, 1979; Sarat, 1976; Leff, 1970).

Most people, however, probably fall somewhere away from the bellicose end of the negotiation style continuum. It is fair to speculate that the majority of persons who reach this stage with their complaints negotiate their problems with at least partial success, though there are few data on what the rates might actually be. The working- and middle-class consumers studied by Ross and Littlefield (1978) not only appeared to negotiate their problems without help, they appeared to prefer doing so (see also Curran, 1977, p. 228), though there is reason to believe that poor and minority-group persons may be much less successful (Best & Andreasen, 1977; Galanter, 1974). On the whole, data will probably show that the majority of grievances or problems are resolved at this stage. But in those grievances that are not resolved, it is important to indicate that it is here that a major escalation of conflict and transformation of the dispute may begin to take place.

If the conflict is not resolved, another choice point is reached. This could involve as many as four options: withdrawal of the grievance; suspension of two-party negotiations with the intention of continuing after time for reflection; resort to illegal means; or involvement of a third party. Again, there are no reliable data to indicate the percentages involved in each of these categories, but it is not unreasonable to speculate that the greatest number of cases fall into the first two categories and that relatively few fall into the third (however, see Merry, 1979). Some additional blowing off of steam, such as a nasty letter, a complaint to a better business bureau, or gossip to neighbors, may accompany withdrawal or suspension, however. But let us turn to the fourth option, the third-party forum.

Resort to a third party can take many forms. It may entail recruiting private citizens (e.g., a religious leader, a higher-status person, or the general community) to intercede. It may involve a forum specifically set up to handle problems, such as a newspaper ombudsman (see Hannigan, 1977; Palen, 1979), a better business bureau, a neighborhood mediation center, or a small claims court. Though costs and risks differ with the types of third-party action, movement to any third party has major psychological implications for a dispute. The simple act of voicing a complaint of a grievance to a third party involves some form of wrongdoing by the complaining party against the second party. Even if it is voiced informally to a private person, accusation is part of the complaint. It places the second party in the role of defending him/herself. And, of course, the more public the airing of the dispute, the greater the potential consequences for the accused party's reputation and image. While public airing may cause the second party to settle or seek compromise to avoid further threat to his/her image and reputation, it may, alternatively, cause the party to respond with indignation and to escalate the conflict.

But the third-party step also has direct consequences for the complaining party as well. Crowe's (1978) recent study of complaints to the Massachusetts Commission against Discrimination lists a number of problems that may be representative of small claims costs: out-of-pocket costs, loss of time, disrespectful treatment, and retaliation by the opposing party. There are some potential social costs as well. The relationship itself and the benefits that accrue from it may be ruptured by confrontation with for example, the otherwise good neighbor who has unruly pets or children. In addition, a number of important psychological costs can potentially be incurred. Public commitment makes it tougher to back down. The complaining person's own reputation and self-image may be affected by the other person's countercharges. In the case of a lawsuit, there may be the realization that the accusation will ultimately involve making charges under oath. And in any third-party action, the complaining person incurs the risk of the third party's taking sides with the second party, of losing, of being shown to be wrong, of incurring the negative personal and social stigma that accompanies losing.

While some persons may take the third-party step largely out of emotion, there is little doubt that many others sober and begin to calculate these various factors, to rethink the strength of their own case at this point. There are almost always a number of unknown factors about the other person and his/her evidence and resources, or about attitudes held by the third party. Thus, for some persons or in some situations, the social-psychological and other costs—or, at least, the perceived risk of incurring them—may outweigh the goals to be gained. For other persons, or in other instances, these potential costs may be ignored or assessed as less important than the possibility of winning. Thus, a move to a third party reflects the fact that at least one of the disputing persons feels the grievance strongly enough to risk the potential negative consequences, or that she/he has failed to calculate them. But not only does this mean that the person's state is psychologically different from that of someone who decides not to press a grievance, it probably also means that attitudes and commitment on the part of one or both parties will harden once the step is taken. In short, further escalation of the conflict may take place simply from the act of going public, regardless of the type of third party to whom the dispute is appealed.

The data from the small claims court study support this hypothesis. For example, we have observed a number of cases where a customer–entrepreneur disagreement resulted in the customer's registering a complaint with the better business bureau. The interviews revealed that up until that step was taken, the entrepreneur had viewed the dispute as a matter to be further negotiated with the customer, with an eye toward a compromise settlement. Involvement of the better business bureau was a threat to the entrepreneur's integrity; he/she then escalated the dispute to force the customer into a court confrontation in the belief that he/she would win and thereby remove the perceived stain on his/her reputation. While we should not generalize to all cases, it seems plausible that the act of seeking help from a neighborhood mediation center might be perceived as a public act that challenges the dog owner's integrity and causes him/her to escalate the dispute with the gardener. In brief, public acts have major implications in a dispute.

Finally, associated with the decision to involve a third party is the choice of that party. The choice, of course, depends on availability. The number of options may differ considerably, depending on the nature of the dispute itself or the parties to the dispute. Businesspeople, for example, often do have a series of communications networks whereby they can apply pressure in clear-cut cases of default or other misdealings, for example, putting out the word that Jones doesn't pay his bills (see Leff, 1970). Additionally, contracts between businesspeople frequently specify that any disagreement about contract obligations will be submitted to arbitration (see Leff, 1970; Macaulay, 1963). Obviously, when a party has these channels available but instead resorts to, say, a legal forum, it implies something about the attitudes of the party or of the seriousness of the dispute.

In the case of the individual citizen—the gardener, the consumer, or the owner of the damaged automobile—the options may be limited. The gardener may not even know his other neighbors, and it is unlikely that they would want to get involved. A similar state would hold for the automobile owner. A nonlegal third-party forum would not have the power to force the dispute into a settlement arena. The disgruntled consumer may have the options of a complaint to the better business bureau, or to a newspaper "action line" or ombudsman. These two options may be perceived as having little chance of success (see Hannigan, 1977; Palen, 1979). Therefore, the only realistic forum for redress may be the legal forum.

4.2. Theoretical and Practical Implications

This analysis of choice points in the development of a public dispute has a number of implications. First, there is more to the issue of the utilization of third-party forums than simple access. Felstiner (1974, 1975), in response to Danzig's (1973) proposals for moot-type forums, argued that avoidance may be an adaptive mechanism in disputes—at least, when alternative mechanisms are available. Though he conceded the fact that we do not have data on the social and psychological costs of avoidance, his analysis seems to imply that there are some important costs; if some alternative third-party forums could be successful in our modern Western society, they would be used. Our analysis, however, suggests that nondisputers may be psychologically different than disputers in a number of ways. Some persons who perceive grievances and voice them are probably skilled in negotiating them—at least, most of the time.

For other persons, and perhaps for even the successful negotiators who occasionally meet failure, the social and psychological costs of confrontation may be greater than any justice needs, such as compensation or retribution. Confrontation has potential costs whether it takes place in a third-party forum or in two-person negotiations, and these costs may take many forms. They may disrupt the relationship and the benefits which accrue from it. Admitting victimization may imply incompetence. Additionally, there is the potential of losing, of being shown to be wrong, which is another potential indication of incompetence. One can also be concerned about being seen as disagreeable, unpleasant, a complainer. In some persons, there may simply be an aversion to the negative emotions evoked by an argument, and the anticipation of a potential argument may inhibit the inclination to voice a complaint directly to the other party or to a third party. In brief, various forms of avoidance may be preferable to confrontation. How avoidance is handled is an interesting question requiring more thought and research. Perhaps, some people develop psychological mechanisms to compartmentalize and rationalize the little injustices in the world. They may still believe in "justice" but merely recognize that it cannot be attained in all cases. Perhaps they save their indignation or their spite for "more important" issues. In brief, avoidance does not necessarily imply cynicism, maladjustment, or other negative consequences for the person.

We have proposed some complicated thoughts here, and it is best to pause to make some further observations about the issue of dispositions versus situations. On the one hand, our analysis of choice points does imply that some persons may be more prone to perceive a grievance and pursue a dispute than others. Certainly, persons who pursue a small dispute in a public forum have overcome hurdles that other people would consider formidable. Studies in legal anthropology suggest that there are differences between cultures in attitudes toward and rates of legal disputes (see Galanter, 1974, pp. 104–108; Hayden & Anderson, 1979). Psychological research as well indicates cross-cultural differences in approaches to conflict (Kelley, Shure, Deutsch, Faucheux, Lanzetta, Moscovici, Nutten, Rabbie, & Thibaut, 1970). Though there are no reliable data regarding specific attitudes toward real-world disputes, there is psychological research showing individual differences in two-person game behavior, the most intriguing being the notion of trust elaborated by Deutsch (1973) and the cooperator–competitor hypothesis set forth by Kelley and Stahelski (1970). But while we can talk about dispositions toward either dispute or nondispute behavior, the generality of the proposition and the interaction with situational factors are empirical issues. For some persons, the tendency to exercise justice motives may be limited to, say, business dealings and may not extend to other areas of behavior. On the other hand, some persons may be inclined to view a problem in almost any sphere as potential grounds for a dispute. Our initial observations of small claims litigants suggest that some persons are seemingly continuously involved in public disputes of various sorts. We should also take clear cognizance of the fact that some persons inclined toward avoidance can, depending on circumstances, be provoked into a public dispute. In short, situational factors are important, though it remains the task of future research to determine the relationships between dispositions and situations.

Tracing the dispute through the various choice points also has implications for the debate about the extent to which legal procedure transforms disputes. While the legal forum undoubtedly does have its effect on the dispute, our analysis also suggests that much escalation and transformation of the dispute may take place prior to entry into the legal arena. In fact, that entry is probably often the culmination of a long escalation and transformation rather than its immediate cause. The legal forum itself should not be held solely accountable for the high levels of affect and conflict that the parties exhibit.

One final caveat concerns the fact that our choice-point analysis does not imply that the development of a dispute is necessarily a rational process. Though we have argued that individuals do engage in calculation and odds assessment to some degree, there are many other factors that make it a less-than-rational process. Leff (1970) argued persuasively that "ignorance" of the other side's position, the economic costs, and the social costs is, more often than not, responsible for disputes' moving into a legal arena. This same inability to calculate the other person's reaction is true of the earlier stages of a dispute as well. Moreover, disputants frequently become so emotionally involved that they may miscalculate their own position and/or the other party's reactions even when some veridical information about the other side is available. But even though the decision making in a dispute is subject to such non-rational factors, the important thing is that the decisions to pursue or not to pursue the dispute into the next stage are made by some calculation of costs and gains.

5. IMPLICATIONS AND RESEARCH NEEDS

I have used the debate over the need for alternative resolution forums as a vehicle for exploring some of the social-psychological factors involved in disputes and their resolution. It should be mentioned again that disputes are embedded in legal and sociostructural contexts that have independent effects on disputes as well as interaction effects with these psychological factors. I have tended to ignore or simply overlook some of these elements so as to put emphasis on the disputants' motives and behaviors; the latter have been given insufficient attention in previous analyses of the dispute process. Regardless of its limitations, the present analysis yields some thoughts directly germane to the policy debate.

By focusing on the types of motives that people carry forward in disputes, we can raise questions about procedural justice in third-party forums. For persons pursuing a dispute primarily on equity or restitution grounds, a nonadjudicatory format may be a preferable procedure, all other things being equal. But for persons pursuing a dispute out of motives involving forms of retribution, prior procedural injustice, or behavior control, the adjudicatory forum may be preferable. For disputes motivated by status or power inequality, predictions are more complicated, but at least some of them might be better suited to adjudication. Other psychological factors, such as the disputants' knowledge, perceived control, risk-taking propensities, and willingness to compromise, may act as mediators of these motives and may influence the preference for procedure.

Consideration of the dynamics of disputes in the stages prior to entry into a third-party forum also raises questions about another argument regarding the advantages of alternative forums. It suggests that the high degree of emotion and other negative effects observed in adjudicatory hearings should not be ascribed solely to the tranforming effects of legal procedure. Rather, it suggests that much of the transformation that does occur in disputes occurs at earlier stages of the dispute process. This observation has direct implications for the argument that nonadjudicatory forums would, on the whole, avoid the transformation problem or even ameliorate the conflict underlying the dispute.

By highlighting differences between persons with regard to behaviors and attitudes at various choice-points in the dispute, the analysis suggests that for many people, confrontation may be more psychologically costly than various avoidance mechanisms. Thus, regardless of the existence of minor grievances, these persons would not be inclined to use an alternative resolution forum.

Some of the more interesting theoretical implications, while far from irrelevant to policy considerations, are those that pertain to individual differences in justice motives. Prior research in this area, even when it has recognized that disputes should be viewed from the perspective of disputants, has generally ignored the types of justice motives that give rise to disputes. The type, salience, and strength of such motives probably differ from person to person. In turn, these differences manifest themselves from the very inception of the dispute onward.

Anthropologists and legal scholars have acknowledged differences in dispute behavior between cultures and between social classes (see, e.g., Galanter, 1974), and without question, culture and class are important variables. But even among the less-privileged classes within our society, or within cultures that discourage litigation, there are individuals who do take legal action (e.g., see Merry, 1979). Cross-cultural research, for example, suggests that aside from cultural attitudes toward the legal system, people from different cultures differ in the type and strength of justice motives (e.g., Peristiany, 1966). Galanter (1974) has made reference to these differences in the context of disputes. Yugoslavs, for example, appear to be a dispute-prone people. Major disputes arise over what many of us would consider trivial incidents; the Serbian language has many expressions for and about spite; legal suits based on spite grievances are common (*New York Times*, 1966). Research on conflict in the field of social psychology suggests that such tendencies are differentially held by individuals within a culture (see generally, Austin & Worschel, 1979; Brickman, 1974). The competitor–cooperator hypothesis of Kelley and Stahelski (1970), for instance, indicates important differences between persons in their tendency to perceive grounds for conflict versus cooperation. And in our own preliminary study of the small claims court, we have been struck by the apparent similarity of the attitudes of many plaintiffs with the competitor syndrome sketched by Kelley and Stahelski.

An allied set of questions is raised by the Danzig and Lowey (1975) versus Felstiner (1974, 1975) debate. How do people deal psychologically with unresolved grievances, and are the effects of these grievances as harmful as Danzig and Lowey asserted? The psychological work on consistency theory (see Abelson et al., 1968) and justice theory (see Lerner, 1975, 1977; Lerner & Miller, 1978), to take just two examples, suggests that people have capacities to develop mechanisms to deal with psychologically troubling events. At the extreme, one might even hypothesize that persons who avoid disputes may be more psychologically healthy than persons who engage in legal or other forums, a hypothesis somewhat contradictory to that proffered by Danzig and Lowey. The truth probably lies somewhere in between. In any event, there is a clear need to determine how people adapt to unresolved grievances, to determine the effects of being involved in a public dispute, and especially to learn the extent to which there are long-term consequences arising from both avoidance and dispute behavior. We are on the edge of an uncharted field.

Finally, let us reverse our focus from how psychological motives influence the choice of resolutions forums and consider the potential influence of institutionalized resolution forums on psychological motives. Specifically, the question is, might the creation of more third-party forums and greater access increase levels of dispute behavior—at least, in the long run? For the most part, advocates of more justice centers and easy access have recognized that such centers will become institutionalized. Their main concern has been whether such institutions will promise more than they can deliver and whether they will eventually become dysfunctional through bureaucratization (see, e.g., McGillis & Mullen, 1977). But there is another consideration.

I have speculated that the majority of persons in our society engage in forms of self-help with their grievances, and even without adequate data, this seems a reasonable assumption. But is it possible that the publicity from, and official or semiofficial sanctioning of, these additional third-party forums might affect both grievance behavior and self-help? For example, people might become sensitized to the notion of conflict and more prone to perceive grievances. While such sensitization might be seen as having some beneficial consequences with respect to, say, consumer behavior among the poor, it could clearly be deleterious if it affected relationships among friends or neighbors or even the levels of trust between small entrepreneurs. Rather than seeing the grievances as problems requiring self-help and negotiation, people may seek confrontation and, further, come to believe that the appropriate response to confrontation is the utilization of a third-party forum. The phenomenological logic might be roughly as follows: there are many forums to handle public disputes; perhaps disputes are an inevitable consequence of social life; the world is full of competitors, not compromisers, and therefore I should be prepared to look for encroachment on my interests and prepared to defend them; the socially acceptable and safest course of action in conflict is to voice my complaints to a third party instead of engaging in self-help.

This general hypothesis is not so farfetched as it might seem at first blush. Platt (1973) has elaborated on the concept of social traps, that is, "situations in society that contain traps formally like a fish trap, where men or organizations or whole societies get themselves started in some direction or some set of relationships that later prove to be unpleasant or lethal and that they see no easy way to back out of or avoid" (p. 641).

In this book Peachey and Lerner utilize the social-trap concept with reference to law, arguing that short-term legal solutions may in the long run condition psychological attitudes in a way that is socially harmful. And law is a two-way street: it socializes as well as controls behavior and creates opportunity (e.g., Andennaes, 1977; Friedman, 1976; Morton, 1962). Indeed, empirical social-psychological research demonstrates that law may affect human judgments (Berkowitz & Walker, 1967; Kaufman, 1970; Walker & Argyle, 1964). The creation of a network of new institutions for dispute resolution, viewed from this perspective, could therefore have long-term negative consequences for social attitudes. Specifically, people's attitudes might be conditioned to operate on a conflict rather than a cooperation model of human behavior and to avoid self-help when conflict does arise.

In concluding, let me again take cognizance of the fact that this essay has covered much ground and has sometimes oversimplified some issues in an attempt to emphasize the importance of justice and other psychological motives and behaviors in the development and resolution of disputes. Such factors have been given insufficient attention by policymakers and by scholars in other disciplines. Social psychologists, on the other hand, have given attention to the psychodynamics of conflict and conflict resolution, but their theoretical focus has been narrowly circumscribed and has ignored the complexities of real-world disputes. Hopefully, this essay has drawn attention to the theoretical and policy richness of the area and has suggested avenues for both laboratory and field research. Disputes are an unavoidable part of society and will become even more salient as we face an uncertain future. The understanding of justice motives and other psychological factors in the dispute process is necessary to help us plan for that future.

ACKNOWLEDGMENTS

I am indebted to Cheryl Lawrence, David Saunders, and Blair Sheppard for comments and suggestions.

REFERENCES

Abel, R. A comparative theory of dispute institutions in society. *Law & Society Review*, 1973, *8*, 217–347.
Abel, R. From the editor. *Law & Society Review*, 1977, *11*. Unnumbered introduction pages to Issue Number 3.
Abelson, R., Aronson, E., McGuire, W., Newcomb, T., Rosenberg, M., & Tannenbaum, P. (Eds.). *Theories of cognitive consistency: A source book.* Chicago: Rand-McNally, 1968.

Andennaes, J. The moral or educative influence of criminal law. In J. Tapp & F. Levine (Eds.), *Law, justice and the individual in society*. New York: Holt, Rinehart & Winston, 1977.

Aubert, V. Competition and dissensus: Two types of conflict and of conflict resolution. *Journal of Conflict Resolution*, 1963, 7, 26–42.

Aubert, V. Courts and conflict resolution. *Journal of Conflict Resolution*, 1967, 11, 40–51.

Austin, W. G., & Worschel, S. *The social psychology of intergroup relations*. Monterey, Calif.: Brooks/Cole, 1979.

Berkowitz, L., & Walker, N. Laws and moral judgments. *Sociometry*, 1967, 30, 410–422.

Best, A., & Andreasen, A. Consumer response to unsatisfactory purchases: A survey of perceiving defects, voicing complaints, and obtaining results. *Law & Society Review*, 1977, 11, 701–742.

Brickman, P. *Social conflict: Readings in rule structures and conflict relationships*. Lexington, Mass.: D. C. Heath, 1974.

Crowe, B. Complainant reactions to the Massachusetts Commission Against Discrimination. *Law & Society Review*, 1978, 12, 217–236.

Curran, B. *The legal needs of the public*. Chicago: American Bar Foundation, 1977.

Danzig, R. Toward the creation of a complementary decentralized system of criminal justice. *Stanford Law Review*, 1973, 26, 1–103.

Danzig, R., & Lowey, M. Everyday disputes and mediation in the United States: A reply to Professor Felstiner. *Law & Society Review*, 1975, 9, 675–684.

Deutsch, M. *The resolution of conflict*. New Haven, Conn.: Yale University Press, 1973.

Deutsch, M. Equity, equality, and need: What determines which value will be used as the basis of distributive justice? *The Journal of Social Issues*, 1975, 31, 137–150.

Felstiner, W. The influence of social organization on dispute processing. *Law & Society Review*, 1974, 9, 63–94.

Felstiner, W. L. Avoidance as dispute processing: An elaboration. *Law & Society Review*, 1975, 9, 695–706.

Friedman, L. The idea of right as a social and legal concept. *Journal of Social Issues*, 1971, 27, 189–198.

Friedman, L. *The legal system: A social science perspective*. New York: Russell Sage Foundation, 1976.

Fuller, L. Mediation—Its forms and functions. *Southern California Law Review*, 1971, 44, 305–339.

Galanter, M. Why the "haves" come out ahead: Speculations on the limits of legal change. *Law & Society Review*, 1974, 9, 95–160.

Galanter, M. Afterward: Explaining litigation. *Law & Society Review*, 1975, 9, 347–368.

Hannigan, J. The newspaper ombudsman and consumer complaints: An empirical assessment. *Law & Society Review*, 1977, 11, 679–700.

Hayden, R. M., & Anderson, J. K. On the evaluation of procedural systems in laboratory experiments: A critique of Thibaut and Walker. *Law and Human Behavior*, 1979, 3, 21–38.

Hirschman, A. *Exit, voice and loyalty: Responses to decline in firms, organizations, and states*. Cambridge, Mass.: Harvard University Press, 1970.

Houlden, P., La Tour, S., Walker, L., & Thibaut, J. Preferences for modes of dispute resolution as a function of process and decision control. *Jounal of Experimental Social Psychology*, 1978, 14, 13–30.

Johnson, E. A preliminary analysis of alternative strategies for processing civil disputes. *A report submitted to the National Institute for Law Enforcement and Criminal Justice*. University of Southern California, February 1977.

Johnson, E. Toward a responsive justice system. In T. Fetter (Ed.), *State courts: A blueprint for the future.* Washington: U.S. Department of Justice, 1978.

Kaufman, H. Legality and harmfulness of a bystander's failure to intervene as determinants of moral judgment. In J. Macaulay & L. Berkowitz (Eds.), *Altruism and helping behavior.* New York: Academic Press, 1970.

Kelley, H., Shure, G., Deutsch, M., Faucheux, C., Lanzetta, J., Moscovici, S., Nutten, J., Rabbie, J., & Thibaut, J. A comparative experimental study of negotiation behavior. *Journal of Personality and Social Psychology,* 1970, *16,* 411–438.

Kelley, H., & Stahelski, A. Errors in perception of intentions in a mixed motive game. *Journal of Experimental Social Psychology,* 1970, *6,* 379–400.

Leff, A. Injury, ignorance, and spite—The dynamics of coercive collection. *Yale Law Journal,* 1970, *80,* 1–46.

Lerner, M. (Ed.). The justice motive in social behavior (whole issue). *Journal of Social Issues,* 1975, *31.*

Lerner, M. The justice motive: Some hypotheses as to its origins and forms. *Journal of Personality,* 1977, *45,* 1–50.

Lerner, M. The law as a social trap. Paper presented at the Bi-annual meeting of the Law and Society Association, Minneapolis, Minn., May 1978.

Lerner, M., & Miller, D. Just world research and the attribution process: Looking back and ahead. *Psychological Bulletin,* 1978, *85,* 1030–1051.

Leventhal, G. S. Fairness in social relationships. In J. Thibaut, J. Spence, & R. C. Carson (Eds.), *Contemporary topics in social psychology.* Morristown, N.J.: General Press, 1976.

Macaulay, S. Non-contractual relations in business: A preliminary study. *American Sociological Review,* 1963, *28,* 55–67.

McGillis, D., & Mullen, J. *Neighborhood justice centers: An analysis of potential models.* Washington: U.S. Department of Justice, 1977.

Merry, S. Going to court: strategies of dispute management in an American urban neighborhood. *Law & Society Review,* 1979, *13,* 891–926.

Morton, J. D. *The function of the criminal law.* Toronto: Canadian Broadcasting Corporation, 1962.

Nader, L. Styles in court procedure: To make the balance. In L. Nader (Ed.), *Law in culture and society.* Chicago: Aldine, 1969.

Nader, L. Forums for justice: A cross-cultural perspective. *Journal of Social Issues,* 1975, *31,* 151–170.

Nader, L., & Singer, L. Dispute resolution . . . what are the choices? *California State Bar Journal,* 1976, *51,* 281–320.

Nader, L., & Todd, H. *The disputing process—Law in ten societies.* New York: Columbia University Press, 1978.

New York Times. Spite feuds fill Yugoslavia's courts. Sunday, October 16, 1966.

Palen, F. Media ombudsmen: A critical review. *Law & Society Review,* 1979, *13,* 799–850.

Peristiany, J. (Ed.). *Honor and shame.* Chicago: The University of Chicago Press, 1966.

Platt, J. Social traps. *American Psychologist,* 1973, *28,* 641–651.

Ross, L., & Littlefield, N. Complaint as a problem-solving mechanism. *Law & Society Review,* 1978, *12,* 199–216.

Sander, F. *Report on the National Conference on Minor Disputes Resolution.* Chicago: American Bar Foundation, 1977.

Sarat, A. Alternatives in dispute processing: Litigation in a small claims court. *Law & Society Review,* 1976, *10,* 339–376.

Starr, J., & Yngvesson, B. Scarcity and disputing: Zeroing-in on compromise decisions. *American Ethnologist,* 1975, *2,* 553–566.

Steele, E. Two approaches to contemporary dispute behavior. *Law & Society Review*, 1977, *11*, 667–678.

Thibaut, J., & Walker, L. *Procedural justice: A psychological analysis.* Hillsdale, N.J.: Erlbaum, 1975.

Thibaut, J., & Walker, L. A theory of procedure. *California Law Review*, 1978, *66*, 541–566.

Vidmar, N. *Creating access: Alternative dispute forums, conflict, and justice.* Paper presented at the American Psychological Association Convention, Toronto, August 28, 1978.

Vidmar, N. *Social psychological aspects of disputes and their transformation in a small claims court.* Paper presented at the Meetings of the Law and Society Association, San Francisco, May 1979.

Vidmar, N., & Miller, D. T. Social psychological processes underlying attitudes toward legal punishment. *Law & Society Review*, 1980, *14*, 565–602.

Walker, N., & Argyle, M. Does the law affect moral judgments? *British Journal of Criminology*, 1964, *4*, 570–581.

Yngvesson, B. Responses to grievance behavior: Extended cases in a fishing community. *American Ethnologist*, 1976, *3*, 353–373.

18

Down-to-Earth Justice

Pitfalls on the Road to Legal Decentralization

ROBERT L. KIDDER

1. INTRODUCTION

"Less is more." This slogan, reminiscent of Earth Days and the ecology movement, contains the essence of two popular positions that support the simplification of law in the United States. One position is that decreasing resources and environmental crisis mandate simplification. If we can no longer count on an ever-expanding supply of those resources that made advanced industrial society possible, then our complex institutions will become increasingly inappropriate. Terms like *appropriate technology* and *intermediate technology* describe proposed adjustments at the economic level (Schumacher, 1975). By implication, all of our institutions, including law, require adjustment to a shrinking scale of operations.

The second position is summarized in the term *deinstitutionalization*. It stems from what Nonet and Selznick (1978) call the "crisis of authority" (p. 4). Promoters of deinstitutionalization share the belief that regardless of environmental considerations, our professionals have failed us. Their failure lies in their highly specialized institutions, which often cost more in social dislocation and psychic damage than they pay out in viable solutions to society's problems. Specialization ignores or destroys the holistic character of human relationships. This leads to eventual failure of institutions despite sophisticated theories (e.g., see Illich, 1973, 1976).

These two positions together form a contemporary context of social

ROBERT L. KIDDER • Department of Sociology, Temple University, Philadelphia, Pennsylvania 19122 criticism congenial to a rebirth of proposals for simplifying American methods of producing justice. Critics charge that existing legal institutions sacrifice justice because they are too slow, too expensive, too burdened with mystifying details, and too professionalized to provide a sense of involvement and efficacy to those with only occasional legal needs.

Decentralization of law might teach people to use their own resources and give them back a sense of self-control. Especially in an era of declining resources, ordinary citizens might find more equitable formulas of distributive justice because legal procedures would be more accessible, understandable, and affordable. Simplification might thus increase the *feeling* that justice is being done.

Several alternatives to conventional legal procedures and institutions have been debated. These include the ombudsman, the mediator, the arbitrator, and the professional negotiator. They also include the primary focus of this paper: neighborhood courts and neighborhood-level action for collective crime prevention. Lay participants in existing legal institutions often feel cheated by the slow, expensive, lawyer-dominated mysteries of the court or the law office. Justice does not *seem* to be a product of the system. So, proponents of the neighborhood solution say, bring the process back to the neighborhood. Eliminate the role of lawyers and related professionals. Rely on local wisdom. Then justice will be swift, understandable, cheap, and more equitable. Then a *sense* of justice stands a chance of emerging.

In this chapter, I argue that we know enough from previous attempts at legal simplification and decentralization to be skeptical of the promises offered in the name of neighborhood justice. I also argue that the problems that have plagued decentralization policies in the past may actually be accentuated by conditions of increasing scarcity.

In what follows, I use the term *neighborhood justice* to refer to all those programs, proposals, and practices that substitute neighborhood-level, lay-operated procedures for professionally operated, bureaucratized legal procedures of the official legal system. These programs and proposals include neighborhood "courts," community mediators, arbitrators, and the mobilization of neighborhood residents to combat crime in their communities.

2. SCARCITY, CENTRALIZATION, AND NEIGHBORHOOD JUSTICE

Past attempts at legal simplification or decentralization do not offer great hope for future success in an environment of increasing scarcity, despite the presence in our literature of many descriptions of working alternatives to our heavily professionalized system. Anthropologists, for example, report on the justice systems of small villages, tribal groups, nomads, and caste groups (see, e.g., Nader, 1969; Bohannan, 1967). Their reports often demonstrate the capacity of nonprofessionals to develop effective, appropriate procedures of local justice. Peaceable group life seems possible without the complexities and expenses of modern legal systems.

Even industrial societies seem viable with little law. We are told that the Japanese successfully rely on alternatives to legal institutions because of their strong aversion to public confrontations (Kawashima, 1963). Their tiny legal profession and low rates of litigation testify to the strength of their localized alternatives. Even their police are organized on a decentralized basis (Bayley, 1977). Assigned to small neighborhood police stations, they become familiar figures, relied on for numerous, only occasionally safety-related, services. How sharp the contrast with American police, who are remote strangers seen infrequently cruising past in expensive patrol cars.

We see an even more dramatic example in the People's Republic of China. There, local committees of neighbors, fellow factory workers, and commune members handle both crime and disputes through public debate and group consensus (Li, 1977). Divorce, vandalism, industrial sabotage, theft, and even choice of career are left in the hands of such committees. Formal law and lawyers play no role in most such actions. Local conditions, interpreted through constantly debated "correct political thinking," are weighed, and ad hoc solutions are developed by the local committees, who, despite lack of professional training, also have the authority to enforce their judgments.

Thus, our literature repeatedly shows that society is possible without lawyers and legal complexity. Some analysts even argue that legal complexity has been a tool in the development of totalitarianism because it contributes to the breakdown of customary, local forms of social control (Diamond, 1971).

But showing that social life is viable without legal complexity is not the same as showing that where legal complexity has developed, it can be surgically excised from the social body in which it grew. Attempts to reverse the legal–social clock may actually accelerate the pace of legal formalization and centralization.

Consider, for example, that "simplifying" reform known as the small claims court. These courts were supposed to bring simple justice to people with "small" legal problems, problems they could not afford to bring to ordinary courts (Yngvesson & Hennessey, 1975). They were to operate without the obscurities loved by lawyers. Wherever possible, lawyers were to be barred from participating. These courts were to be inexpensive and swift. They were for ordinary people.

But as they became established, their actual function was transformed. They have evolved into routinized tools dominated by business operators and landlords, who use them as collection agencies. As Yngvesson and Hennessey demonstrated, small claims court policymakers wrongly equated *small* claims (involving little wealth) with case simplicity. They therefore created procedures incapable of handling the typically complex relationships and emotions hidden by the relative insignificance of the monetary claim. As in the case of the examples I describe below, the reform missed its mark because of the inadequate sociological model held by the professionals who drew up the plans.

In India, following its independence in 1948, there was strong political pressure to replace the "alien justice" of British colonial courts with a form of justice that the masses could understand and afford (Kidder, 1977, pp. 172–173; Galanter, 1972). The legal system given by the British masters was said to be corrupt, expensive, unconscionably slow, and incomprehensible to the vast majority of illiterate peasants who make up most of the population. So the cry went out for village "panchayats," quasi-judicial groups of village elders who would hear dispute cases and criminal charges and resolve them on the spot. This was thought to be the method used in the golden age before colonial rule was imposed.

A system of panchayats was established. But it failed to achieve the stated purpose of simplification and decentralization. The legislation setting up the system was so laden with restrictions, and the response of legal and administrative bureaucrats was so intrusive, that the panchayats never achieved the promised autonomy (Baxi & Galanter, 1979). Instead, panchayat elders became tools in the plans of government developers who saw local leaders as useful conduits for government policy.

In Allende's Chile, reformers advocated popular justice in the form of neighborhood courts that would allow the masses to seek equitable solutions to persistent problems of slum living (Spence, 1978). Spence reported that two different models were simultaneously instituted one conducted by professional judges and the other by lay persons chosen from squatters' neighborhoods. Both were supposed to provide an alternative to the lawyer- and business-dominated regular courts. Again, as in India and the American small claims courts, the objective was swift, cheap, understandable justice.

Yet, despite different origins and different kinds of participants, neither system delivered the promised service. In the program using regular judges, the judges entered the program reluctantly, refused to move their courts to the neighborhoods, and devoted almost none of the courts' resources to neighborhood cases. The sessions that they did hold actually simply channeled people into ordinary administrative procedures. Their minimal compliance with the objectives of the program stemmed from the pressure they felt from ordinary legal circles to devote the courts' resources to cases involving large sums, complex legal issues, and influential institutions in Chilean society.

The squatters' courts, on the other hand, failed because they were the fruit of a nationally mobilized, ideologically dominated anticapitalist political movement. These courts were constantly concerned with expressing general ideological issues. With this external orientation, they tended to ignore the particular structural and dispute-processing needs of the neighborhood. As Spence (1978) concluded, in both experiments, "Their development was held back by a lack of organizational resources and a relatively low priority accorded their activities by the larger organizational context in which they were situated." (p. 175)

Even in China, Li (1977) wrote, the purity of the localized "internal" system is constantly challenged by pressures from the more professionalized, bureaucratized sector (Spence's "larger organizational context"). Local committees must have trained cadres who are responsible for ensuring that "correct political thinking" will prevail. "Correct political thinking" is, of course, thinking that accords with the dominant view held in Peking. Furthermore, there has been constant pressure to standardize procedures and rules so that ad hoc decision making would be eliminated. Li described this "internal–external" tension as a basic feature of postrevolutionary China and argued that the vitality of the internal, or local, side and its relative immunity to external interference are mostly the result of economic necessity rather than specific governmental policy. The implication is that as China's economic strength increases, external legal personnel and doctrine will increasingly temper, modify, or eliminate the authority of the internal system, making its operations serve the purposes of central institutions.

In the United States, one of the latest expressions of the simplification ideal has come in the form of federal assistance for neighborhood-level crime-stopping and dispute-processing programs. On the crime side, the Law Enforcement Assistance Administration has tried to encourage programs of citizen involvement. Town-watch programs and similar civilian-operated surveillance activities put neighbors on the streets in patrol vans with two-way radios. Block organizing has been initiated in those neighborhoods that lack organization. In these programs, "facilitators" call neighbors together to discuss the local crime situation and propose group actions to reduce crime. These actions include "target hardening" (installation of locks, alarms, dogs, and the engraving of valuables), the development of warning systems for individuals (whistles or freon horns, which are expected to frighten would-be attackers and alert neighbors), and the development of neighborhood cohesion so that "eyes on the street" will reduce the free flow of strangers and eliminate the anomic climate within which crime is thought to thrive.

I argue here that these programs have been plagued by the same problems and for the same reasons as those in India, Chile, China, and our small claims courts.

The model on which these federal programs have been developed comes from sociological studies showing "urban villages," cohesive urban communities, in which crime is low because of strong communal solidarity and routine patterns of interaction (e.g., Gans, 1962; Jacobs, 1961; Stack, 1974). Government planners, assisted by academics, hold these neighborhoods up as evidence that urban living need not lead to anomie and the reliance on professionals. They argue that government resources should be able to stimulate neighborhood self-reliance even where it does not exist, and that if this happens, inequities that now stem from inaccessible legal institutions will be eliminated. This reasoning leads them to propose neighborhood crime-fighting programs, sometimes combined with experiments in neighborhood councils for dispute settlement.

Research I have conducted on urban neighborhood responses to crime indicates that federal sponsorship of decentralized programs may actually increase the intrusive influence of centralized authority. This research is part of the Reactions to Crime project conducted by Northwestern University's Center for Urban Affairs. It involved surveys and extended participant observation in 20 urban neighborhoods in three cities.

The general pattern that I see in our results matches the cases I have discussed above; that is, where programs are initiated from outside the community, they express the interests and the generalized perspectives of various parties seeking to influence centralized governmental policy. These parties include economic institutions, politicians, government bureaucrats, and professionals acting as advisers. Because these concerns underlie the establishment of "community crime prevention" or "neighborhood dispute settlement" programs, they express from the outset the ideologies and the evolving relationships of external sponsors, rather than the existing conditions and spontaneous inclinations of local residents.

The programs then have a contrived quality. We found, for example, that funded programs typically have participants who devote much time and energy not only to stopping crime but to figuring out how to keep the sponsoring agency interested enough to continue funding. We also found that outsiders, with their global bureaucratic definitions of the crime problem, are reluctant or unable to tailor their programs to the special concerns of specific neighborhoods. As a result, they impose one or more remotely framed definitions of the problem and then engage in "educational" activities (films, speeches by police officers, group sessions with "facilitators") designed to eliminate the "irrelevant," "unfocused" concerns of the residents.

As an example of the distortions produced by sponsored neighborhood programs, consider the problem of defining crime at the local level. We found that federal programs assume a legalistic notion of "street crime." Policymakers accept the premise that the most serious problem facing urban residents is their victimization by stranger-predators, that is, muggers, robbers, burglars, and rapists whose origins and identities are shrouded in mystery.

We found that such crimes are only a small part of the total range of problems that residents seek to address when they think about crime. One important shortcoming of the policymakers' view is that many technically criminal acts that residents experience are committed by people they know—their own neighbors. These "crimes" are sometimes manifestations of conflicts that have split neighborhoods into competing or warring camps. Such conflicts may have old and well-established roots, or they may be of more recent vintage. Burglaries, purse snatchings, and assaults are only occasionally perpetrated by "outsiders." Organizing against them, then, often means organizing some neighbors against others. Furthermore, many noncriminal acts (verbal harassment, "stealing" customary parking spaces, failing to maintain abandoned buildings, or keeping junked cars parked on the street) that upset some residents may also express and exacerbate ongoing conflicts between people who have continuing relationships.

When a federally sponsored program is brought into such a community, its enthusiastic adoption by local residents may actually signal the beginning of an escalated attack by one faction against another within the neighborhood. Federal aid then becomes a means of magnifying rifts that predated the funding, because federal policy was developed around an isolated, abstracted, professionally specialized definition of crime, which ignored the relationship between crime and interpersonal conflict. The rift between adults and "kids" is often aggravated in exactly this way, as the kids sense themselves to be the target of new neighborhood patrol programs. Another rift that we found exaggerated by federal funding was that between old-timers and newcomers to neighborhoods, especially where immigration is associated with racial transition.

But what about the "urban villages" where people look out for each other? Is the federal program founded on a myth? The answer is a partial "no." Not all neighborhoods in our study contradicted the assumptions underlying the federal programs. We did find "urban villages" where there was indeed a great deal of informal crime control and dispute handling. "Eyes on the street" were in abundance in one neighborhood, for example, where our bearded research assistant was pursued for two hours by several police cars. When they found him casually dictating notes into his tape recorder, they told him that their switchboard had been "lit up like a Christmas tree" all morning with calls from worried residents.

This "nosey" response is the kind that government programs are supposed to promote. But we found these patterns only in neighborhoods with characteristics that were not the result of government-funded crime-fighting programs. Specifically, they were *stable* working-class areas inhabited by people whose economic circumstances were such that they had no alternative to staying in the neighborhood.

Ericksen and Yancey (1979; Yancey & Ericksen 1979) found in their studies of "ethnicity" and residential stability that certain populations were, in a sense, "captured" by housing. transportation, and employment patterns associated with industrialization in the United States. Migrants to urban centers were selectively drawn into different patterns of employment and housing. Ethnic whites went into the factories and blacks were left with domestic jobs. Those in industrial jobs had to live close to the factories or to public transportation. They thus became the original inhabitants of large industrial housing tracts in what are now the "old neighborhoods" (often also thought of as the "ethnic neighborhoods" or "urban villages") of our cities. Their low incomes permitted purchase of inexpensive homes in industrial areas but were inadequate to support a move to the suburbs and higher-priced housing when that became the dominant trend. Thus, the "old neighborhoods" are now populated by those who have remained in industrial manufacturing jobs and cannot afford to sell their homes and move.

So the "urban villages," which serve government policymakers as a model of home-grown crime fighting and dispute resolving, are actually the product of a combination of circumstances, including economic hardship, that have resulted in isolated pockets of long-term residential stability. The most important point here is that successful examples of localized "justice" are not produced by external sponsorship. Rather, they emerge from a combination of external socioeconomic conditions and processes internal to the neighborhood that produce residential stability, long-term relationships, and neighborhood-level economic interdependence (see Stack, 1974).

Our research, then, combined with studies that show simplified legal alternatives in other societies, suggests that legal simplicity works only where certain circumstances prevail. These include residential stability, ongoing interdependent relationships that cannot be simply avoided (see Felstiner, 1974), and conditions that allow local residents to exercise authority not mediated by external authority. Furthermore, we must recognize that such simplicity "works" only in the limited sense that it produces some level of internal stability without reliance on outsiders.

But are we really willing to pay the price of such stability? Its costs include the potential for localized tyranny, development of parochial values and rules, heretical belief systems, and the promotion of the idea that insiders are different from (usually superior to) outsiders. This last idea is also likely to be associated with the rise of conflict between localized pockets of autonomy. In Japan, for example, while custom seems adequate for the control of conflicts within the generally self-sufficient villages and towns, legal institutions tend to become involved whenever conflict between villages and/or towns occurs.

The control of local deviation has been the target of government attempts to promote societywide civility in the United States. Civil rights legislation, guarantees of free speech and freedom of religion, the development of programs to eliminate disease, poverty, and discrimination are all founded on the premise that a higher authority based on higher principles should prevail over local authority based on "backward" (parochial, redneck, podunk) principles. Our discomfort with localized "justice" is summed up in the connotations of the word *vigilantism*.

The dilemma is thus before us. We recoil from the abuses we know to have resulted from the free play of local variation in the past. Our concepts of freedom and the legitimate right to privacy have developed as a product of struggle between local tyrannies and the federal-level protectors of our constitutional rights. But we know that outside control of neighborhood justice programs typically distorts them in the direction of the nonlocal concerns of the programs' initiators and administrators. If we insist that neighborhood justice be accountable to general values, we undercut the authority that makes the neighborhood system a credible mechanism of control.

Having seen what we can learn of proposals to simplify based on existing experience, let us consider next what we may expect in a society of increasing scarcity.

3. SCARCITY, TECHNOLOGY, AND DECENTRALIZATION

It is a mistake to assume that the depletion of resources will drive our modern industrialized societies backward in a uniform regression toward earlier forms of production and social organization. While certain forms of modern activity may become obsolete (e.g., heavy reliance on the automobile), other technological breakthroughs will become even more dominant. For example, no foreseeable decline in natural resources is likely to eliminate reliance on computers and computer-controlled cybernetic methods of production (Theobald, 1964). As resource scarcity intensifies, we may reasonably expect even heavier reliance on computers because they use very little energy compared with the efficiencies they can effect.

A key feature of this increased use of computers is that fewer and fewer people are involved in decision making (see Braverman, 1974). Computers increase the power of those who control them and eliminate jobs involving repetitive tasks. Resource scarcity is not likely to reverse this trend toward managerial centralization, though it may reverse the move toward capital-intensive production as labor becomes an increasingly competitive alternative. But the labor-intensive result would be a largely unskilled work force taking orders from computer-assisted managers. If anything, resource scarcity should increase the pressure for the adoption of rationalized, cost-efficient production systems. Computers and cybernetic production techniques have become part of what is probably an irreversible development.

A second trend that is not likely to regress is the technology of mass communications. Advances in electronics make broader communication possible at lower levels of energy expenditure.

Both of these features of modern society contribute in important ways to the system of economic and political power. The efficiency imperative has, in the past, contributed to the development of centralized bureaucratic control that now typifies postindustrial society. Resource scarcity, rather than returning us to the comparative simplicity of frontier life, will increase the pressure of the efficiency imperative. This pressure can be expected to provide further support for the continuing trend toward centralization. Thus, instead of expanding the layman's self-reliance, resource scarcity in the computer–mass-communication age may well increase the power imbalance between the professional–managerial class and everyone else. In a land of increasing scarcity, then, we may find strong processes that increase the inequality of distribution.

Ironically, however, we may also see the increase of conditions that encourage the growth of neighborhood justice. As I pointed out above, we found instances of viable local alternatives in those neighborhoods where conditions fostered residential stability and economic interdependence. If increasing scarcity curtails long-distance commuting patterns, we may see a return to patterns of locating residences near workplaces. If, simultaneously, the processes of managerial centralization continue to develop, as I suggested above, then we may experience a growth in the prevalence of "urban villages." These would again be inhabited by people tied to workplace and residence by low income. In such neighborhoods, there would develop a strong sense of belonging-ness, an ease of identifying "outsiders" and strangers, a reluctance to become involved with outside officials such as police and courts, and indigenous ways of resolving disputes and punishing deviance.

Thus might the dreams of neighborhood justice proponents be re-alized by the press of scarcity.

But before celebrating this escape from professional dominance, we should examine the broader issue of its implications for the equitable distribution of resources in an economy of growing scarcity. One of the key features of the neighborhoods, we found, that do develop local alternatives is that they are poor, though not necessarily destitute. One of the major reasons that they develop alternatives is that the residents of these neighborhoods lack the alternatives available to wealthier citi-zens, including the wherewithal to mobilize supportive lawyers and legal processes, or to choose the option of avoiding problem-causing situations and people by going elsewhere. If you cannot move or afford legal remedies, you are more likely to develop alternatives. Hence, lo-calized alternatives to formal law may become more common because of the *increasing* inequality associated with increased scarcity. Thus, in-stead of seeing proposals for neighborhood justice bearing fruit with a more equitable distribution of diminishing wealth, we are likely to see its opposite: a spontaneous upsurge of neighborhood justice as a *con-sequence* of spreading proverty and diminishing alternatives.

But there is an even more fundamental criticism of the neighborhood justice option, namely, that the tensions of inequitable distribution in our society are not created at the neighborhood level and therefore cannot be adequately attacked by modifications at that level. Over the past 15 years, we have experienced a dramatic increase of a kind of "rights consciousness," which stems from a recognition of shared eco-nomic fate. It perhaps began with the black struggle for equal rights. But it has spread well beyond those origins. The women's movement, the gray rights movement (for "senior citizens"), political organizations of handicapped persons, gay rights activists, and mental patients' rights activists are but the better-known tip of a potentially huge iceberg of rights advocacy that cuts across all neighborhood, community, and even regional lines. Similarly, occupational groups, having experienced the proletarianization of their once respected "professions" (Sarfatti-Larsen, 1977), have begun reacting with strikes and other collective actions as adversarial relations have become a common stance between employers and employees. Thus, we now see strikes by police, fire fighters, social workers, teachers, and even college professors. Medical personnel, once thought of as the epitome of the autonomous professionals, increasingly find themselves in institutional contexts where only collective action can promote their interests. Consumers' rights groups at various levels of organization have reified yet another once-abstract sociological category.

The inadequacy of the anthropological model for neighborhood justice stems from the fact that neighborhoods in most of urban America are neither producing nor consuming units as are the small tribes and villages where anthropologists have found effective legal simplicity. People in all but the poor neighborhoods do not depend extensively on each other for economic survival. Their interdependence is limited to relations with city government (on some issues, such as street lighting and trash collection) and isolated services (such as phone and electricity and natural gas). With industrial and corporate centralization, people become dependent on highly specialized patterns of division of labor that are organized by remote corporate decision-makers operating at national and international levels of policymaking. Only where poverty forces mutual interdependence do we find conditions that begin to approximate the anthropological model.

The new coalitions of activists join people with similar problems in our centralized, bureaucratized, postindustrial economy. If decision-making centralization does continue, as I predict, even in the face of increasing scarcity, then we may expect to see an ever-growing number of such specialized coalitions. Thus far, these groups have not shown much sign of preference for decentralized alternatives to law. On the contrary, they typically combine attention-getting action with legal strategies at the most sophisticated levels of legal action, providing new sources of work for lawyers and new legal theories to be pondered by judges.

This is not to say that there are no nonlegal alternatives operating at this level of interest aggregation. Many cities, for example, now have newspaper columnists or radio or TV programs that are actively involved in "solving" the problems of subscribers or listeners. Consumer advocates, "action lines," financial advisers, and counselors of various kinds try to use mediation, publicity, and shame, along with the spreading of specialized information, as an alternative to legal action. Organizations like Women Organized against Rape in Philadelphia supplement the legal assistance they provide to rape victims with psychological support and the opportunity to find solace in a "community" of sympathetic comrades. Canada and the Scandinavian nations have pioneered the role of the ombudsman who acts to mediate between government agencies and aggrieved citizens. Many states in the United States maintain lists of professional mediators who intervene in the labor disputes that arise between government and public sector employees.

In none of these cases, however, are the alternative mechanisms expected to replace legal institutions. Rather, they are conceived of as additions to a system of problem solving that has become overburdened. The legal system continues to be treated as the core around which these accretions grow in piecemeal fashion. Most importantly, though they are alternatives to traditional legal action, they have not significantly weakened the position of legal professionals in the United States. The number of lawyers has actually surged dramatically during the period of development of many of these alternatives. And, in some cases, as in medical malpractice lawsuits for example, lawyers have actually been at the center of the development of new "rights consciousness," creating a kind of social movement around the individualized activity of litigation. These developments, thus, do not provide a substitute for legal complexity. They leave intact the legalistic view of the world shared by legal professionals, and they leave the profession in a decisive position to determine their parameters.

4. CONCLUSION

It is tempting to interpret systems of law and economic distribution in nomadic, tribal, and agrarian societies as evidence that the simplification and decentralization of American law can adequately insulate us from disruptive inequities that may be produced by increasing scarcity. Since we find cases of "neighborhood justice" already operating in our urban centers, we are tempted to assume that careful planning for reorganization can spread this simpler social "technology" throughout American society. We have ample evidence that the "haves" come out ahead under our existing legal system (Galanter, 1974). Can't we short-circuit this inequitable system by empowering the "have-nots" in their own communities?

The fly in this ointment is the probable continuation of the trend toward centralization in other significant technologies. As long as these technologies increase the monopolization of policymaking in such major institutions as education, health care, communications, banking, housing, insurance, and welfare even as scarcity increases, neighborhood "justice" is likely to be either a spontaneous local response to the resulting impoverishment and powerlessness or a disguised element in strategies that actually further reduce local autonomy.

Because the major sources of inequities that may result from resource scarcity are not found within neighborhood-level processes, effective responses to those inequities will not be found in neighborhood justice. Rather, I think we can expect to see an accelerating pace of broadly based coalition formation. The law's role in this acceleration is likely to be that of moderator, attempting to deflect the potentially explosive effects of direct political action. This confrontation will probably involve lawyers on both sides, creating challenges and new responses to rapidly shifting definitions of inequity. Lawyers will thus continue to be central to both the drive for the redistribution of scarce wealth and the resistance to that drive.

REFERENCES

Baxi, U. & Galanter, M. "Panchayat Justice": An Indian experiment in legal access. In M. Cappelletti & B. Garth (Eds), *Access to justice.* Vol. III: *Emerging issues and perspectives.* Milan: Guiffre; Alphen aan den Rijn: Sijthoff and Noordhoff, 1979.

Bayley, D. H. Modes and mores of policing the community in Japan. *Law and Society: Culture learning through law.* Honolulu: East-West Center, 1977, pp. 71–82.

Bohannan, P. *Law and warfare.* New York: Natural History Press, 1967.

Braverman, H. *Labor and monopoly capital.* New York: Monthly Review Press, 1974.

Diamond, S. The rule of law versus the order of custom. In R. P. Wolff (Ed.), *The rule of law.* New York: Simon & Schuster, 1971.

Ericksen, E., & Yancey, W. Work and residence in industrial Philadelphia. *Journal of Urban Studies,* 1979, 5, 147–182.

Felstiner, W. L. F. Influences of social organization on dispute processing. *Law & Society Review,* 1974, 9, 63–94.

Galanter, M. Why the haves come out ahead: Speculations on the limits of legal change. *Law & Society Review,* 1974, 9, 95–160.

Galanter, M. The aborted restoration of indigenous law in India. *Comparative Studies in Society and History,* 1972, 14, 53–70.

Gans, H. *Urban villagers: Group and class in the life of Italian-Americans.* New York: Free Press, 1962.

Illich, I. *Medical nemesis.* New York: Pantheon, 1976.

Illich, I. *Tools for conviviality.* New York: Harper & Row, 1973.

Jacobs, J. *Death and life of Great American cities.* New York: Vintage, 1961.

Kawashima, T. Dispute resolution in contemporary Japan. In A. T. von Mehren (Ed.), *Law in Japan: The legal order of a changing society.* Cambridge, Mass.: Harvard University Press, 1963, pp. 41–72.

Kidder, R. L. Western law in India. In H. M. Johnson (Ed.), *Social system and legal process.* Special Issue of Sociological Inquiry, 1977, 47, San Francisco: Jossey-Bass.

Li, V. *Law without lawyers.* Stanford, Calif.: Stanford Alumni Association, 1977.

Nader, L. (Ed.), *Law in culture and society.* Chicago: Aldine, 1969.

Nonet, P., & Selznick, P. *Law and society in transition: Toward responsive law.* New York: Harper & Row, 1978.

Sarfatti-Larson, M. *The rise of professionalism: A sociological analysis.* Berkeley: University of California Press, 1977.

Schumacher, E. F. *Small is beautiful.* New York: Harper & Row 1975.

Spence, J. Institutionalizing neighborhood courts: Two Chilean experiences. *Law & Society Review,* 1978, 13, 139–182.

Stack, C. B. *All our kin: Strategies for survival in a black community.* New York: Harper & Row, 1974.

Theobald, R. Cybernation and human rights, In *Liberation*, August 1964. Reprinted in H. Hodges (Ed.), *Conflict and consensus*. New York: Harper & Row, 1973.

Yancey, W., & Ericksen, E. The antecedents of community: The economics and institutional structure of urban neighborhoods. *American Sociological Review*, 1979, 44, 252–263.

Yngvesson, B., & Hennessey, P. Small claims, complex disputes: A review of the small claims literature. *Law & Society Review*, 1975, 9, 219–274.

19

Law as a Social Trap

Problems and Possibilities for the Future

DEAN E. PEACHEY and MELVIN J. LERNER

1. INTRODUCTION

When we think about the law, we are faced with a curious set of contrasts. We have come to rely heavily on the legal structures in our everyday lives. Ideally, the law guarantees a wide array of personal and civil liberties. It ensures education for our children, provides a structure for our business transactions, and serves to curb the power and ambitions of our governments. In addition, the law prescribes what is right and good in our society, and it offers sanctions for those who would trangress those values and pose a threat to us or our security. Of course, we recognize that the law is not perfect and that courts sometimes make mistakes, but few of us would dispute the claim that, all things considered, our system of law and justice is a relatively good one and that "if the law were abolished today, it would have to be reinvented tomorrow."

On the other hand, there are enormous recurring doubts when we think about the law, and particularly our system of criminal justice. As

DEAN E. PEACHEY and MELVIN J. LERNER • Department of Psychology, University of Waterloo, Waterloo, Ontario, Canada N2L 3G1. The first author received partial support from a Mennonite Central Committe (Canada) Peace Scholarship. The initial work on the "Law as a Social Trap" was made possible by a Leave Fellowship Grant from the Canada Council and a Senior Fellowship from the Culture Learning Institute of the East-West Center, 1975–76, to Melvin J. Lerner.

indicated in this chapter, the American crime rate has increased dramatically in recent years. And more and more people are perturbed at the proliferation of regulations and "red tape" that inconvenience them or prevent them from living, working, or playing as they wish. The courts and efforts at criminal rehabilitation have come under heavy criticism. In short, something appears to be seriously awry in our legal system.

In this chapter, the argument is advanced that such problems are more than just the miscarriage of a good system of law and justice— that they are predictable, if not actually inevitable, results of our legal structures.

To present such a seemingly one-sided position on the law is not to deny the remarkable achievements of our legal systems, or to claim that they do not retain positive functions in our corporate life as a society. Rather, the argument draws on the results of social-psychological research to suggest that while the legal system initially provides numerous benefits, it also poses long-term hazards. Therefore, it represents a kind of social trap (Hardin, 1968; Platt, 1973).

The present discussion examines some results of the rapidly expanding role of law in modern society. Negative side effects of creating and enforcing laws are identified, such as less trust among people, a heightened sense of injustice, and the perceived need for more surveillance in society. To the extent that these occur, legal institutions become self-contained feedback mechanisms that call for a even greater exercise of law.

2. TOWARD AN UNDERSTANDING OF THE LAW

2.1. A Functional Approach

This analysis begins with a look at a common functional understanding of the role of law.[1] According to some social analysts, the codified law, like other human institutions, arose to fulfill certain needs or functions in society. (Friedman, 1977; Black, 1976). The historical events of prime interest here are the emergence of the nation-state and

---

[1] The term *law* is used in reference to those rules of society that are enforceable directly by the police and where interpretation and administration of the rules are carried out by the court system. Some phenomena that are enforced by quasi-judicial organizations, such as governmental regulatory and licensing agencies, are also included in the discussion. This is presented as a working definition for the purposes of the present discussion; it is not intended as a definitive statement.

centralized government, along with the rapid industrialization process of the last two centuries under a free-enterprise economy. The changing state of affairs resulted in a need for a much greater degree of management or control within a given population. While this can be broadly thought of as social control and social integration, it can also be analyzed in terms of more specific functions (Friedman, 1977). An overview of the primary functions follows.

In an era of rationalism and libertarianism, the law emerged as a *guarantor of rights*. The emphasis was strongly on individual rights, and particularly those freedoms that would allow the individual to determine his private destiny. And in the American system, there was a particular thrust toward protecting the individual from the tyranny of an abusive government.

Guaranteeing economic rights and privileges also became important. Protection of land and capital were required for the success of the free market, and with the rapid appearance of corporations during the last 150 years, which made possible the long-term integration of activities required for economic development, the legal system received its greatest impetus toward the codification and professionalization of the law (Friedman, 1977).

There is also the role of *"social engineering"*—determining the shape and direction of a changing society. Who will have access to education, how will employees be treated, and what will be the rights of minority groups? In general, what will be the goals and methods of the nation? In a scientific, secularizing society, the role increasingly fell to the state not only to establish these policies but also to articulate an emerging morality and make judgments of right and wrong. By and large, the state did this through its usual vehicles, laws and courts.

Coupled with social engineering, there is also an element of *secondary social control*. Execution of the law provides a multitude of opportunities for vicarious learning in the society. The law acts as both deterrent and moral instructor. We learn (often unconsciously) such truisms as "People eventually get what they deserve," or "The Mounties always get their man."

Thus, along with the traditional responsibilities for *control of deviants* and *dispute settlement*, law has assumed other major functions in modern society. These innovations are linked to changes in society. While it may lack universal truth, Black's (1976) proposition that the "[amount of] law varies inversely with other kinds of social control" (p. 107) is a useful observation. Where direct social bonds among the people are lacking, laws and legal mechanisms tend to be more prevalent (Nader & Metzger, 1973: Schwartz, 1973).

Put simply then, the increasing rule of law can be viewed as a function of the increasing complexity of society coupled with weakening social bonds.

> Over the centuries, social life has been drifting in two great patterns. On the one hand, the intimate life of earlier times has been coming apart. On the other hand, what was separate has been drawing together . . . all this time, law has been increasing. (Black, 1976, p. 132)

In summary, the functional understanding of law recognizes a rapidly increasing level of legalization in contemporary society and views that increase as being rooted in the needs of the society.

2.2. Law as an Effort to Avoid Social Traps

An alternate way to understand the function of law has emerged recently, based on the concept of social traps (M. J. Lerner, 1976). Traps exist when the results of an act are initially rewarding, but they later become increasingly destructive—so much so that they create the need for even greater use of the seemingly effective device (Platt, 1973; Hardin, 1968).

A common social trap involves the construction of superhighways to eliminate the traffic congestion created by urban commuters from suburbs. Initially, the problem and the solution seemed obvious; the old country road was not designed for and cannot handle the traffic created by the suburb. It "makes sense," then, to spend the funds to acquire land and build wider, better highways. And for a relatively brief period after construction, that worked. But with the easy access to the city, developers created new, larger suburbs, which generated heavier traffic using the highway. Before long, the new freeway was as congested and probably more dangerous than the old country road had been before the "obvious" solution was adopted. Meanwhile, train service and other forms of public transportation have deteriorated from disuse, to the point where they are even less attractive options.

Platt (1973) analyzed the social-trap phenomena in terms of reinforcement theory. People engage in a behavior that offers immediate rewards: the opportunity is just too good to let pass by. But over a period of time, the reward structure changes. The actors now begin to incur substantial costs for their behavior; yet, for any one individual to pursue an alternate course of behavior would result in an even greater cost for that person. Thus, the cycle perpetuates and escalates.

The social-trap analysis has been applied to a wide variety of social problems in recent years, particularly in regard to natural resources, population, and environmental concerns (Hardin, 1968; Hardin & Baden, 1977; Stern, 1978).

Proposals for avoiding or escaping the traps have varied, but many of them (e.g., Baden, 1977; Ostrom & Ostrom, 1977) have adopted some form of Platt's (1973) argument for the democratic creation of *superordinate authorities* that are "able to manage and correct social traps . . . [that lead] to collective bads" (p. 650).

The argument presented by Platt provides the current framework for understanding the increasing legalization of society. It is possible to view the law as a responsible attempt by society to prevent a variety of social traps from occurring by altering the rewards and costs of social behavior (e.g., through the use of fines, imprisonment and tax incentives).

At first glance, it would appear that the legal system is functioning well in this role. Deviants are arrested and placed safely in jail, damage claims are awarded in court, and contractual agreements allow the wheels of business to turn. One can see immediately the value of enforcing laws, regulations, and contracts. This evidence of successful social control and integration is the "immediate reinforcement" that provides the inducement to continue or increase the use of the legal mechanism.

But the question being raised here is whether relying on the law as a superordinate authority, rather than solving the problem, does not actually pose yet another collective trap (M. J. Lerner, 1976). What is the evidence for this position? How well are our legal institutions performing the functions that they supposedly were designed to fulfill?

3. CURRENT PROBLEMS IN THE LEGAL SYSTEM

In recent times there has been a proliferation of research and writing on the problems facing the criminal justice system in North America (see Menninger, 1969; Casper, 1972; Bayley & Mendelsohn, 1969; Quinney, 1974, for some illustrative examples). We will not attempt a comprehensive review of this literature but will focus instead on two crucial questions: (a) To what extent is social pathology a product of the legal mechanisms themselves? (b) Can the problems that exist be remedied through a greater exercise of the law?

3.1. Crime Control

3.1.1. *Effective Sanctions versus Criminal Labeling*

What is the effect of our corrections systems on criminal behavior? Is punishment effective in preventing future crime, or does the application of criminal sanctions actually tend to encourage criminal behavior through people's becoming labeled as deviant by themselves and society?

While there is evidence that certain kinds of punishment have a deterrent value in some situations (e.g., Blumstein, Cohen, & Nagin, 1978), it appears that the labeling process can also be dominant.

According to their own reports, 40% of a sample of high-school students continued to shoplift after being apprehended for that offense. Those who were reported to the police were *more likely* to continue shoplifting than the youth who were dealt with only by store personnel (Klemke, 1978). However, there is some evidence that the detrimental effects of labeling may be limited primarily to persons with relatively high social status and/or self-esteem (Mahoney, 1974).

3.1.2. Police as a Deterrent to Crime

Correlational studies have shown a high positive correspondence between crime rates and the size of the community's police force (Kidder, Kidder, & Snyderman, 1976). While the lack of an inverse relationship is significant, whether the police ranks swell in response to climbing crime rates or the reported crime index increases because of larger police deployment has generally not been clear. More compelling evidence comes from the Kansas City Preventive Patrol Experiment (Kelling, Pate, Dieckman, & Brown, 1974). Routine police patrols were virtually eliminated in selected districts and were increased in others for a period of one year. The crime rates did not change, and the residents reported feeling neither more nor less secure in their neighborhoods as a result of the presence or absence of police patrols.

3.1.3. Disillusionment with Justice

After an extensive review of survey evidence on American attitudes about police, lawyers, and courts, Sarat (1977) identified three recurring themes:

1. There is a recurring demand for equal treatment; violation of equality is the most persistent source of dissatisfaction.
2. Yet, for the most part, there is overall approval of the legal system: "people seem to be dissatisfied without being detached" (p. 441).
3. Persons having firsthand experience with police, lawyers, and courts tend to be the most dissatisfied with the legal system.

While people's exact understanding of "justice" is illusive, it is clear that there is not a faith in the judicial *procedure* itself, as guaranteeing a fair outcome. The proverbial opportunity to "have one's day in court" is not as assurance of fair treatment. For example, Casper (1977) found that defendants who had gone through a trial viewed their sentence as less fair than did those who forfeited trial by entering a guilty plea. More defendants reported that they did *not* have a fair chance to present their side when they were sentenced by trial rather than by guilty plea. In addition, they were more likely to feel that the judge and/or the prosecutor was biased against them when they were sentenced by trial. (The defendants also tended to evaluate the fairness of their sentence by comparing it with the sentences received by others convicted of the same offense.) That direct experience with legal proceedings is related to enhanced feelings of injustice reflects a serious failure on the part of our present legal system.

3.2. Dispute Settlement

Courts provide a means for handling disagreements and seeking redress for grievance in a socially approved, "civilized" manner, without resort to physical force or violence. While this function is performed successfully at times, there are many occasions when taking a case to court intensifies the conflict between the disputants and leads to the further deterioration of social relationships. The increasing alienation is a consequence of both the process (adversary) and the final outcome (win–lose), as civil courts have often employed the same methodology used in criminal proceedings. Thus, the civil suit is geared more toward a winner-take-all, right–wrong approach than toward problem solving (McGillis & Mullen, 1977).

While this may be an appropriate methodology for some cases, there are many conflicts where a problem-solving approach to justice is more appropriate. Persons with a conflict often have to choose between simply enduring their problem or resorting to the law and destroying whatever relationship they may have. If the issue goes to court, one party usually feels unjustly treated and resentful, and may appeal the decision or seek some other type of retaliation (Nader, 1974, 1975). This concern is currently spawning a variety of alternative modes of dispute resolution (see, for example, McGillis & Mullen, 1977).

3.3. Ensuring Civil Rights

The notion that all men are endowed with certain inalienable rights has permeated our legal and political thinking for the past two centuries, and the task of assuring that this "self-evident truth" is practiced has fallen on the law.

Unfortunately, it appears that the law may actually be detrimental to the welfare of various groups that exist on the fringes of our society. While racial minorities are the groups that most frequently have their conditions worsened by the law, this may also be the case for political and other minority groups.

The manner in which cultural bias is built into the legal system has been assessed by an anthropologist, Daniel Swett (1969), who made his observations while working as a reserve police officer. Swett first described the police recruiting procedures, which favored persons who subscribed to the dominant values of the society. The police were subsequently trained to maintain an alert for suspicious persons and incidents. The officers on patrol knew from their department's statistics and training program that the crime-prone persons were those who were poor or were members of a minority group, and the "high-risk" areas were those districts inhabited by such persons. Thus, persons who were visibly poor or members of minority groups became the object of greater surveillance by the police, and thus subject to more frequent interrogation and arrest.

The courts have received widespread attention as sources of injustice through the inability of many persons to hire expensive legal counsel (Casper, 1977; Clarke & Koch 1976), but there are other aspects of court procedures that also merit scrutiny. The rigidity and formality of courtroom proceedings, while formidable to the average citizen, can be completely overwhelming to the disadvantaged person. The legal language is incomprehensible or confusing to the uneducated, or to persons from a subculture that relies on a well-developed slang. Witnesses may consequently appear to be uncooperative, unsure of the facts of the case, self-contradictory, or even perjurious (Swett, 1969; Nader, 1974).

Using a path analysis on data from homicide cases, Farrell and Swigert (1978) concluded that race, occupational status, and sex influence the likelihood of defendants' having a prior record, which in turn influences the disposition of the case. (Persons having prior records were more likely to be convicted and to receive longer sentences on conviction.)

Chiricos, Jackson, and Waldo (1972) studied felony cases in which the judge invoked an option to *withhold adjudication*. While the defendant's guilt had been established in court, a Florida law made it possible for the judge to withhold passing sentence in those cases that he felt would benefit from such action. The individual would simply be placed on probation and thereby spared the label of *convicted felon*, and the accompanying stigma in community life, on employment and license applications, etc. The researchers found that for each level of age or occupation, and regardless of type of attorney, plea, court, or prior record, blacks were more likely than whites to receive the stigma of a felony conviction.

It appears then that while the law may sometimes act to correct injustices, it often serves to perpetuate and institutionalize societal biases and inequities.

3.4. Social Engineering: The Law as an Instrument of Social Change and Adaptation

When times are changing, a society needs to have the means to be adaptive. The law frequently has been the vehicle for such change. Emancipation, suffrage, and desegregation are notable social innovations achieved through the exercise of the law. (But the fact that desegregation laws were necessary after the Emancipation Proclamation also testifies to the limitations of laws that do not receive popular support.)

The law also supplies a link with the past that assures continuity and stability in society. It is, after all, a means of perpetuating dominant values and ideology (Swett, 1969). Those who write the law, enforce it, and administer it tend to be successful members of the society who have a definite interest in seeing things continue without major upheavals. Law is therefore a limited means of change. As Lempert (1976) cautioned:

> Law is ultimately the power of the state. People mobilize law so that they can enlist that power, actually or potentially, to advance their own interests. It is not surprising then, that law has its limits as an instrument of social change. . . . This is important to any general theory of legal mobilization, for such a theory must also specify its limits. (pp. 184–185)

Thus, even in a democracy, law as an instrument of social change lends itself more to "top-down" social changes than to "bottom-up" innovations (Rogers, 1971). It is more amenable to moving society ahead in the current direction than it is to changing directions.

The traditional virtues of our legal mechanisms are being called into question in the present discussion. We have seen some evidence that more laws generate more enforcement, and that more enforcement leads to more perceived injustice, less faith in the system, and new laws to place limits on the enforcement. Social ties deteriorate, mistrust is generated, and those who have little receive less. As Sarat (1977) concluded, "contemporary survey research indicates that, for the average citizen, familiarity (with the legal system) breeds contempt" (p. 441).

Are the problems reviewed in this section simply minor flaws in what is basically a good system? Can they be remedied by "more of the same"—passing more or better laws, employing more police and judges, or improving access to the law for various subgroups in society? The suggestion has been made that relying on elaborate and complex superordinate authorities to avoid social traps may in itself prove to be yet another trap. Thus, there is an implication that miscarriages of justice are not due simply to understaffing or imperfect and biased persons working in our legal institutions, but that these negative effects are products of the structures themselves.

Let us assume for a moment that this analysis has merit. Can we begin to understand *how* the trap operates? Are there known psychological processes that could mediate the metamorphosis of a reinforcing outcome to a negative one? Evidence that such processes exist comes from several lines of current social-psychological research.

4. SOCIAL STRUCTURES AS CREATORS OF SOCIAL REALITY

Of course, many of us are law-abiding citizens not out of fear of punishment, but because we have internalized the values of the legal system. Law serves an important function in our society of defining what is right, good, and just (Chein, 1975; Andenaes, 1977). Indeed, it can be shown that public attitudes and values can change to conform to modifications in the law (Sheppard, 1968). The law conveys messages to us that we learn, often without being aware of it at the time. But what additional implicit messages are communicated by the presence of *any* law, regardless of its content?

4.1. Effect of Surveillance and Rules on Perceived Motivation

Our responses to the actions of another person can vary greatly, depending on the intention we attribute to the person's behavior (Heider, 1958; Kelley, 1973).

For example, when we receive a gift, we are sometimes uncertain of the giver's motivation. He may have felt obligated to present a gift because of the occasion (birthday, Christmas, etc.); it may be that he is lonely and is trying to "buy" friendship and approval; he could be trying to ingratiate himself and secure a future favor; or the gift might be a genuine expression of love and friendship.

We use various means to decide which cause or causes are operating, but to the extent that we believe there is more than one cause that would be "sufficient" to explain someone's behavior, we will not be able to place much confidence in any one explanation (Kelley, 1973). (For instance, as long as there remains more than one plausible reason for receiving the gift, we will have difficulty discerning the giver's "true" motivation.)

Applying these attribution principles to the law produces a rather startling hypothesis: to the extent that there are laws that define how we must treat one another in order to avoid sanction (fine, imprisonment, public scandal, etc.) and there are corresponding legal institutions designed to enforce these rules, then we are less able to trust other people. Behavior that is in accord with the laws cannot be confidently interpreted as indicating an intention or a motivation other than the desire to avoid the punishment for failure to comply.

In other words, the fact that we might live and work together quite amicably tells me very little about what you think of me or would do if there were not a contract and bylaws that governed how we act. I will never know what kind of person you "really" are or know your "true" feelings until I see what you do when the external incentives and constraints are removed. This is not unlike the dilemma of the very rich or the very beautiful in trying to find someone who likes them for "themselves" and not for the other gratifications they can offer.

A related line of research has indicated that the act of keeping a person under surveillance can reduce the observer's trust in the other person and the confidence that he will perform appropriately without surveillance (Strickland, 1958; Kruglanski, 1970).

Such phenomena are not confined to the psychological laboratory. When an ongoing relationship is valued, even the supposedly "crass" business world, which ostensibly is a prime benefactor of civil law, hesitates to utilize the law. Businessmen frequently fail to write their contracts in a way that will include legally enforceable sanctions (Macauley, 1963). When dealing with a known party, they prefer to say and act on the assumption that "We can trust old Max." Detailed contracts are drawn up only when there is a likelihood that problems will arise, and filing a suit in court usually means the end of a business relationship (Macauley, 1963).

The businessmen interviewed by Macauley suggested that to be too concerned about exercising the provisions of the law would damage their reputation in the business community; they would be viewed as people who did not trust others, and who perhaps could not be trusted in return.

Not only do attributions of causation affect our perceptions of other people's motivations, but there is good reason to believe that our understanding of the reasons for our own behavior is similarly influenced (Bem, 1972). Although the theoretical reasons have been under debate, two research findings have been consistently reported: (a) behaviors that are initially self-rewarding will decrease in frequency when subjected to external rewards, especially if the additional rewards are emphasized or "highly salient" in the situation (Lepper & Greene, 1978; Ross, 1975); and (b) threats of severe punishment result in less internalization of a rule than do much milder punishers (Aronson & Carlsmith, 1963).

Thus, social-psychological theories of attribution, self-perception, and dissonance have far-reaching implications for our understanding of legal processes. They suggest that the exercise of the law may limit our capacity to trust other people's intentions in a variety of ways. Or the disproportionate amount of police surveillance devoted to minority groups (Swett, 1969) can be expected to result in even less trust in those groups by the officers. Commonly accepted views on the benefits of contracts can be questioned. And finally, we may even come to understand our own behavior as being less motivated by intrinsic factors and more directed by the threat of sanctions.

4.2. Loss of a Sense of Personal Control

That people like to have some voice in determining what happens to them is common knowledge, but the extent to which the loss of control over one's environment affects the individual is relatively new as an area of systematic study. In one controlled study, institutionalized senior citizens who were given enhanced choice and personal responsibility over mundane behaviors were found to be happier, more alert and active, and in better overall health than similar persons who had less opportunity for control over their lives (Langer & Rodin, 1976). From a sociological perspective, Reisman (1950) warned about the debilitating consequences of large bureaucracies that remove from the individual control over his own destiny.

Is it possible that the expansion of the law has a similar effect? As laws and law-enforcing agencies multiply, control is readily eroded from the individual or the local group and is placed in the hands of a centralized authority. Gradually people lose a cherished possession: their perception of freedom and choice.

4.3. The Illusion of Autonomy and "Diffused Interdependence"

With the social and geographic mobility associated with industrialization, an increasing number of people live in settings with few stable social ties to one another (Singer, 1973). As a result, there appear to be few if any common bonds among members of society that can be employed to cope with conflicts or to deter abuse and exploitation. The intergrating influence of neighbors, community, or religious leaders is minimal, if present at all.

Despite the weakened social and emotional bonds, the *actual interdependence* in contemporary society is greatly heightened. Virtually every citizen interacts with many others in highly interdependent ways during the course of normal activities, with an elaborate exchange of goods and services (Greer, 1962). We are directly affected by the behavior of many persons whom we may never even meet. (Consider for example, the far-reaching ramifications of power failures, price increases by those who control oil and gas supplies, or strikes by teachers, truckers, and transportation or communication workers.)

In the midst of this complex web of relationships, people cling to the notion that they are independent and free, when in actuality, there has simply been a shift from dependence on specific individuals to dependence on various classes or groups of people. These relationships might be described as *diffused interdependence.*

Because of its advanced state of interdependence and technological effectiveness, this intricate web is extremely vulnerable to the vagaries of human performance. Our technology ties us together in many ways, while at the same time providing us with extraordinary means whereby we can harm one another (or merely cause inconvenience and stress).

Presumably, the widespread use of law is a product of weakened social bonds (Black, 1976; Greer, 1962). However, the current analysis suggests that the exercise of law may in fact increase the emotional distance between people and may contribute to relationships' being viewed in purely functional terms. To the extent that this is true, we need to understand the role that legal structures play in maintaining diffused interdependence and the illusion of autonomy.

4.4. The Model of Man Implicit in the Law: Man as a "Rational Animal"

A system of legal sanctions assumes that people can and will rationally assess a range of alternative actions and elect the one that promises the greatest benefits over an extended period of time.

Research of the past decade has raised serious questions about this assumption. Platt (1973) has described how people may ignore long-term costs in order to achieve immediate gains (and, likewise, avoid incurring an initial cost, even when it would lead to eventual pleasure). Langer (1978) has highlighted the observation that people often perform routine activities in what is essentially a nonrational, nonthinking manner.

Game theorists have shown that people often lock themselves into escalating conflicts, while ignoring or remaining oblivious to alternative choices of cooperation that would yield greater gains for all parties (Deutsch & Krauss, 1960).

In light of these findings, is it possible that the system of legal sanctions may rest on an unwarranted confidence in the rationality of man's behavior? The problem may go even deeper, however. To the extent that the law is accepted as being "right" or "good," individuals are likely to learn and accept its assumptions about human nature; that is, they may come to believe in the rational, cost-accounting image of man. When viewed in this way, people are understood to be essentially self-seeking, balancing in some cognitive ledger the personal gains or costs of a given behavior. The danger is that if we view others in a certain way and act toward them accordingly, they are likely to confirm our expectations (Jones, 1977).

It appears plausible that there has been an overemphasis on man as a self-seeking, cost-accounting mechanism.[2] There is no direct proof that this phenomenon results from the ubiquitous nature of legal structures, but the possibility is of such social importance that it should be the object of vigorous study and research. It seems entirely possible that the indirect effects of our legal institutions on how we view and live with one another are to create or to alter social reality. They establish perceptions that lead to behaviors. The behaviors in turn "prove" the correctness of the perceptions. Thus, our legal institutions can create events that provide social validation for an essentially self-seeking view of human relations.

While the discussion in this section has been brief and somewhat preliminary, it perhaps provides a glimmer of how legal structures can become a source of social breakdown. Specifically, the following hypotheses have been presented:

1. Pervasive legal structures can serve to reduce our confidence in the motivations of one another.
2. Our perceptions of the motivations of our own behavior may likewise be altered.
3. Large-scale legal institutions may reduce one's sense of control and may contribute to the depersonalization of modern society.
4. Through the pervasiveness of laws, people learn that they need rules to tell them how to behave, and that external sanctions are necessary to overcome their motivations to do otherwise.

---

[2] There is, of course, considerable debate on this issue. See Lerner (1977) for a further development of this theme. Walster *et al.* (1978) present the case for a divergent view.

Before passing final judgment on our present law, however, it is imperative that we look at the future. In a final sense, the legal system, with its entrapping potential, must be evaluated not on its past performance but on its capability to serve in the times that lie ahead.

5. SCENARIOS OF THE FUTURE: THE EFFECTS OF SCARCITY

Since the *Limits to Growth* (Meadows, Meadows, Randers, & Behrens, 1972) was published by the Club of Rome in 1972, there has been widespread recognition that whatever else the future might hold, it will bring critical shortages of important resources.

The world population will continue to rise (United Nations, 1974), and the supply of essential commodities will not keep pace with demand (Heilbroner, 1974). It is now recognized that the "green revolution" of the 1950s and the early 1960s will not miraculously provide perpetual increases in food production. There are limited supplies of minerals and other nonrenewable resources (Brown, McGrath, & Stokes, 1976; Ehrlich & Ehrlich, 1970).

The effects of scarcity have been studied extensively by social psychologists (S. C. Lerner, 1979). Not only does the potential for conflict increase dramatically under conditions of real or threatened scarcity, but the resultant competition readily comes to be viewed as legitimate. It is considered acceptable, if not admirable, in such circumstances to use extralegal means to advance one's end if they can be used effectively and successfully. It is assumed in these circumstances that everyone will do his utmost to win and to get as much as he can for himself with little or no regard for the consequences to others (Lerner & Lichtman, 1968; M. J. Lerner, 1971).

The individual does not remain a neutral strategist in this context; rather, he engages in cognitive changes to adapt to the competitive encounter he anticipates. This process was demonstrated in an experimental situation by Lerner, Dillehay, and Scherer (1967). Adults were given the opportunity to meet another person briefly. Those who had been led to believe that they would later be working with the stranger in a cooperative project found the person to be interesting, desirable, and similar to themselves, while those who expected to be competing with the same stranger considered him to be relatively undesirable, inferior, and "just different."

Ralph White (1966) has described the tendency for contesting nation-states to generate an image of the "enemy" that portrays him as untrustworthy and clearly malevolent in intention. As a result, any action taken by the other side is perceived as an aggressive strategy—either as overt hostility or a devious ploy. Because of the closed system of distrust, nothing that either party can do will provide reliable evidence to counter the ominous, threatening image. Eventually, any action toward the enemy is justified, including "preemptive" aggression.

This escalating cycle between nations parallels what goes on among individuals who see themselves in "zero-sum" competitive relations. Wrightsman (1966) has shown, for example, that people who are distrustful in orientation will, if given the opportunity, engage in self-serving behavior rather than acting cooperatively. Similar are many examples of how the perception of a competitive act elicits retaliation, which leads to an escalation of the conflict (Deutsch & Krauss, 1960; Kelley & Stahelski, 1970). Since the competitive behavior often reduces the likelihood that the person can obtain the desired outcomes, the resources become increasingly scarce, and thus, they become more valuable and less likely to be shared. Even if he did not perceive others' intentions as malevolent, each person would increase his efforts on his own behalf, which in turn causes the others' harm.

Will these processes of competition and conflict actually be manifested in our daily lives? As indicated above, the probability of future shortages has been known and predicted for years. But always the knowledge was "theoretical" for most of us; we knew the problems existed, but they were projected to "sometime in the future" or "over there" in some other country or social stratum.

The issue that first made these predictions immediate and concrete for most North Americans was *energy*. With rising energy consumption in the westernized world, and political tensions in the Middle East during the 1970s, the shortage became "real"; that is, it affected our daily activities.

Several changes in human relationships that are most likely to result from scarcity have already been identified: heightened distrust and suspicion, the complete justification of what would otherwise be considered selfish or immoral behavior, competition perceived in "zero-sum" terms, and derogation of the perceived competitor. Thus, it is hardly surprising that gasoline shortages in the summer of 1979 produced service station violence (*Newsweek*, 1979).

What will be the role of law in the decades of dwindling resources that lie ahead? Can it be used to lessen the problems just described, or will it worsen them?

Oddly enough, many people who view contemporary and future problems as collective traps regard legal devices as the means for establishing the cooperative efforts required to end destructive conflict and to prevent the self-defeating consequences of social traps. (See, for example, Platt, 1973; Baden, 1977.)

In the course of this chapter, the argument has been presented that the prevailing legal structures frequently exacerbate rather than curtail elements of distrust, zero-sum competition, and "justifiable" selfish motivation. The evidence presented indicates that superordinate authorities must be used much more judiciously if we are to respond to these pressures in a manner that is just and responsible to both the human race and the intricate ecology of our planet.

Are there ways to respond to crises other than an increase in "law and order"? And more importantly, is our society capable of making the changes that will be required to escape the tightening trap?

6. ALTERNATIVES

The suggestions presented here fall roughly into two categories: changes in the legal system *per se*, and modifications in the social system that would be necessary to implement and sustain those changes.

6.1. Delegalizing Portions of the Law

In some cases, an effective approach may be to change the law so as to encourage *less* legal activity rather than more of it. Significant innovations in dealing with civil disputes are being implemented in many areas (McGillis & Mullen, 1977; Stulberg, 1975). Using various styles of third-party mediation and/or arbitration, these services are designed to divert disputes away from slower and more expensive court action or to prevent minor conflicts from escalating into criminal behavior.

Programs that delegalize the sentence for minor criminal offenses, while they are less common than civil dispute services, have also proved to be successful. One such program in Ontario utilizes third-party volunteers to mediate an encounter between the victim of a crime and the offender, in which the offender agrees to make restitution to his victim during a period of supervised probation. This program addresses the often-neglected needs of the victims of crime, as well as encouraging the offender to maintain responsible employment and avoiding the detrimental effects of imprisonment (VORP, 1977). An experimental project in New York City has even achieved success in dealing with felony cases through a mediation–arbitration process rather than criminal prosecution (Davis, 1979).

A number of other proposals for modifying the judicial processes have been discussed by Galanter (1976). He suggested that we

change the character of courts by creating tribunals that are more "popular,"
responsive and participatory, less professional and alien, thereby reducing
the cultural and psychological distance between tribunal and parties. (p. 227)

He also proposed making existing judicial institutions more "active,"
with the idea that for processing numerous complaints involving small
amounts, an inquisitorial adjudication may be preferable to an adversary
process where the judge plays a more passive role in fact finding. In
addition, Galanter encourages further development of tribunals in the
private sector (such as labor relations boards).

The use of "champions," such as ombudsmen and media ombudsmen, may also prove to be effective and less costly methods of redressing
grievances (Hannigan, 1977; Johnson, Kantor, & Schwartz, 1977).

These proposals are presented with a view toward greater citizen
participation in redressing grievances and toward enhanced personal
contact among members of the community. Many of them also contain
the potential for giving people a greater sense of personal control of and
responsibility in their environment.

It may also be worthwhile to consider altering the rewards and costs
associated with court action in order to reduce court congestion and
abuse. For example, the "contingency fee" arrangement, whereby a
client pays his attorney only if he wins the case, is permitted in the
United States but not in Canada. Canada appears to have fewer and
shorter civil suits than does the United States.

6.2. Changes in the Larger Social System

The incapability of the current legal and political system to effect
the changes necessary for adapting to the future, as well as the frustration of political leadership, was reflected by President Jimmy Carter
when he addressed the nation on energy needs on July 15, 1979:

> The gap between our citizens and our government has never been so
> wide. . . . What you see too often in Washington and elsewhere around the
> country is a system of government that seems incapable of action. (New York
> Times, July 9, 1979)

Carter then prefaced his legislative proposals by calling on the citizenry
to evaluate their life goals, renew their sense of patriotism, and rekindle
their allegiance to one another.

Black (1976) suggested, and it has been argued in this chapter, that
law tends to be most prevalent when other types of social control resulting from recognized social and emotional bonds among people are
lacking. Given this assumption, if any significant "delegalization" is to
take place, it will be necessary to highlight and strengthen the social

bonds within society. Obstacles of mobility and urbanization notwith-standing, the authors believe that some progress can be made in this direction.

Encouraging organization and interaction within neighborhoods should be given more attention. Small-scale projects offer possibilities for personal participation, and they contribute toward the sense of per-sonal control that was discussed above. Ironically, the problems asso-ciated with resource scarcity could also provide the impetus for more cooperative problem-solving at a local level. Recycling drives, car pools, and neighborhood solar-heating installations are examples of responses that are well suited to community participation.

As Stern (1977) argued in the context of energy conservation, col-lective traps require collective action. The effectiveness of individual incentives tends to be short-lived. However, incentives for change that are directed at groups offer the possibility for sustained behavior change as social influences gradually replace the external incentives. For ex-ample, the federal government could subsidize communities to develop renewable energy resources, or it could return tax money to local gov-ernments or other units that successfully meet conservation goals.

Stern suggested that such a plan would motivate local groups to rally around common projects, and perhaps even to engage in healthy competition for energy management grants:

> These incentives would coalesce local groups which would be of an appro-priate size to develop renewable energy resources and which would also, not incidentally, develop civic pride and group norms concerning energy con-servation. Individuals would become more intent on the collective welfare and less likely to fall into the trap of individualistic consumption. (p. 15)

While the initial cause of such changed behavior might lie in the eco-nomic incentives, the conservation efforts could soon be maintained by a variety of social influences. Stern's example is presented here because it illustrates a more appropriate role for superordinate authorities and legal mechanisms. Government action should enhance social organi-zation through positive processes that emphasize human relationships. There are many areas in addition to energy conservation where the responsibility for meeting resource needs might be focused at the local level rather than on impersonal hierarchical programs that foster ano-nymity.

A variety of proposals for implementing such changes in an indus-trialized society have already been articulated and will not be repeated here. (See, for example, Lakey, 1973; Schumacher, 1974; Morris & Hess, 1975.) The point of this chapter is to suggest that such proposals are not simply the product of wild-eyed visionaries and social malcontents, but that widespread social reform may actually be in the best interests of most members of society.

While encouraging localism, there can also be a complementary emphasis on larger human ties—national and international—in order to avert destructive conflict (see Sherif, 1966). There needs to be a greater recognition of the diffused interdependence of modern society to dispel illusions of autonomy and minimize "us" versus "them" perceptions. Traditional approaches to reducing tension and fostering competition may be effective, but only when these efforts are motivated by a recognition that we are truly linked together "in common predicament" (Sherif, 1966).

These suggestions are only a beginning. The analysis on which they rest must be viewed as tentative, but its embryonic nature only serves to highlight the need for further development. While much of the work that lies ahead falls outside the customary realm of the social psychologist, there is an acute need for ongoing programs of careful research. An extensive classification of questions for research has been complied by Lempert (1976). Crucial areas of inquiry include the following:

1. What are the conditions under which law will instill prosocial values rather than eroding them?
2. What would be the impact of increased access to legal services on individuals and society? How would enhanced access to legal services affect the way in which social and legal problems are resolved? Does expanding the availability of legal services increase or decrease the amount of contentiousness and violence in a culture?
3. What would promote self-help remedies? How does the desired outcome affect the ability of people to solve their problems without using the law?

REFERENCES

Andenaes, J. The moral or educative influence of criminal law. In J. L. Tapp & F. J. Levine (Eds.), *Law, justice, and the individual in society.* New York: Holt, Rinehart & Winston, 1977.

Aronson, E., & Carlsmith, J. M. Effect of the severity of threat on the devaluation of forbidden behavior. *Journal of Abnormal and Social Behavior,* 1963, *1,* 145–155.

Baden, J. A primer for the management of common pool resources. In G. Hardin & J. Baden (Eds.), *Managing the commons.* San Francisco: W. H. Freeman, 1977.

Bayley, D., & Mendelsohn, H. *Minorities and the police.* New York: Free Press, 1969.

Bem, D. Self-perception theory. In L. Berkowitz (Ed.), *Advances in experimental social psychology,* Vol. 6. New York: Academic Press, 1972.

Black, D. J. *The behavior of law.* New York: Academic Press, 1976.

Blumstein, A., Cohen, J., & Nagin, D. *Deterrence and incapacition: Estimating the effects of criminal sanctions on crimes.* Washington: National Academy of Sciences, 1978.

Boorstin, D. The perils of indwelling law, in R. P. Wolfe (Ed.), *The rule of law.* New York: Simon & Schuster, 1971.

Bordua, D. J., & Tifft, L. L. Citizen interviews, organizational feeeback, and police-community relations decisions. *Law & Society Review,* 1971, *6,* 155–182.

Brown, L. R., McGrath, P. L., & Stokes, B. *Twenty-two dimensions of the population problem.* Washington: Worldwatch Institute, 1976.

Casper, J. D. *American criminal justice: The defendant's perspective.* Englewood Cliffs, N.J.: Prentice-Hall, 1972.

Casper, J. D. Having their day in court: Defendant evaluations of the fairness of their treatment. *Law & Society Review,* 1977, *12,* p. 237–251.

Chein, I. There ought to be a law: But why? *Journal of Social Issues,* 1975, *31,* 221–244.

Chiricos, T. G., Jackson, P. D., & Waldo, G. P. Inequality in the imposition of a criminal label. *Social Problems,* 1972, *19,* 553.

Clarke, S. H., & Koch, G. C. The influence of income and other factors on whether criminal defendants go to prison. *Law & Society Review,* 1976, *11,* 57–92.

Davis, R. Evaluation of the use of mediation in resolving felony offenses. Paper presented at American Psychological Association, New York, 1979.

Deutsch, M., & Krauss, R. M. The effect of threat on interpersonal bargaining. *Journal of Abnormal and Social Psychology,* 1960, *61,* 181–189.

Ehrlich, P. R., & Ehrlich, A. H. *Population, resources, environment: Issues in human ecology.* San Francisco: W. H. Freeman, 1970.

Farrell, R. A., & Swigert, V. L. Prior offense record as a self-fulfilling prophecy. *Law & Society Review,* 1978, *12,* 437–453.

Friedman, L. M. *Law and society.* Englewood Cliffs, N.J.: Prentice-Hall, 1977.

Galanter, M. Delivering legality: Some proposals for the direction of research. *Law & Society Review,* 1976, *11,* 225–246.

Greer, S. A. *The emerging city.* New York: Free Press 1962.

Hannigan, J. A. The newspaper ombudsman and consumer complaints: An empirical assessment. *Law & Society Review,* 1977, *11,* 679–699.

Hardin, G. The tragedy of the commons. *Science,* 1968, *162,* 1243–1248.

Hardin, G., & Baden, J. *Managing the commons.* San Francisco: W. H. Freeman, 1977.

Heider, F. *The psychology of interpersonal relations.* New York: Wiley, 1958.

Heilbroner, R. L. *An inquiry into the human prospect.* New York: W. W. Norton, 1974.

Johnson, E., Kantor, V., & Schwartz, E. *Outside the courts: A survey of diversion alternatives court cases.* Denver: National Center for the State Courts, 1977.

Jones, R. A. *Self-fulfilling prophecies.* Hillsdale, N.J.: Erlbaum, 1977.

Kelley, H. H. The processes of causal attribution. *American Psychologist,* 1973, *28,* 107–128.

Kelley, H. H., & Stahelski, A. J. Social interaction basis of cooperators' and competitors' beliefs about others. *Journal of Personality and Social Psychology,* 1970, *16,* 66–91.

Kelling, G. L., Pate, T., Dieckman, D., & Brown, C. E. *The Kansas City preventive patrol experiment: A technical report.* Washington D.C.: Police Foundation, 1974.

Kidder, L. H., Kidder, R. L., & Snyderman, P. A cross-lagged correlational analysis of the causal relationship between police employment and crime rates. Paper presented at the annual meeting of the American Psychological Association, 1976.

Klemke, L. W. Does apprehension for shoplifting amplify or terminate shoplifting activity? *Law & Society Review,* 1978, *12,* 391–403.

Kruglanski, A. W. Attributing trustworthiness in supervisor-worker relations. *Journal of Experimental Social Psychology,* 1970, *6,* 214–232.

Lakey, G. *Strategy for a living revolution.* New York: Grossman, 1973.

Langer, E. J. Rethinking the role of thought in social interactions. In J. H. Harvey, W. T. Ickes, & R. F. Kidd (Eds.), *New directions in attribution research*, Vol. 2. Potomac, Md.: Erlbaum, 1978.

Langer, E. J., & Rodin, J. The effects of choice and enhanced personal responsibility for the aged: A field experiment in an institutional setting. *Journal of Personality and Social Psychology*, 1976, *34*, 191–198.

Lempert, R. O. Mobilizing private law: An introductory essay. *Law & Society Review*, 1976, *11*, 173–189.

Lepper, M. R., & Greene, D. Overjustification research and beyond: Toward a mean–ends analysis of intrinsic and extrinsic motivation. In M. Lepper & D. Greene (Eds.), *The hidden costs of reward*. Hillsdale, N.J.: Erlbaum, 1978.

Lerner, M. J. Justified self-interest: A replication and extension. *Journal of Human Relations*, 1971, *20*, 127–135.

Lerner, M. J. The law as a social trap. *Culture Learning Institute Report*, August 1976.

Lerner, M. J. The justice motive: Some hypotheses as to its origin and forms. *Journal of Personality*, 1977, *45*, 1–52.

Lerner, M. J., Dillehay, R. C., & Sherer, W. C. Similarity and attraction in social contexts. *Journal of Personality and Social Psychology*, 1967, *51*, 481–486.

Lerner, M. J., & Lichtman, R. R. Effects of perceived norms on attitudes and altruistic behavior toward a dependent other. *Journal of Personality and Social Psychology*, 1968, *9*, 226–232.

Lerner, S. C. Behavior in the crunch. *Alternatives*, 1979, 8–2, 5–11.

Macaulay, S. Non-contractual relations in business: A preliminary study. *American Sociological Review*, 1963, *28*, 55.

Mahoney, A. R. The effect of labeling upon youths in the juvenile justice system: A review of the evidence. *Law & Society Review*, 1974, *8*, 583–614.

McGillis, D., & Mullen, J. *Neighborhood justice centers: An analysis of potential models*. Washington: U.S. Government Printing Office, 1977.

Meadows, D. H., Meadows, D. L., Randers, J., & Behrens, W., III. *The limits to growth*. New York: Universe, 1972.

Menninger, K. *The crime of punishment*. New York: Viking, 1969.

Morris, D., & Hess, K. *Neighborhood power: The new localism*. Boston: Beacon, 1975.

Nader, L. Perspectives on the law and order problem. In M. J. Lerner & M. Ross (Eds.), *The quest for justice: Myth, reality and ideal*. Toronto: Holt, Rinehart & Winston of Canada, 1974.

Nader, L. Forums for justice: A cross-cultural perspective. *The Journal of Social Issues*, 1975, *31*, 151–170.

Nader, L., & Metzger, D. Conflict resolution in two Mexican communities. In D. Black & M. Mileski (Eds.), *The social organization of law*. New York: Seminar Press, 1973.

Nader, R. Consumerism and legal services: The merging of movements. *Law & Society Review*, 1976, *11*, 247–256.

Nemeth, C. Effects of free versus constrained behavior on attraction between people. *Journal of Personality and Social Psychology*, 1970, *15*, 302–311.

Newsweek, Over the oil barrel. July 9, 1979, 18–26.

New York Times, Transcript of President's address to country on energy problems. July 17, 1979.

Ostrom, V., & Ostrom, E. A theory for institutional analysis of common pool problems. In G. Hardin & J. Baden (Eds.), *Managing the commons*. San Francisco: W. H. Freeman, 1977.

Platt, J. Social traps. *American Psychologist*, 1973, *28*, 641–651.

Quinney, R. *Critique of the legal order*. Boston: Little, Brown, 1974.

Reisman, D. *The lonely crowd.* New Haven, Conn.: Yale University Press, 1950.

Rogers, E. M. Social structure and social change. *American Behavioral Scientist*, 1971, *14*, 691–718.

Ross, M. Salience of reward and intrinsic motivation. *Journal of Personality and Social Psychology*, 1975, *25*, 245–254.

Sarat, A. Studying American legal culture: An assessment of survey evidence. *Law & Society Review*, 1977, *11*, 427–488.

Schumacher, E. F. *Small is beautiful: A study of economics as if people mattered.* London: Sphere Books, 1974.

Schwartz, R. D. Social control in two Israeli settlements. In D. Black & M. Mileski (Eds.), *The social organization of law.* New York: Seminar Press, 1973.

Schwartz, R. D., & Miller, J. C. Legal evolution and societal complexity. In D. Black & M. Mileski (Eds.), *The social organization of law,* New York: Seminar Press, 1973.

Sheppard, D. *The 1967 drink and driving campaign: A survey among drivers.* Crowthorne, Eng.: Road Research Laboratory, 1968.

Sherif, M. *In Common predicament: Social psychology of intergroup conflict and cooperation.* Boston: Houghton Mifflin, 1966.

Singer, B. D. *Feedback and society: A study of the uses of mass channels for coping.* Lexington, Mass.: Lexington Books, 1973.

Smith, P., & Hawkins, R. Victimization, types of police contact, and attitudes toward police. *Law & Society Review*, 1973, *8*, 135.

Stern, P. C. When do people act to maintain common resources? Reformulated psychological question for our times. *International Journal of Psychology*, 1978, *13*, 149–158.

Stern, P. C., & Kirpatrick, E. M. Energy behavior: Conservation without coercion. *Environment*, 1977, *19*, 10–15.

Strickland, L. H. Surveillance and trust. *Journal of Personality*, 1958, *26*, 200–215.

Stulberg, J. A. Civil alternative to criminal prosecution. *Albany Law Review*, 1975, *39*, 359–376.

Swett, D. H. Cultural bias in the American legal system, *Law and Society Review*, 1969, *4*, 79.

United Nations. *Assessment of the world food situation, present and future world food conference.* Rome: United Nations, 1974.

VORP. The developmental steps of the Victim/Offender Reconciliation Project, Kitchener, Ont. Unpublished paper, on file with the authors, 1977.

Walster, E. G., Walster, W., & Berscheid, E. *Equity theory and research.* Boston: Allyn & Bacon, 1978.

White, R. K. Misperception and the Vietnam War. *Journal of Social Issues*, 1966, *22*, 1–164.

Wrightsman, L. S. Personality and attitude correlates of trusting and untrustworthy behaviors in a two-person game. *Journal of Personality and Social Psychology*, 1966, *41*, 328–332.

ENDNOTE

20

Adapting to Scarcity and Change (II)

Constructive Alternatives

SALLY C. LERNER

1. THE ROLE OF JUSTICE IN AN UNCERTAIN FUTURE

The question that must be addressed here, at the conclusion of the volume, is this: How can we use our better understanding of the development and functioning of justice concerns so as both to minimize their contribution to negative societal consequences and to foster their mobilization in the service of cooperative problem-solving? This is essentially the challenge we face in trying to deal with the tragedies of the commons. Such tragedies, we now realize, arise from the difficulties of maintaining the viability of any finite system to which there is unlimited access by those who believe that they stand to gain much and lose little by exercising that right of access. Historically, of course, societies have sometimes ignored or been unaware of such situations, with resulting serious degradation of resources and consequent episodes of human suffering, embitterment, warfare and societal upheaval. There are also the countless examples of ruling groups that have exercised rigid regulation of access based on criteria that reflect the variety of power relations, as well as caste and class structures, throughout human history.

SALLY C. LERNER ● Environmental Studies, University of Waterloo, Waterloo, Ontario, Canada N2L 3G1.

Indeed, much of what we know as human history consists of accounts of power struggles over access to scarce resources—and the resultant human suffering, embitterment, warfare, and societal upheaval.

In our own recent history, the Western industrialized nations have experienced a period of continually increasing material affluence. As a consequence, we had begun to believe that we might be on the threshold of escape from the problems of scarce or finite resources. "Technology" would assure an ample supply of existing goods for nearly everyone and would find replacement for whatever we ran out of—thus dramatically reducing human suffering, embitterment, etc. Certainly, this promise has been realized to some extent—for some groups—in some parts of the world. Perhaps that is why it has been so difficult for both haves and have-lesses to accept the diagnoses and prescriptions of the "limits to growth" and "tragedy of the commons" analyses. No one wants to believe that life will be "worse" tomorrow than it is today, whether in terms of absolute deprivation, loss of relative status, or loss of hope.

As we noted in the "Introduction," justice concerns are inextricably bound up in the complex process of how people perceive, define, evaluate, and respond to what is happening to them. Thus, a better understanding of how justice concerns develop and function in people's lives should enable us to plan more effectively for institutional and other social change to deal with the problems that confront humankind. The articles in this volume have approached the question of how justice concerns function in social behavior from a number of quarters: the origins and development of justice concerns; their role and functioning in situations of resource allocation, close interpersonal relations, and dispute resolution; and the problematic nature of their involvement at the interactive levels of culture, social unit and individual psyche. What, then, can we conclude—or even suggest—about the most desirable paths to follow in trying to recruit justice concerns to the service of a more humane society?

A standard assertion about justice concerns is that they become salient in situations where there is seen to be some conflict of interest. This generalization covers situations ranging from the allocation of available resources among family members to the necessity of dealing with the individual who inflicts bodily harm or death on a stranger. Whether "justice concerns" involve careful attention to the question of what principle, (e.g., equity, parity, need) should govern the amount of some resource that different people receive (the implication being that the interest of those involved are in conflict due to the finite nature of the resource) or focus on deciding what negative sanctions to apply to someone defined as a wrongdoer (the implication being that the wrongdoer's

actions conflict with the interests of others in the social system), justice
concerns do appear to arise only in conflict-of-interest situations. Presumably, if no conflict of interest is perceived, such concerns do not
arise, emerge, or become salient; other motives and interest predominate.

And yet, there is considerable reason to believe that such a clear-
cut distinction fails to capture the actual state of affairs. As a number
of the authors in this volume suggest, a variety of justice concerns seem
to be pervasive in our culture if not intrinsic to the process of becoming
human. Perhaps a more useful model—at least, for Westerners, is the
one that postulates a constant monitoring system that develops in a
complex way as the individual interacts with the natural and social
environment—not necessarily identical in nature for everyone but present to some extent—which "keeps tabs" on the just–unjust quality of
all events in the ongoing life experience. This perspective identifies justice concerns as a powerful motivating force in social behavior (as is
apparent from the lengths to which people have gone to redress what
they feel are injustices); and it also recognizes that the extent to which
people respond to events in terms of the just–unjust dimension is a
function of a number of variables, beginning with their perception of
"what is happening"—whether they flag the situation as one involving
a potential violation of a relevant justice principle. This view speaks to
the point raised by a number of these authors that justice concerns are
most usefully conceptualized as emergent and negotiated rather than
as invariant, and it opens the way for a consideration of the influence
on justice-related behavior of individual differences (e.g., in role-taking
ability and experience); objectively different situational contexts (e.g.,
scarcity, abundance, economic uncertainty); and variations in cultural
values, social norms, and interpersonal bonds. It should be noted, of
course, that in pluralistic cultures such as our own, which stress an
individualistic, competitive view of social relations, we would expect a
high proportion of situations to be perceived as involving conflict of
interest for the person as individual or subgroup member.

2. TOWARD INSTITUTIONAL CHANGE

Several very general principles that have emerged from the chapters
in this volume are these:

1. Justice concerns can take a variety of forms, depending on what
a person perceives and feels in a particular situation.

2. These perceptions and feelings are shaped by the person's socialization and prior experiences; by the characteristics of others in the

situation and the person's relationships with them; by the rules, norms, and values that are seen to pertain to the situation; by certain characteristics of the situation, such as number of participants and duration; by the extent to which the person has a controlling influence in the situation; by the amount of information that the person has about the situation and about the others involved; and by the complex interaction of these and other variables.

3. Certain rules, norms, and relationships may be made more-or-less salient in a number of ways, including:

(a) By deliberate instruction (as in a game or an experiment)
(b) By the way an institution is structured, as may be seen in the variety of institutional approaches to formal education, completion of work necessary to society, dealing with deviant behavior, and other basic social needs
(c) By implicitly or explicitly invoking some central cultural value, such as "Share and share alike" or "To the victor belong the spoils"

In short, the usual conditions and processes of constructing reality are present in the emergence and specification of justice concerns. What sorts of justice concerns do our present cultural values and institutions encourage? To answer this question, and to chart desirable alternatives, we must consider the nature of the "realities" that our society promotes: what models of human nature are articulated or implicit; what goods and goals are stressed as paramount; and what modes of interpersonal relating are permitted, required, engendered, or discouraged by institutional forms? Brief examinations of two central institutional areas—education and work—suggest not only that our institutional forms have often promoted less than humane justice concerns, but also that rational attempts to alter this state of affairs are quite possible.

2.1. Education

The problems of our mass public-education experiment are legion and well-publicized, as have been attempts to ameliorate them. Perhaps the most telling criticism, in the context of justice concerns, is the assertion that the schools are designed to create and identify winners and losers, adumbrating outcomes in the adult world. To the extent that we set up the learning process as a contest among individuals for individually beneficial rewards (grades, approbation), we establish with one stroke a world view (zero-sum game), a required way of relating to others (competition), and the predominance of one type of justice concern (equity). It is no accident that many passionate attempts to reform the schools, including those of the 1960s and their current versions, have been initiated by people who realized that any move to a more cooperative, compassionate society must be based on a different way of socializing the young. Along with the desire to foster "no-lose" motivation and learning processes, many of the parents and teachers who established what they called alternate or free schools planned explicitly to structure learning as a cooperative enterprise and to alter radically the hierarchical, authority-oriented nature of the school. Many beneficial effects were hoped for from these changes, including less anxious, more creative children. Equally important have been such goals as fostering relationships that revolve around mutual respect, sharing of resources, and pooling of effort; and providing for self-initiated learning and a view of adults as helpful resource people rather than as authorities to whose ideas one must conform or risk punishment (cf. Kohl, 1969). Again, the implication of these reforms, which have been implemented with varying degrees of success, is that children's perceptions of what is just, fair, and desirable in a situation can be influenced by changing the rules and goals of the game and making salient different values and norms, particularly regarding how people relate to one another. Ideally, in terms of the objectives of many of the new schools, children develop a problem-solving approach to learning that leads them to seek input from others—peers and adults—and to reciprocate in kind, with the result that a group orientation develops, based on mutual help, common goals, and identification with one another. The justice concerns fostered by this type of school situation would almost certainly focus less exclusively on equity and more on parity, group welfare, and individual need. It is also likely that reducing the pressure to conform to authority figures in a reward–punishment game would lessen to some degree children's preoccupation with punitive justice—at least Hogan and Miller and Vidmar suggest that possibility in the articles here.

It is interesting to note that in the highly competitive institutions of higher education, including graduate and professional schools, students have increasingly turned to cheating, a time-honored way of dealing with competitive situations that submerges all justice concerns in a *realpolitik* of self-interest. More suggestive, however, is the student-initiated study group, a survival technique that is based on principles of mutual dependence, sharing, and cooperative problem-solving. While the ultimate objectives of the study group can be characterized as self-interested, in that the group serves to improve the chances of the members in a larger competition, it would be valuable to identify the variables that lead to one or the other of these responses in competitive situations, since it is far from certain that periods of diminishing resources will ever be perceived as anything except highly competitive. However, it is also clear that in such forms as the spontaneous study group, there is the seed of a will and a means to redefine situations, to opt for different, less competitive games and relationships.

2.2. Work

All well and good, says the agreeable reader. But how can we prepare youngsters for a real world that is highly competitive if not in an identical microcosm? The world of work, for which the schools supposedly provide basic training, has traditionally, in industrialized capitalist society, been predicated on competition—for resources, buyers, jobs, promotions, and economic and emotional survival. Conflict of interest was not a regrettable side effect, it was the name of the game. And a meritocracy based, at least publicly, on individual achievement has been the North American version of the game (Rinehart, 1975). If you prepare yourself, most recently through education, to make valued inputs in a valued sector, your efforts will be rewarded with a well-paying job, the respect of the community, stock options, and a comfortable retirement. If you don't, someone else will beat you out, and you'll have to settle for second best, tenth best, or no job at all. And so forth. While conditions of employment, particularly in terms of job security have improved considerably in the last 50 years, it is still accurate to say that for many in our society, the process of "making a living" has in fact been accurately mirrored in the traditional competitive classroom, particularly in terms of the kinds of justice concerns that come to be paramount. There are few surprises, then, in the case studies and analyses that detail the malaise and alienation of large segments of the work force at all levels. What is of more immediate interest are contemporary efforts to restructure work (French & French, 1975; Hunnias, Garson, & Case, 1973), initiated with the objectives of increasing job satisfaction and thus "productivity" by those who control the work of others, and on a small scale by people who have developed or are seeking to develop what is often called an alternate lifestyle. We will focus on the former, since at present it could potentially involve more people.

In both industrial and white-collar settings, there are currently some interesting experiments under way that hold promise for sharpening the more benign and supportive justice concerns, that is, those that could motivate people to plan their way out of commons dilemmas. Called *quality-of-work-life* or *job-enrichment* programs, they have in common several characteristics. These include redesigning work processes so that people are part of small teams with control over and responsibility for an entire product or process; providing for some kind of job rotation within the teams so that each member can perform or at least understand all operations; and introducing more decision-making latitude for employees at all levels, both in organizing their own tasks and in larger issues that affect the organization. There is heated debate over a number of matters involved in these developments: whether "worker control" (as ownership) necessarily guarantees better working conditions; whether the Japanese and various European models of worker involvement with management damage union effectiveness; whether the entire initiative is a management ploy to get more work for the same money. Yet, there is evidence that for some people in some work situations, redesigning the rules, norms, and relationships of the situation does lead to heightened job satisfaction (O'Toole, 1974; Sanderson, 1979). This satisfaction seems to stem from the opportunity to work cooperatively as part of a small team (rather than as one person doing some part of a larger task) and from the sense of having more knowledge about and control over the work process (rather than mindlessly following orders). From one point of view, it might be said that workers appreciate being treated less like children in the traditional classroom. In fact, it has been hypothesized that the steady increase of dissatisfaction and unrest among workers in recent years is at least partially due to the failure of workplace arrangements to keep pace with the changes toward more "permissiveness" in both family and school settings.

In any event, we would expect that changes in the work situation such as those described would promote a sense of group identity, concern for group goals, appreciation of individual differences, awareness of interdependence, and the like. And from what we have begun to learn about the development and shaping of justice concerns, there should follow a shift in perspective away from a narrow focus on equity tied largely to self-interest and toward more reliance on the principles of parity, need, and group welfare, at least within the work team. Further research is needed on the question of how changes in justice concerns engendered by one set of institutional arrangements would generalize to other relations and situations in a person's life. At the moment, though, it does seem reasonable to suggest that while North American society is only beginning to experiment intelligently and creatively with work arrangements, the restructuring of such situations—"social engineering," if you like—holds promise as a way of influencing people to develop the kinds of human bonds and thus the justice concerns that are more socially constructive and relevant to future needs.

In concluding this comment on the place of justice concerns in planning for social change, it seems appropriate to suggest, as this author has suggested elsewhere (Lerner, 1979, 1980), that social and policy impact assessments should routinely include an examination of immediate or eventual effects of any proposed initiative on the constellation of justice concerns among involved segments of the population. This would not be an easy task, but it seems a necessary one if we hope to see justice develop as a positive force in social relations.

REFERENCES

French, D., & French, E. *Working communally: Patterns and possibilities.* New York: Russell Sage, 1975.

Hunnius, G., Garson, G. D., & Case, J. *Workers' control.* New York: Random House, 1973.

Kohl, H. R. *The open classroom: A practical guide to a new way of teaching.* New York: Random House, 1969.

Lerner, S. C. Environmental impact assessment in an era of relative scarcities. *Contact,* 1979, *2*, 61–67.

Lerner, S. C. *Energy Policy: A potential source of possitive social impacts.* Paper presented at the First National Energy Policy Conference, Morgantown, West Virginia, May 1980.

O'Toole, J. (Ed.). *Work and the quality of life.* Cambridge, Mass.: M.I.T. Press, 1974.

Rinehart, J. W. *The tyranny of work.* Don Mills, Ontario: Longman Canada, Ltd., 1975.

Sanderson, G. F. (Ed.). Adapting to a changing world: Readings on the quality of working life. Ottawa: *The Labour Gazette*(Special Edition), 1979.

Author Index

Subject Index